BIOPHYSICAL CHEMISTRY

Readings from
**SCIENTIFIC
AMERICAN**

BIOPHYSICAL CHEMISTRY

PHYSICAL CHEMISTRY IN THE BIOLOGICAL SCIENCES

With Introductions by
Victor A. Bloomfield
University of Minnesota

and

Rodney E. Harrington
University of Nevada, Reno

W. H. Freeman and Company
San Francisco

Cover: Schematic illustration of band patterns of
interacting collagen molecules (see p. 77). [From
"Giant Molecules in Cells and Tissues," by F. O.
Schmitt. Copyright© 1957 by Scientific American,
Inc. All rights reserved.]

Most of the SCIENTIFIC AMERICAN articles in
BIOPHYSICAL CHEMISTRY are available as separate
Offprints. For a complete list of more than 950 articles
now available as Offprints, write to W. H. Freeman
and Company, 660 Market Street, San Francisco,
California 94104.

Library of Congress Cataloging in Publication Data.

Main entry under title:

Biophysical chemistry: physical chemistry in the
   biological sciences.

   Bibliography: p.
   Includes index.
   1. Biological chemistry—Addresses, essays, lectures.
   2. Chemistry, Physical and theoretical—Addresses,
essays, lectures. I. Bloomfield, Victor A.   II. Harrington, Rodney E.   III. Scientific American. [DNLM:
1. Biochemistry—Collected works.   2. Chemistry,
Physical—Collected works. QU4 B617] QH345.B544
574.1′9283   75–8748
ISBN 0–7167–0513–3
ISBN 0–7167–0512–5 (pbk.)

Printed in the United States of America.

9 8 7 6 5 4 3 2 1

PREFACE

Biophysical chemistry is the application of the concepts and techniques of physical chemistry to the study of biological systems. Physical chemistry is concerned with a quantitative understanding of the structure, energetics, and dynamics of molecular systems. Starting in the middle 1930s, but particularly since the early 1950s, great progress has been made in biology by the application of physico-chemical methods to the study of biological molecules and molecular aggregates. Physical chemistry has led to a comprehension of such basic problems as the three-dimensional structure of an enzyme molecule, the mechanisms that enable the conversion of metabolic energy into muscular work, and many important architectural aspects of aggregated structures. The physico-chemical approach affords a *unified* understanding of the structure and properties of atoms and small molecules, such biological macromolecules as proteins and nucleic acids, and such functional aggregates of macromolecules as chromosomes, membranes, and viruses. This unification is satisfying not only scientifically, but also aesthetically and philosophically.

As quantitative chemical understanding has become more important in biology and biochemistry, more and more students in these areas have taken courses in physical chemistry. However, the textbooks generally do not adequately indicate pertinent biological applications of the subject; those of us who teach physical chemistry have had to formulate our own examples. This collection of articles from *Scientific American* is intended to supplement the textbooks, and to show students of the biological sciences how physical chemistry has been useful in biology.

Physical chemists have become increasingly aware that biochemistry and biology offer an almost unlimited range of interesting and important problems on which their special skills may be exercised. The articles in this volume indicate those areas in which the physico-chemical approach has been successful, and those in which significant problems remain. Finally, for the general reader whose expertise lies neither in biology nor in physical chemistry, a survey of the interactions between chemistry, physics, and biology will be found to have its own fascination.

The articles in this book are grouped into four sections. The first includes five articles on the structure of biological macromolecules, especially the proteins and nucleic acids, whose structural-functional relationships lie at the heart of modern biochemistry and molecular biology. The second section deals with the structures of macromolecular aggregates, typified by viruses and membranes, which often constitute the true functional units of cells. This section includes two articles on important physical phenomena in these systems: active transport, and the remarkably efficient mechanochemical properties of muscle. Because of the fundamental importance of enzymatic catalysis and its regulation in living systems, the third section is devoted exclusively to enzymes and their unusual behavior with respect to biochemical control. The fourth section deals with the experimental tools—such as ultracentrifugation and electrophoresis—that have made the brilliant work described in the first three sections possible.

In choosing the articles for this book, we necessarily had to omit many that are pertinent to the areas covered here. For the reader who wishes to explore further the topics in this volume, we have cited several *Scientific American* articles at the end of each introduction. In addition, a field so enormous and diverse as modern biophysical chemistry simply cannot be covered in a single volume. References to additional topics occur throughout the book, and the reader should not be misled into thinking that these are less important in biophysical chemistry. Those wishing to pursue additional topics will find a vast number of important articles in *Scientific American*.

Among the topics excluded from this volume is water, that most fundamental of all life substances. Important papers on the structure and physical and biological properties of water certainly include "Water" by Arthur M. Buswell and Worth H. Rodebush, April, 1956 (Offprint 262); "Ice" by L. K. Runnels, December, 1966 (Offprint 307); "pH" by Duncan A. MacInnes, January, 1951; and "The State of Water in Red Cells" by Arthur K. Solomon, February, 1971 (Offprint 1213).

Another area of importance, only indirectly included in this volume, is the relationship between molecular energetics and dynamics and biochemical processes. Articles on this topic include "What is Heat?" by Freeman J. Dyson, September, 1954; "Light" by Gerald Feinberg, September, 1968; "Life and Light," by George Wald, October, 1959 (Offprint 61); "How Light Interacts with Matter" by Victor F. Weisskopf, September, 1968; "The Chemical Effects of Light" by Gerald Oster, September, 1968; "How Light Interacts with Living Matter" by Sterling B. Hendricks, September, 1968; "The Mechanism of Photosynthesis" by R. P. Levine, December, 1969 (Offprint 1163); "Molecular Isomers in Vision" by Ruth Hubbard and Allen Kropf, June, 1967 (Offprint 1075); "Biological Luminescence" by William D. McElroy and Howard H. Seliger, December, 1962 (Offprint 141); "Heat and Life" by Frank H. Johnson, September, 1954; "Free Radicals" by Paul D. Bartlett, December, 1953; "Free Radicals in Biological Systems" by William A. Pryor, August, 1970 (Offprint 335); "The Solvated Electron" by James L. Dye, February, 1967; "The Processes of Vision" by Ulric Neisser, September, 1968 (Offprint 519); "Energy and Information" by Edward C. McIrvine and Myron Tribus, September, 1971 (Offprint 670); "How Cells Transform Energy" by Albert L. Lehninger, September, 1961 (Offprint 91); "The Flow of Matter" by Marcus Reiner, December, 1959 (Offprint 268); "The Mechanical Properties of Polymers" by Arthur V. Tobolsky, September, 1957; "How is Muscle Turned On and Off?" by Graham Hoyle, April, 1970 (Offprint 1175); "The Cooperative Action of Muscle Proteins" by John M. Murray and Annemarie Weber, February, 1974 (Offprint 1290); "ATP" by Paul K. Stumpf, April, 1953 (Offprint 41); and "Muscle as a Machine" by A. Katchalsky and S. Lifson, March, 1954.

Some articles that deal with pertinent but more general topics are: "The Fundamental Physical Constants" by Donald N. Langenberg, Barry N. Taylor and William H. Parker, October, 1970 (Offprint 337); "Mathematics in the Physical Sciences" by Freeman J. Dyson, September, 1964; "Mathematics in the Biological Sciences" by Edward F. Moore, September, 1964; "The Principle of Uncertainty" by George Gamow, January, 1958 (Offprint 212); "The Quantum Theory" by Karl K. Darrow, March, 1952 (Offprint 205); "What is Matter?" by Erwin Schrödinger, September, 1953 (Offprint 241); "Organic Chemical Reactions" by John D. Roberts, November, 1957 (Offprint 85).

March, 1975   Victor A. Bloomfield
                                                                     Rodney Harrington

CONTENTS

Note on cross-references to SCIENTIFIC AMERICAN *articles:* Articles included in this book are referred to by title and the beginning page in this volume. Articles not included here, but available as Offprints, are referred to by title and offprint number; articles that are not available as Offprints are referred to by title and date of publication.

I

BASIC
BIOMOLECULAR
STRUCTURE

BASIC BIOMOLECULAR STRUCTURE

I

INTRODUCTION

Biological systems consist of both small and very large molecules. The operation of the systems is determined in general by the way the small molecules interact chemically with each other and with macromolecules, and by the physical properties of the macromolecules themselves. Structural molecules tend to be highly elongated, which promotes intermolecular aggregation; enzymes, whose catalytic activity requires close juxtaposition of numerous functional groups, tend to be more globular in shape.

These examples may be thought of as illustrations of the dictum, first articulated by the architect Louis Sullivan early in the twentieth century, that form follows function. Clearly, it is essential to understand the structures of large molecules if one is to comprehend their modes of action. Consequently, an immense effort has been put into determining the structure of biopolymers.

Determination of the structure of biological macromolecules is a principal goal of biophysical chemistry, which is both a branch of molecular biology and an outgrowth of polymer chemistry. Particular attention has been paid to the structure of the proteins, which function as enzymatic catalysts and as structural elements in the cell, and to the structure of the nucleic acids, which carry genetic information and which provide much of the machinery for the expression of this information in the form of proteins and other cell constituents. The articles in this section describe some of the structural principles of biophysical chemistry and report in detail on such important macromolecules as the proteins and nucleic acids.

The first article, "The Structure of Protein Molecules" by Linus Pauling, Robert B. Corey, and Roger Hayward, introduces several themes that play an important role in nearly all discussions of biopolymer structure. The work described here, done in the early 1950s, has been enormously influential. According to Pauling and his co-workers, in order to understand the function of a biopolymer, one must elucidate the arrangement of its thousands of atoms. For this task, X-ray diffraction is of paramount importance; even by this technique, however, it is not easy to get enough data to define atomic coordinates precisely. Therefore, it is necessary to make the assumption that the properties of the monomers—the amino acid subunits that link together to form long protein chains—carry over into the polymer. By X-ray crystallographic studies of small model compounds, Pauling and co-workers found that two such properties were of prime importance: the planarity and the *trans*-conformation of the amide group. Further, hydrogen bonds between amino and carboxyl groups in these compounds were noted, and hydrogen bonding was postulated to be an important factor in stabilizing polypeptide conformations. Finally, the assumption of symmetry—the equivalence of the amino acid subunits of a polypeptide—led to the prediction of helical structures. The alpha helix, and several other types of helixes defined by Pauling and other workers, have since been identified as prominent structural elements in proteins and other biopolymers.

Paul Doty, in the article "Proteins," develops additional themes that have become common in protein structural research. He points out that most proteins are globular, and thus do not give the relatively easily interpreted diffraction patterns of fibrous, helical proteins. Using synthetic polypeptides of appropriate chemical composition, it is possible to develop models for some aspects of the behavior of naturally occurring proteins. In particular, by proper manipulation of their temperature or chemical environment, synthetic polypeptides may be induced to undergo a structural transition from a helical to a randomly coiled conformation. This transition bears some relation to the denaturation of proteins by heat (for example, the coagulation of egg protein by boiling) and chemicals. Doty also discusses the use of optical rotation to measure the fraction of amino acids that may be incorporated in a helical region of a protein or polypeptide. Finally, he describes the interaction of separate polypeptide chains to form triple-stranded collagen fibers.

This brings up an important idea in the study of biopolymer structure: hierarchies of structure. The most commonly accepted classification is due to the great Danish protein chemist, K. U. Linderstrøm-Lang. *Primary structure* is the linear sequence of amino acids along the polypeptide backbone. *Secondary structure* is the folding of certain regions of the protein into helical arrays. *Tertiary structure* is the three-dimensional arrangement of helical and nonhelical sections within a single polypeptide chain. *Quaternary structure* is the relative arrangement of polypeptide chains in a higher-order aggregate. These aggregates may contain just a few chains—there are only four in hemoglobin (see "The Hemoglobin Molecule" by M. F. Perutz, p. 40)—or several thousand (see "The Structure of Viruses" by R. W. Horne, p. 81). Investigations of quaternary structure are of central importance in modern biophysical chemistry.

In the article "The Three-Dimensional Structure of a Protein Molecule," John C. Kendrew reports on the first high-resolution determination of the structure of a globular protein. The protein is myoglobin, an oxygen-storage protein found in muscle. A single polypeptide chain, its tertiary structure is remarkably similar to that of one of the chains of hemoglobin, the oxygen-transport protein of blood. In both, oxygen binds to iron at the center of the heme group, around which the globin part of the protein is arrayed. Kendrew's article, along with those of Perutz ("The Hemoglobin Molecule," p. 40) and Phillips ("The Three-Dimensional Structure of an Enzyme Molecule," p. 141), is a useful introduction to important ideas in the use of X-ray crystallography in the determination of protein structure. Two points are particularly important. First, the very fact that proteins crystallize is evidence that they are definite chemical species, not indefinite, amorphous aggregates as was once thought. Second, in order to determine the positions of thousands of atoms in a protein, thousands of diffraction spots on film have to be analyzed. This great mass of data puts severe demands on data-analysis techniques, and high-speed computers have become essential tools. Some of the structural features characteristic of myoglobin have been found to occur in other proteins as well, and have guided thought on the general nature of protein conformation. Among these are a substantial degree of helical structure, either alpha helix (of which myoglobin has an unusually high amount) or some other kind; stabilization of the structure by hydrogen-bonding interactions; the location of charged and polar amino acid side chains on the exterior of the molecule and hydrocarbon side chains clustered in the interior; and the helix-interrupting property of the amino acid proline.

No protein has been studied more devotedly than hemoglobin, the oxygen transport protein of blood. In the article "The Hemoglobin Molecule," Max Perutz describes the results of studies of this protein, again based mainly on X-ray crystallography. Hemoglobin has four polypeptide chains. Its structure, both tertiary and quaternary, and its oxygen affinity are affected by the state of oxygenation of the molecule. This change in conformation and properties attendant on ligand binding, known as allosteric binding, also underlies much fine-tuned regulation of enzyme-catalyzed processes. Hemoglobin has served so effectively as a model for allosteric effects in enzymes that it is affectionately called an honorary enzyme. Perutz discusses several points of technical interest. He provides a lucid description of the "phase problem" in X-ray crystallography, and shows how the method of isomorphous substitution of heavy atoms, which he was the first to apply to proteins, can be used to solve the phase problem. All subsequent determinations of protein structure have relied for their success on this method. Perutz also discusses in detail hydrophobic bonding, which denotes the tendency of amino acids with oily, hydrocarbon side chains to cluster together in the interior of protein molecules, away from the solvent water. Hydrophobic bonding, perhaps even more than hydrogen bonding, has emerged as a major determinant of protein structure and stability. Finally, it is interesting to note that in hemoglobin, perhaps more than in any other protein, it has been possible to correlate structure with function. Various diseases, such as sickle-cell anemia, have been related to amino acid substitutions that modify interactions between the chains, and other properties, of hemoglobin. Once we know where these mutant amino acids are located in the three-dimensional structure of the protein, it becomes possible to connect molecular defects with specific disease states.

With the article by F. H. C. Crick, "The Structure of the Hereditary Material," we turn our attention to the nucleic acids. Twenty years ago, when this article was written, the fact that DNA is the genetic material, and the basic aspects of its composition, had only recently been established. Today, this fact and the structure and mode of replication of DNA proposed by Crick in his 1954 article are taken as dogma and are the intellectual foundation of the enormously successful field of molecular biology. Many of the themes that are now familiar to us from the study of proteins appear in this article: helical structures (here the famous double helix suggested by Watson and Crick), the importance of hydrogen bonding in providing structural specificity and stability, and the use of X-ray diffraction techniques. In addition, the idea of complementarity—of A with T and G with C—assumes central importance and is used to explain the mode of DNA replication.

Another aspect of complementarity is discussed by Sol Spiegelman in "Hybrid Nucleic Acids." Here, complementarity leads to "recognition" between nucleic acid chains, which then form hybrid structures that are used in studies of the mechanism of protein synthesis. In these studies the important nucleic acid is RNA, which is transcribed from DNA. The experiments described by Spiegelman helped to demonstrate that information flows from DNA to RNA to protein, which constitutes the so-called "central dogma" of molecular biology. (Recently, certain viruses have been found containing an enzyme that enables the transcription of RNA into DNA, thereby reversing the first step in the central dogma.) The experimental techniques used in the work described by Spiegelman—electrophoresis, ultracentrifugation, density gradients, radioactive labeling, and hybridization—are major tools in molecular biology. Several of these techniques are described in detail in articles in Section IV.

Other *Scientific American* articles relevant to the topic of biomolecular structure are "The Shapes of Organic Molecules" by Joseph B. Lambert, January, 1970 (Offprint 331); "X-Ray Crystallography" by Sir Lawrence

Bragg, July, 1968 (Offprint 325); "Giant Molecules" by Herman F. Mark, September, 1967 (Offprint 314); "The Nature of Polymeric Materials" by Herman F. Mark, September, 1967; "The Chemical Structure of Proteins" by Stanford Moore and William H. Stein, February, 1961 (Offprint 80); "The Structure and History of an Ancient Protein" by Richard E. Dickerson, April, 1972 (Offprint 1245); "The Specificity of Antibodies" by S. J. Singer, October, 1957; "The Structure and Function of Antibodies" by Gerald M. Edelman, August, 1970 (Offprint 1185); "Nucleic Acids" by F. H. C. Crick, September, 1957 (Offprint 54); "The Nucleotide Sequence of a Nucleic Acid" by Robert W. Holley, February, 1966 (Offprint 1033); "The Synthesis of DNA" by Arthur Kornberg, October, 1968 (Offprint 1124); "Information Transfer in the Living Cell" by George Gamow, October, 1955; "The Genetic Code: II" by Marshall W. Nirenberg, March, 1963 (Offprint 153); "Genetic Repressors" by Walter Gilbert and Mark Ptashne, June, 1970 (Offprint 1179); "The Isolation of Genes" by Donald D. Brown, August, 1973 (Offprint 1278); "The Molecule of Infectious Drug Resistance" by Royston C. Clowes, April, 1973 (Offprint 1269); "Monomolecular Films" by Herman E. Ries, Jr., March, 1961; "The Membrane of the Living Cell" by J. David Robertson, April, 1962 (Offprint 151); and a final article on biomolecular structure, which may well point the way to particularly exciting future advances, "Molecular Model-Building by Computer" by Cyrus Levinthal, June, 1966 (Offprint 1043).

The Structure of Protein Molecules

by Linus Pauling, Robert B. Corey and Roger Hayward
July 1954

A protein molecule is composed of thousands of atoms. How are they arranged? Progress has been made toward answering the question for fibrous proteins such as hair, horn, fingernail and porcupine quill

The human body is about 65 per cent water, 15 per cent proteins, 15 per cent fatty materials, 5 per cent inorganic materials and less than 1 per cent carbohydrates. A molecule of water consists of three atoms, two of hydrogen and one of oxygen. The structure of this molecule has been determined in recent years: each of the two hydrogen atoms is 0.96 Angstrom unit from the oxygen atom (an Angstrom unit is one ten-millionth of a millimeter), and the angle formed by the lines from the oxygen atom to the hydrogen atoms is about 106 degrees. Compared to this simple molecule, a protein molecule is gigantic. It consists of thousands of atoms, mostly of hydrogen, oxygen, carbon and nitrogen. The problem of how these atoms are arranged in a protein molecule is one of the most interesting and challenging now being attacked by workers in the physical and biological sciences.

The proteins are of especial interest not only because of their complexity of structure but also because of their variety and versatility. There are tens of thousands, perhaps as many as 100,000, different kinds of proteins in a single human body. They serve a multitude of purposes: collagen, a constituent of tendons, bones and skin, seems to have the main purpose of providing a framework which has suitable mechanical properties; hemoglobin, found inside of the red blood cells, has the primary function of combining with oxygen in the lungs and liberating it in the tissues; keratin, in the hair and in the epidermis, provides pro-

tection for the body, and in the fingernails it functions as a tool; pepsin, trypsin and many similar enzymes are involved in the digestion of food; cytochrome c and other oxidation-reduction enzymes catalyze the oxidation of foodstuffs within the cells; the muscle protein myosin plays an important part in the process of converting chemical energy into mechanical work. The tabulation could be continued almost indefinitely.

The Amino Acids

During the second half of the 19th century it was found that proteins can be broken down by boiling them in water for a long time or by treating them with acid or alkali, and that simple chemical substances, called amino acids, can be obtained as the products of this treatment.

Just 50 years ago it was discovered by the German chemist Emil Fischer that proteins consist of long chains of amino-acid residues. (An amino-acid residue is the group of atoms that remains after a molecule of water has been removed from a molecule of an amino acid.) Long chains of amino-acid residues are called polypeptide chains. The chains are usually very large; for example, in the molecule of ovalbumin, the principal protein of egg white, about 400 amino-acid residues form a single polypeptide chain. The number of residues of amino acids of different kinds in a protein molecule can be determined by chemical analysis of the protein; each

molecule of ovalbumin has been found to contain about 19 glycine residues, about 35 alanine residues, about 9 tyrosine residues, and so on for the 17 other kinds of amino acids that are represented.

In the study of the structure of a protein there are two questions to be answered. What is the sequence of amino acids in the polypeptide chain? What is the way in which the polypeptide chain is folded back and forth in the space occupied by the molecule? Significant progress has been made toward answering both of these questions during recent years.

In this article we shall consider only the second question. The experimental technique of greatest value in the attack on this problem is that of X-ray diffraction. It was this technique that in 1914 enabled the Braggs (the late Sir William and his son Sir Lawrence, who is now director of the Davy Faraday Laboratory of the Royal Institution in London) to determine the structure of sodium chloride and other simple substances, and then of more complex inorganic substances, such as silicate minerals and metals, and of organic substances. Only half a dozen years after X-rays were first used for this purpose they were applied to proteins. At the Kaiser Wilhelm Institute in Berlin-Dahlem, R. O. Herzog and W. Jancke made X-ray diffraction photographs of hair, horn, muscle, silk and tendon. The results were disappointing—the definition of the photographs was so poor that it seemed a hopeless job to attempt to determine from them the positions of atoms.

Proteins and X-Ray Diffraction

The situation the investigator had to face is suggested by the two photographs at the bottom of these two pages. The photograph on the opposite page shows the diffraction pattern obtained by passing a beam of X-rays through a small crystal of the simple substance glycylglycine which was being rotated about a vertical axis. The photograph on this page shows the diffraction pattern of a fibrous protein, a horsehair that has been descaled (to remove a protein of a different sort that is present on the surface of the hair). The axis of the hair is vertical.

The X-ray photograph of glycylglycine shows about 400 spots. Each of these spots represents a direction in which X-rays are strongly scattered by the atoms in the crystal. From several photographs of this sort a collection of about 800 characteristic intensity values can be obtained. The glycylglycine molecule contains nine atoms other than hydrogen atoms; the hydrogen atoms do not scatter X-rays very strongly, and their positions are usually not well indicated by the X-ray method. The structure of the glycylglycine crystal can be described when three coordinates, the x, y and z coordinates, have been determined for each of the nine atoms. There are accordingly 27 atomic coordinates to be determined. The intensity of each X-ray reflection depends upon these coordinates, and it is possible from the 800 intensity values to determine all 27 of them with considerable accuracy. In this way each of the atoms, except the hydrogen atoms, in the glycylglycine crystal has been located to within about 0.02 Angstrom unit. This uncertainty is about 2 per cent of the distance between each atom and its nearest neighbors. The structure of the glycylglycine crystal, found in this way, is shown on page 6.

The problem of determining the structure of a protein such as keratin, which makes up the fibers of hair, is a quite different one. The polypeptide chains in fibrous proteins are about as complicated as those in ovalbumin. Chemical analysis of hair has shown that 18 different amino acids are represented in the keratin molecule, and that the repeating unit in the keratin fiber probably consists of about 300 amino-acid residues. Each amino-acid residue contains on the average about nine atoms other than hydrogen; there are accordingly about

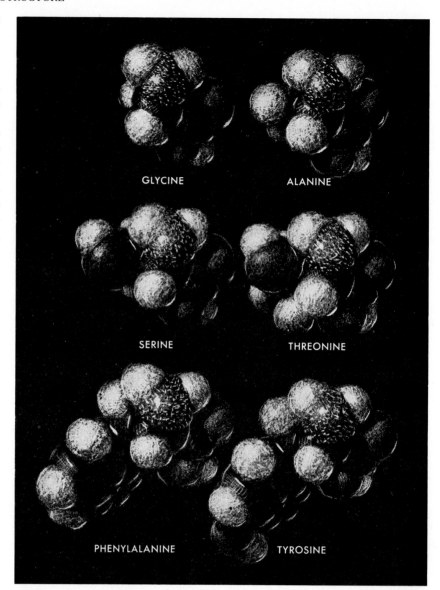

GLYCINE   ALANINE

SERINE   THREONINE

PHENYLALANINE   TYROSINE

AMINO-ACID MOLECULES are typical of those of the 20-odd amino acids now known. These molecules consist of atoms of carbon (*black spheres*), oxygen (*gray*), hydrogen (*white*) and nitrogen (*stippled*). They differ only in side-chains attached at the left.

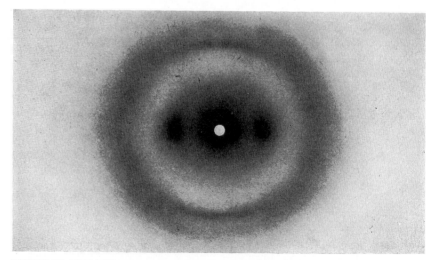

DESCALED HORSEHAIR produced this X-ray diffraction pattern. The X-rays were perpendicular to the plane of the picture; the hair, vertical in a plane parallel to the picture.

POLYPEPTIDE CHAIN is formed when a hydroxyl group (one atom of hydrogen and one of oxygen) attached to a carbon atom in an amino-acid molecule combines with a hydrogen atom attached to a nitrogen atom in another amino-acid molecule (*top*). This forms a molecule of water. As the water is ejected the carbon and nitrogen atoms combine to link the amino-acid residues (*bottom*).

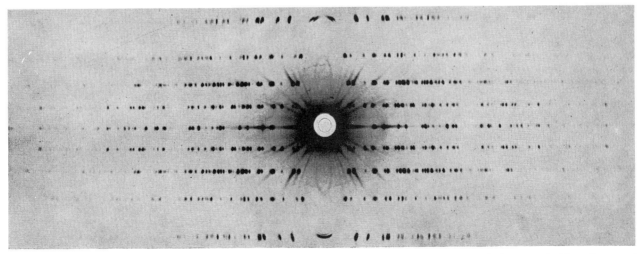

GLYCYLGLYCINE CRYSTAL yielded an X-ray diffraction pattern with a regular array of spots. As the crystal was photographed it was rotated around its *c* axis, which was vertical in a plane parallel to that of the picture. The *c* axis is one of three crystal axes.

10

POSITIONS OF ATOMS in a crystal of glycylglycine have been determined from the intensities of spots on X-ray patterns like that at the bottom of the preceding page. The positions of the carbon, oxygen and nitrogen atoms have been determined within 0.02 Angstrom unit. The hydrogen atoms are not accurately located because they do not scatter X-rays very strongly. On the basis of this information it is possible to calculate the dimensions of the glycylglycine molecule. This is outlined in the drawing above and shown in detail by the diagram at the right. The angles between the bonds that join the atoms are indicated in degrees. The length of each bond is given in Angstrom units.

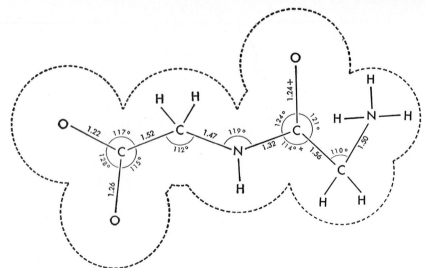

2,700 atoms to be located and 8,100 coordinates to be determined. Such a complex structure obviously cannot be determined from the ill-defined X-ray photograph of horsehair.

Despite this discouraging situation several investigators, most outstanding among them W. T. Astbury of the University of Leeds, continued to use X-rays in the study of the fibrous proteins that occur in plants and animals, and they collected a great deal of valuable information. Herzog and Jancke had observed that the X-ray patterns of hair, silk and tendon are quite different. The later investigators found that almost every one of the many fibrous proteins found in nature gives one or another of these three patterns. Such different proteins as hair, horn, fingernail, porcupine quill, muscle, epidermis, fibrinogen and bacterial flagella give similar X-ray photographs. This similarity strongly suggests that the configuration of the polypeptide chains in all of these proteins is the same; that, regardless of differences in the relative number and sequence of their amino-acid residues, the chains are folded or coiled according to a common pattern. Other fibrous proteins, such as silk and tendon, have different X-ray patterns, and their polypeptide chains must be coiled in different ways.

In 1937 it was decided in the Gates and Crellin Laboratories of Chemistry at the California Institute of Technology to attack the problem of the structure of proteins along an indirect route—by learning enough about the nature of polypeptide chains to permit a good guess as to how the polypeptide chain would naturally fold itself to form a protein molecule or fiber. At that time the X-ray diffraction method had been successfully applied in the determination of the structure of hundreds of crystals, including some very complex ones such as the mineral beryl ($Be_3Al_2Si_6O_{18}$). No structure determination had, however, yet been made of any amino acid or any other simple substance closely related to the proteins.

The attack on these simple substances was begun, and by 1950 precise structure determinations had been made in these laboratories of three amino acids, three simple peptides (short chains of amino-acid residues), and several closely related substances. With the information provided by these structures it was possible to start work on the prediction of likely configurations for polypeptide chains. Since 1950 six more amino-acid and peptide structures have been deter-

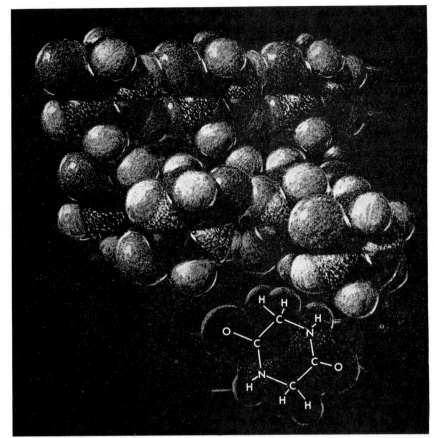

HYDROGEN BONDS join the molecules of diketopiperazine in long laths. In the model of the diketopiperazine crystal at the top the laths are horizontal. At the bottom is the structure of a diketopiperazine molecule. Hydrogen bonds occur at the points N-H and O-C.

mined at the California Institute of Technology, and several have been worked out in other laboratories also.

Much work was needed to learn the distances between the atoms, the angles between the chemical bonds, and other structural features. This had to be done with an accuracy corresponding to errors in atomic position not greater than about 0.02 Angstrom unit. The investigation of one crystal, the amino acid threonine, required the efforts of four post-doctoral research workers for an average of one full-time year apiece.

When several of these structures had been determined it was found that they were strikingly uniform from substance to substance. This uniformity permitted the reliable prediction of the dimensions of a polypeptide chain. The dimensions of the chain, as derived in this work, are given in the illustration at the top of the next page. The distances between atoms are believed to be reliable to about 0.02 Angstrom unit, and the angles between chemical bonds to within about three degrees.

One characteristic feature of the structure is of special importance. The six atoms of the so-called amide group

(CCONHC) are coplanar—they lie within a few hundredths of an Angstrom unit of a common plane. This planar amide group is a rigid part of the polypeptide chain; the amide group can be only slightly distorted from the planar configuration. The rigidity of the amide group greatly simplifies the problem of finding the ways in which the polypeptide chain can be folded.

The planarity of the amide group is explained by its electronic structure. The chemical valence bonds can be represented for this group in two different ways:

In one valence-bond structure there is a double bond between a carbon atom in the group and the adjacent oxygen atom. In the other valence-bond structure there is a double bond between the same carbon atom and the adjacent nitrogen atom. The actual structure of the amide group may be described as

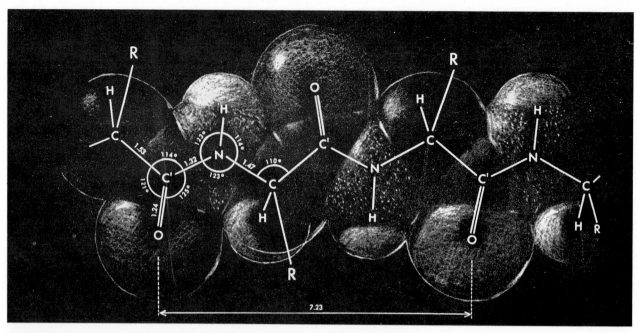

POSITIONS OF ATOMS in a polypeptide chain have been determined by precise X-ray analysis of amino-acid and peptide crystals. The angles between bonds are in degrees. Linear dimensions are in Angstrom units. The symbol R represents atoms in side chains.

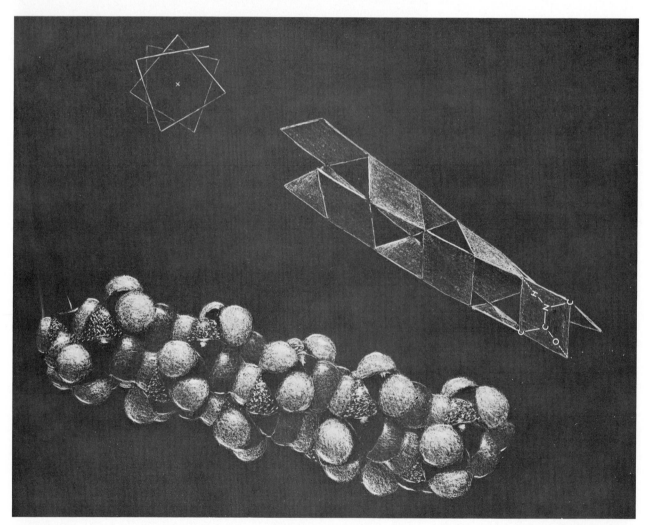

ALPHA HELIX is an arrangement of the polypeptide chain characteristic of hair, horn and related proteins. As shown at the upper right, the flat amide groups of the chain are joined by carbon atoms and held in place by hydrogen bonds. At the upper left this configuration is seen along its axis. At the bottom is a model of the helix. For simplicity the amino-acid residues are those of glycine.

a hybrid of the two valence-bond structures, with the double bond resonating between the two positions.

There are two alternative ways in which the atoms of the amide group can be arranged in a plane. One is called the trans configuration; in it there are carbon atoms at opposite corners of the group. The other is the cis configuration, in which the carbon atoms are at adjacent corners. The trans configuration appears in the structure of a polypeptide chain shown at the left, and the cis configuration in the structure of the substance diketopiperazine on page 7. There is evidence that the trans configuration is considerably more stable than the cis configuration, and the cis configuration is probably rare in the polypeptide chains of proteins.

A polypeptide chain of amide groups with the trans configuration might be folded in a great many ways. The bonds to the corner carbon atoms of the group are single chemical bonds, and the molecule may assume any one of various angles about the axis of each single bond. Of the resulting configurations, the satisfactory ones are those in which each amide group forms so-called hydrogen bonds with other amide groups. The hydrogen bond between two amide groups is a weak bond connecting a hydrogen atom and a nitrogen atom of one amide group with the oxygen atom of the other amide group. In the illustration on page 7, hydrogen bonds join the molecules of diketopiperazine into long laths, which lie side by side in the crystal. The presence of the hydrogen-bonded laths of diketopiperazine molecules is reflected in the physical properties of the crystal. One would expect that it would be rather easy to separate one lath from another, and more difficult to break a lath, which would require that the hydrogen bonds be broken. It is in fact found that the diketopiperazine crystal can easily be cleaved along planes parallel to the long axis of the laths.

By studying this crystal and others it has been shown that the average distance between a nitrogen atom and an oxygen atom connected by a hydrogen bond is 2.79 Angstrom units. It is accordingly reasonable to believe that an acceptable configuration for a polypeptide chain should be one permitting the formation of hydrogen bonds about 2.79 Angstrom units long.

The folding of the polypeptide chain that seems to occur most widely among proteins is shown in the illustration at the bottom of the opposite page. This

SYNTHETIC POLYPEPTIDE poly-gamma-methyl-L-glutamate is depicted in this model. The polypeptide chain is horizontal. Only two of the residues in the chain are shown.

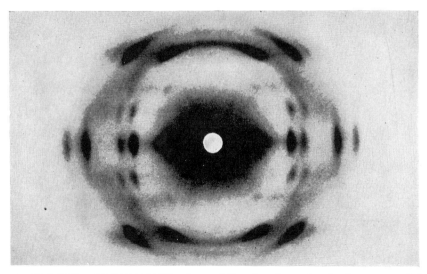

X-RAY DIFFRACTION PATTERN of a specially oriented sample of poly-gamma-methyl-L-glutamate was made by the English investigators C. H. Bamford, W. E. Hanby and F. Happey. The axis of the synthetic fiber was vertical in a plane parallel to that of the picture.

configuration was discovered by analyzing the consequences of a simple assumption—that all of the amino-acid residues in the polypeptide chain are equivalent to one another, except for the difference in the nature of the side chains. Except for glycine the amino acids that occur in proteins are asymmetric; they are described as left-handed molecules. When asymmetric objects in space are joined together in such a way that every one has the same geometrical relationship to its neighbors, a helix is formed. An example is provided by a spiral (properly a helical) staircase; the first step is converted into the second step by moving it along the axis of the staircase and rotating it around the axis. The same operation converts the second step into the third, the third into the fourth and so on. When a search was made for helixes in which each amide group in the polypeptide chain is attached by hydrogen bonds to two others, two structures were found. One of these structures does not seem to occur in proteins. The other structure, which is

called the alpha helix, is believed to be present in many proteins.

The Alpha Helix

The alpha helix has about 3.60 amino-acid residues per turn of the helix. This number may vary by a small amount. The original prediction was that the number of residues per turn would lie between 3.60 and 3.67. The reliable experimental values that have been obtained so far lie between 3.600 and 3.625. The number 3.60 corresponds to 18 residues in 5 turns of the helix. The pitch of the helix—the distance between one turn and the next turn—was predicted to be 5.4 ± 0.1 Angstrom units. This value corresponds to 1.50 ± 0.03 Angstrom units for the axial length per amino-acid residue—the rise from one step of the helical staircase to the next. The diameter of the molecule, including the side chains, was predicted to be about 10.5 Angstrom units.

It was immediately seen that the alpha helix might represent the structure of

hair, fingernail, horn, muscle and other proteins classified as alpha keratin. However, the agreement between the X-ray pattern predicted for molecules of this configuration, lined up side by side in parallel orientation, and the observed X-ray diagram was far from complete. Encouraging support then came from an unexpected quarter. The English investigators C. H. Bamford, W. E. Hanby and F. Happey had prepared X-ray photographs of some synthetic polypeptides in which all of the amino-acid residues in the polypeptide chain were chemically identical, and they published these photographs early in 1951. One of them, together with the structural formula of the substance, is shown on the preceding page. This photograph agrees very well with the calculated X-ray pattern for a bundle of alpha helixes arranged in hexagonal packing. In particular, the positions and intensities of the X-ray reflections correspond very closely to the calculated values for a helix with 18 amino-acid residues in 5 turns.

There is a general similarity between

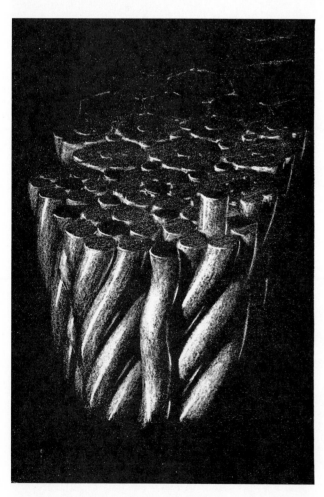

HAIR and similar proteins are probably made up of seven-strand cables. Each consists of an alpha helix surrounded by six compound helixes. In the interstices are compound helixes of different pitch.

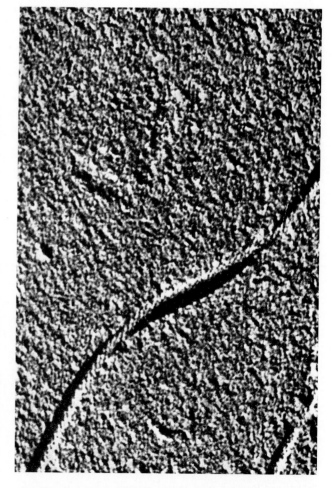

BACTERIAL FLAGELLUM in this electron micrograph is a three-strand cable 180 Angstrom units thick. The diameter of each strand suggests that it is a seven-strand cable of seven-strand cables.

the X-ray photograph of the synthetic polypeptide and the pattern of horsehair on page 4. This lends some support to the idea that alpha helixes are present in hair as well as in synthetic polypeptides, but the similarity is not close enough to be reliable. The convincing proof was provided by Max Perutz of the Cavendish Laboratory at Cambridge University. He pointed out that a strong X-ray reflection corresponding to a spacing of 1.5 Angstrom units should be observed for substances containing the alpha helix; the amino-acid residues, which are separated from one another by a distance of 1.5 Angstrom units along the axis of the helix, would cooperate with one another in scattering X-rays in the direction corresponding to this reflection. Perutz made X-ray photographs of synthetic polypeptides and many proteins, with the fibers oriented at the correct angle for this reflection to appear on the photographic plate, and he found that the reflection was in fact produced by the synthetic polypeptides and by hair, horn, fingernail, epidermis and other proteins which give X-ray photographs of the alpha-keratin type.

One striking difference between the X-ray photograph of horsehair and the expected pattern for an arrangement of alpha helixes in hexagonal packing is the presence of a strong reflection above and below the central image. In the corresponding photograph of the synthetic polypeptide there are strong reflections on either side of these points. About a year ago it was suggested independently by the workers at the California Institute of Technology and by F. H. C. Crick of Cambridge University that the strong vertical reflections result from the presence of molecules with the configuration of the alpha helix which are twisted about one another, to form what Crick calls coiled coils. This twisting of the alpha helix into a compound helix, shown at upper right, might result from a small shortening and lengthening of hydrogen bonds, perhaps within the range 2.7 to 2.9 Angstrom units, in a regular way. A good correlation of all of the evidence is provided by the structure for alpha keratin shown in the illustration at the left on the opposite page. In this structure there are seven-strand cables, each about 30 Angstrom units in diameter, which are made up of six compound alpha helixes twisted about a central alpha helix. These seven-strand cables are arranged in hexagonal packing, and the interstices between them are occupied by additional alpha helixes.

It is of course evident that the X-ray

photograph of horsehair does not provide enough information to settle the question of the structure of alpha keratin in complete detail. There seems to be little doubt, however, that hair and similar proteins are made up of polypeptide chains with the configuration of the alpha helix and that these chains are twisted about one another. It is not unlikely that the way in which the chains are twisted together is the one shown at the right, but there is a possibility that it may be somewhat different.

The simplicity of the helix as a structural element—the fact that it results automatically through the repetition of the most general symmetry operation which does not convert an asymmetric object into its mirror image—tempts us to look for helixes in larger structures. An example is given in the illustration at the right on the opposite page. This is an electron micrograph of a bacterial flagellum, from a bacterium of the diphtheroid class, made by Robley C. Williams and Mortimer P. Starr of the University of California in Berkeley. X-ray photographs of such flagella are of the alpha-keratin type, indicating the presence of alpha helixes. This flagellum is about 180 Angstrom units in diameter. It may be seen in the electron microscope that it consists of three strands, each 90 Angstrom units in diameter, twisted about one another. Although the way in which these strands are built out of alpha helixes is not known, one speculation immediately suggests itself. The seven-strand cables mentioned earlier are 30 Angstrom units in diameter; if a seven-strand cable were similarly made of seven of these seven-strand cables it would be 90 Angstrom units in diameter. Hence it may be found, when these flagella are subjected to more thorough study, that they can be described as three-strand ropes, each strand of which is a seven-strand cable built of seven-strand cables of alpha helixes. It is interesting that the bacterial flagella themselves form a still larger helix.

Through the development during the last quarter-century of the techniques of electron microscopy and X-ray diffraction, the time has now arrived when it is possible to track the structure of living organisms down through successively smaller orders of size, without a gap, from the whole animal through the cell to the atom. We may hope that the knowledge that will be obtained in this way during the coming decades will provide a far more precise and penetrating understanding of life than we now have.

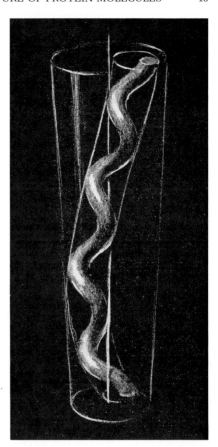

COMPOUND HELIX, or coiled coil, is the probable configuration of the polypeptide chain in hair, horn and related proteins.

SEVEN-STRAND CABLE made up of seven-strand cables, proposed for strand of a bacterial flagellum, might have this structure.

Proteins

by Paul Doty
September 1957

*The principal substance of living cells, these giant
molecules have identical backbones. Each is
adapted to its specific task by a unique combination
of side groups, size, folding and shape*

Thousands of different proteins go into the make-up of a living cell. They perform thousands of different acts in the exact sequence that causes the cell to live. How the proteins manage this exquisitely subtle and enormously involved process will defy our understanding for a long time to come. But in recent years we have begun to make a closer acquaintance with proteins themselves. We know they are giant molecules of great size, complexity and diversity. Each appears to be designed with high specificity for its particular task. We are encouraged by all that we are learning to seek the explanation of the function of proteins in a clearer picture of their structure. For much of this new understanding we are indebted to our experience with the considerably simpler giant molecules synthesized by man. High-polymer chemistry is now coming forward with answers to some of the pressing questions of biology.

Proteins, like synthetic high polymers, are chains of repeating units. The units are peptide groups, made up of the monomers called amino acids [*see diagram below*]. There are more than 20 different amino acids. Each has a distinguishing cluster of atoms as a side group [*see next two pages*], but all amino acids have a certain identical group. The link-

ing of these groups forms the repeating peptide units in a "polypeptide" chain. Proteins are polypeptides of elaborate and very specific construction. Each kind of protein has a unique number and sequence of side groups which give it a particular size and chemical identity. Proteins seem to have a further distinction that sets them apart from other high polymers. The long chain of each protein is apparently folded in a unique configuration which it seems to maintain so long as it evidences biological activity.

We do not yet have a complete picture of the structure of any single protein. The entire sequence of amino acids has been worked out for insulin [*see "The Insulin Molecule," by E. O. P. Thompson; SCIENTIFIC AMERICAN Offprint 42*], the determination of several more is nearing completion. But to locate each group and each atom in the configuration set up by the folded chain is intrinsically a more difficult task; it has resisted the Herculean labors of a generation of X-ray crystallographers and their collaborators. In the early 1930s W. T. Astbury of the University of Leeds succeeded in demonstrating that two X-ray diffraction patterns, which he called alpha and beta, were consistently associated with certain fibers, and he identified a third with collagen, the pro-

tein of skin, tendons and other structural tissues of the body. The beta pattern, found in the fibroin of silk, was soon shown to arise from bundles of nearly straight polypeptide chains held tightly to one another by hydrogen bonds. Nylon and some other synthetic fibers give a similar diffraction pattern. The alpha pattern resisted decoding until 1951, when Linus Pauling and R. B. Corey of the California Institute of Technology advanced the notion, since confirmed by further X-ray diffraction studies, that it is created by the twisting of the chain into a helix. Because it is set up so naturally by the hydrogen bonds available in the backbone of a polypeptide chain [*see top diagram on page 20*], the alpha helix was deduced to be a major structural element in the configuration of most proteins. More recently, in 1954, the Indian X-ray crystallographer G. N. Ramachandran showed that the collagen pattern comes from three polypeptide helixes twisted around one another. The resolution of these master plans was theoretically and esthetically gratifying, especially since the nucleic acids, the substance of genetic chemistry, were concurrently shown to have the structure of a double helix. For all their apparent general validity, however, the master plans did not give us the complete configuration in three dimensions

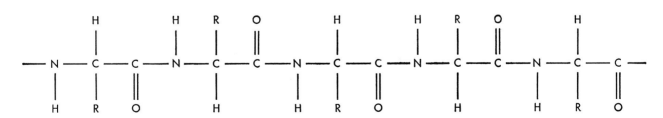

POLYPEPTIDE CHAIN is a repeating structure made up of identical peptide groups (CCONHC). The chain is formed by amino acids, each of which contributes an identical group to the backbone plus a distinguishing radical (R) as a side group.

GLYCINE   ALANINE   VALINE   ISOLEUCINE   LEUCINE

LYSINE   ARGININE   HISTIDINE   PROLINE   HYDROXYPROLINE

AMINO ACIDS, the 20 commonest of which are shown in this chart, have identical atomic groups (*in colored bands*) which react to form polypeptide chains. They are distinguished by their unique side groups. In forming a chain, the amino group (NH_2) of one

of any single protein.

The X-ray diffraction work left a number of other questions up in the air. Since the alpha helix had been observed only in a few fibers, there was no solid experimental evidence for its existence elsewhere. There was even a suspicion that it could occur only in fibers, where it provides an economical way to pack polypeptides together in crystalline structures. Many proteins, especially chemically active ones such as the enzymes and antibodies, are globular, not linear like those involved in fibers and structural tissues. In the watery solutions which are the natural habitat of most proteins, it could be argued, the affinity of water molecules for hydrogen bonds would disrupt the alpha helix and reduce the chain to a random coil. These doubts and suppositions have prompted investigations by our group at Harvard University in collaboration with E. R. Blout of the Children's Cancer Research Foundation in Boston.

In these investigations we have employed synthetic polypeptides as laboratory models for the more complex and sensitive proteins. When Blout and coworkers had learned to polymerize them to sufficient length—100 to 1,000 amino acid units—we proceeded to observe their behavior in solution.

Almost at once we made the gratifying discovery that our synthetic polypeptides could keep their helical coils wound up in solutions. Moreover, we found that we could unwind the helix of some polypeptides by adjusting the acidity of our solutions. Finally, to complete the picture, we discovered that we could reverse the process and make the polypeptides wind up again from random coils into helixes.

The transition from the helix to the random coil occurs within a narrow range as the acidity is reduced; the hydrogen bonds, being equivalent, tend to let go all at once. It is not unlike the melting of an ice crystal, which takes place in a narrow temperature range. The reason is the same, for the ice crystal is held together by hydrogen bonds. To complete the analogy, the transition from the helix to the random coil can also be induced by heat. This is a true melting process, for the helix is a one-dimensional crystal which freezes the otherwise flexible chain into a rodlet.

From these experiments we conclude that polypeptides in solution have two natural configurations and make a reversible transition from one to the other, depending upon conditions. Polypeptides in the solid state appear to prefer the alpha helix, though this is subject to the presence of solvents, especially water. When the helix breaks down here, the transition is to the beta configuration, the hydrogen bonds now linking adjacent chains. Recently Blout and Henri Lenormant have found that fibers of polylysine can be made to undergo the alpha-beta transition reversibly by mere alteration of humidity. It is tempting to speculate that a reversible alpha-beta transition may underlie the process of muscle contraction and other types of

SERINE

H—N—C—C—O—H
(H, H, O backbone)
H—C—H
O—H

THREONINE

H—N—C—C—O—H
H—C—O—H
H—C—H
H

ASPARTIC ACID

H—N—C—C—O—H
H—C—H
C=O
O—H

GLUTAMIC ACID

H—N—C—C—O—H
H—C—H
H—C—H
C=O
O—H

TYROSINE

H—N—C—C—O—H
H—C—H
(benzene ring)
O—H

CYSTEINE

H—N—C—C—O—H
H—C—H
S—H

METHIONINE

H—N—C—C—O—H
H—C—H
H—C—H
S
H—C—H
H

CYSTINE

H—N—C—C—O—H
H—C—H
S
S
H—C—H
H—N—C—C—O—H
H, H, O

TRYPTOPHAN

H—N—C—C—O—H
H—C—H
C=C—H
N—H
(benzene ring)

PHENYLALANINE

H—N—C—C—O—H
H—C—H
(benzene ring)

molecule reacts with the hydroxyl group (OH) of another. This reaction splits one of the amino hydrogens off with the hydroxyl group to form a molecule of water. The nitrogen of the first group then forms the peptide bond with the carbon of the second.

movement in living things.

Having learned to handle the polypeptides in solution we turned our attention to proteins. Two questions had to be answered first: Could we find the alpha helix in proteins in solution, and could we induce it to make the reversible transition to the random coil and back again? If the answer was yes in each case, then we could go on to a third and more interesting question: Could we show experimentally that biological activity depends upon configuration? On this question, our biologically neutral synthetic polypeptides could give no hint.

For the detection of the alpha helix in proteins the techniques which had worked so well on polypeptides were impotent. The polypeptides were either all helix or all random coil and the rodlets of the first could easily be distinguished from the globular forms of the second by use of the light-scattering technique. But we did not expect to find that any of the proteins we were going to investigate were 100 per cent helical in configuration. The helix is invariably disrupted by the presence of one of two types of amino acid units. Proline lacks the hydrogen atom that forms the crucial hydrogen bond; the side groups form a distorting linkage to the chain instead. Cystine is really a double unit, and forms more or less distorting cross-links between chains. These units play an important part in the intricate coiling and folding of the polypeptide chains in globular proteins. But even in globular proteins, we thought, some lengths of the chains might prove to be helical. There was nothing, however, in the over-all shape of a globular protein to tell us whether it had more or less helix in its structure or none at all. We had to find a way to look inside the protein.

One possible way to do this was suggested by the fact that intact, biologically active proteins and denatured proteins give different readings when observed for an effect called optical rotation. In general, the molecules that exhibit this effect are asymmetrical in atomic structure. The side groups give rise to such asymmetry in amino acids and polypeptide chains; they may be attached in either a "left-handed" or a "right-handed" manner. Optical rotation provides a way to distinguish one from the other. When a solution of amino acids is interposed in a beam of polarized light, it will rotate the plane of polarization either to the right or to the left [see diagrams at top of page 22]. Though amino acids may exist in both forms, only left-handed units, thanks to some accident in the chemical phase of evolution, are found in proteins. We used only the left-handed forms, of course, in the synthesis of our polypeptide chains.

Now what about the change in optical rotation that occurs when a protein is denatured? We knew that native protein rotates the plane of the light 30 to 60 degrees to the left, denatured protein 100 degrees or more to the left. If there was some helical structure in the protein, we surmised, this shift in rotation might be induced by the disappearance of the helical structure in the denaturation process. There was reason to believe that the helix, which has to be either left-handed or right-handed, would have optical activity. Further, although it appeared possible for the helix to be wound either way, there were grounds for assuming that nature had chosen to make all of its helixes one way or the other. If it had not, the left-handed and right-handed helixes would mutually cancel out their respective optical rotations. The change in the optical rotation of proteins with denaturation would then have some other explanation entirely, and we would have to invent another way to look for helixes.

To test our surmise we measured the optical rotation of the synthetic poly-peptides. In the random coil state the polypeptides made an excellent fit with the denatured proteins, rotating the light 100 degrees to the left. The rotations in both cases clearly arose from the same cause: the asymmetry of the amino acid units. In the alpha helix configuration the polypeptides showed almost no rotation or none at all. It was evident that the presence of the alpha helix caused a

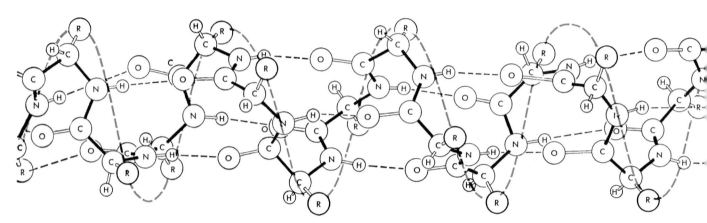

ALPHA HELIX gives a polypeptide chain a linear structure shown here in three-dimensional perspective. The atoms in the repeating unit (CCONHC) lie in a plane; the change in angle between one unit and the next occurs at the carbon to which the side group

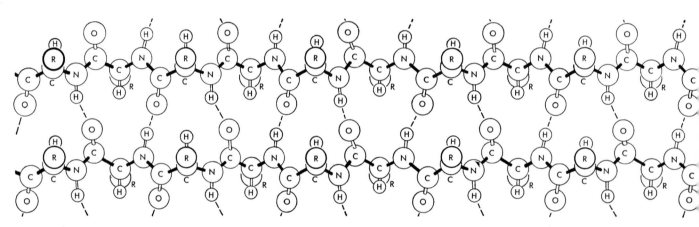

BETA CONFIGURATION ties two or more polypeptide chains to one another in crystalline structures. Here the hydrogen bonds do not contribute to the internal organization of the chain, as in the alpha helix, but link the hydrogen atoms of one chain to the oxygen

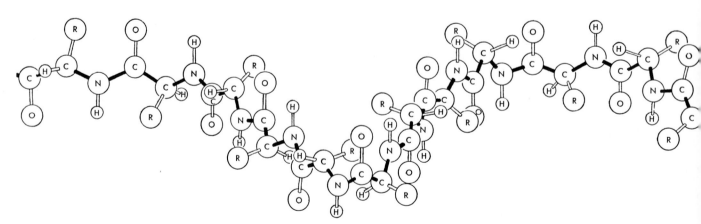

RANDOM CHAIN is the configuration assumed by the polypeptide molecule in solution, when hydrogen bonds are not formed. The flat configuration of the repeating unit remains, but the chain rotates about the carbon atoms to which the side groups are at-

counter-rotation to the right which nearly canceled out the leftward rotation of the amino acid units. The native proteins also had shown evidence of such counter-rotation to the right. The alpha configuration did not completely cancel the leftward rotation of the amino acid units, but this was consistent with the expectation that the protein structures would be helical only in part. The experiment

thus strongly indicated the presence of the alpha helix in the structure of globular proteins in solution. It also, incidentally, seemed to settle the question of nature's choice of symmetry in the alpha helix: it must be right-handed.

When so much hangs on the findings of one set of experiments, it is well to double check them by observa-

tions of another kind. We are indebted to William Moffitt, a theoretical chemist at Harvard, for conceiving of the experiment that provided the necessary confirmation. It is based upon another aspect of the optical rotation effect. For a given substance, rotation varies with the wavelength of the light; the rotations of most substances vary in the same way. Moffitt predicted that the presence of

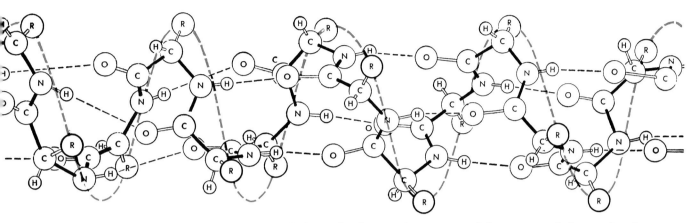

(R) is attached. The helix is held rigid by the hydrogen bond (*broken black lines*) between the hydrogen attached to the nitro-

gen in one group and the oxygen attached to a carbon three groups along the chain. The colored line traces the turns of the helix.

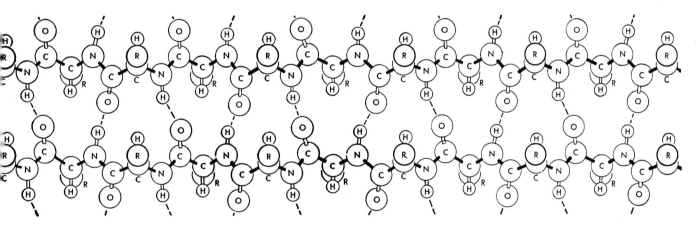

atoms in the adjoining chain. The beta configuration is found in silk and a few other fibers. It is also thought that polypeptide

chains in muscle and other contractile fibers may make reversible transitions from alpha helix to beta configuration when in action.

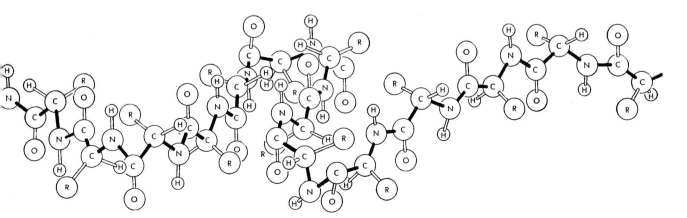

tached. The random chain may be formed from an alpha helix when hydrogen bonds are disrupted in solution. A polypeptide

chain may make a reversible transition from alpha helix to random chain, depending upon the acid-base balance of the solution.

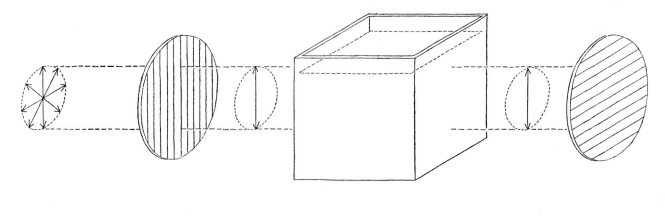

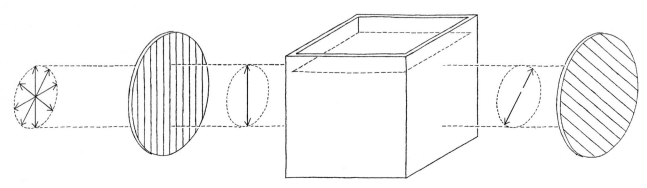

OPTICAL ROTATION is induced in a beam of polarized light by molecules having certain types of structural asymmetry. At top a beam of light is polarized in the vertical plane and transmitted unchanged through a neutral solution. At bottom asymmetrical molecules in the solution cause the beam to rotate from the vertical plane. The degree of rotation may be determined by turning the second polarizing filter (*right*) to the point at which it cuts off the beam. The alpha helix in a molecule causes such rotation.

the alpha helix in a substance would cause its rotation to vary in a different way. His prediction was sustained by observation: randomly coiled polypeptides showed a normal variation while the helical showed abnormal. Denatured and native proteins showed the same contrast. With the two sets of experiments in such good agreement, we could conclude with confidence that the alpha helix has a significant place in the structure of globular proteins. Those amino acid units that are not involved in helical configurations are weakly bonded to each other or to water molecules, probably in a unique but not regular or periodic fashion. Like synthetic high-polymers, proteins are partly crystalline and partly amorphous in structure.

The optical rotation experiments also provided a scale for estimating the helical content of protein. The measurements indicate that, in neutral solutions, the helical structure applies to 15 per cent of the amino acid units in ribonuclease, 50 per cent of the units in serum albumin and 85 per cent in tropomyosin. With the addition of denaturing agents to the solution, the helical content in each case can be reduced to zero. In

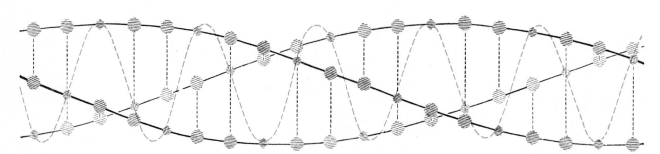

COLLAGEN MOLECULE is a triple helix. The colored broken line indicates hydrogen bonds between glycine units. The black broken lines indicate hydrogen bonds which link hydroxyproline units and give greater stability to collagens in which they are found.

some proteins the transition is abrupt, as it is in the synthetic polypeptides. On the other hand, by the use of certain solvents we have been able to increase the helical content of some proteins—in the case of ribonuclease from 15 to 70 per cent. As in the polypeptides, the transition from helix to random coil is reversible. The percentage of helical structure in proteins is thus clearly a variable. In their natural environment, it appears, the percentage at any given time represents the equilibrium between the inherent stability of the helix and the tendency of water to break it down.

In a number of enzymes we have been able to show that biological activity falls off and increases with helical content. Denaturation is now clearly identified with breakdown of configuration, certainly insofar as it involves the integrity of the alpha helix. This is not surprising. It is known that catalysts in general must have rigid geometrical configurations. The catalytic activity of an enzyme may well require that its structure meet similar specifications. If this is so, the rigidity that the alpha helix imposes on the otherwise flexible polypeptide chain must play a decisive part in establishing the biological activity of an enzyme. It seems also that adjustability of the stiffness of structure in larger or smaller regions of the polypeptide chain may modify the activity of proteins in response to their environment. Among other things, it could account for the versatility of the gamma globulins; without any apparent change in their amino acid make-up, they are able somehow to adapt themselves as antibodies to a succession of different infectious agents.

The next step toward a complete anatomy of the protein molecule is to determine which amino acid units are in the helical and which in the nonhelical regions. Beyond that we shall want to know which units are near one another as the result of folding and cross-linking, and a myriad of other details which will supply the hues and colorings appropriate to a portrait of an entity as intricate as protein. Many such details will undoubtedly be supplied by experiments that relate change in structure to change in function, like those described here.

In the course of our experiments with proteins in solution we have also looked into the triple-strand structure of collagen. That structure had not yet been resolved when we began our work, so we did not know how well it was designed for the function it serves in structural tissues. Collagen makes up one third of the proteins in the body and 5 per cent of its total weight; it occurs as tiny fibers or fibrils with bonds that repeat at intervals of about 700 Angstroms. It had been known for a long time that these fibrils could be dissolved in mild solvents such as acetic acid and then reconstituted, by simple precipitation, into their original form with their bandings restored. This remarkable capacity naturally suggested that the behavior of collagen in solution was a subject worth exploring.

Starting from the groundwork of other investigators, Helga Boedtker and I were able to demonstrate that the collagen molecule is an extremely long and thin rodlet, the most asymmetric molecule yet isolated. A lead pencil of comparable proportions would be a yard long. When a solution of collagen is just slightly warmed, these rodlets are irreversibly broken down. The solution will gel, but the product is gelatin, as is well known to French chefs and commercial producers of gelatin. The reason the dissolution cannot be reversed was made clear when we found that the molecules in the warmed-up solution had a weight about one third that of collagen. It appeared that the big molecule of collagen had broken down into three polypeptide chains.

At about the same time Ramachandran proposed the three-strand helix as the collagen structure. Not long afterward F. H. C. Crick and Alexander Rich at the University of Cambridge and Pauline M. Cowan and her collaborators at King's College, London, worked out the structure in detail. It consists of three polypeptide chains, each incorporating three different amino acid units—proline, hydroxyproline and glycine. The key to the design is the occurrence of glycine, the smallest amino acid unit, at every third position on each chain. This makes it possible for the bulky proline or hydroxyproline groups to fit into the links of the triple strand, two of these nesting in each link with the smaller glycine unit [see diagram on page 22].

One question, however, was left open in the original model. Hydroxyproline has surplus hydrogen bonds, which, the model showed, might be employed to reinforce the molecule itself or to tie it more firmly to neighboring molecules in a fibril. Independent evidence seemed to favor the second possibility. Collagen in the skin is irreversibly broken down in a first degree burn, for example, at a temperature of about 145 degrees Fahrenheit. This is about 60 degrees higher than the dissolution temperature of the collagen molecule in solution. The obvious inference was that hydroxyproline lends its additional bonding power to the tissue structure. Moreover, tissues with a high hydroxyproline content withstand higher temperatures than those with lower; the skin of codfish, with a low hydroxyproline content, shrivels up at about 100 degrees. Tomio Nishihara in

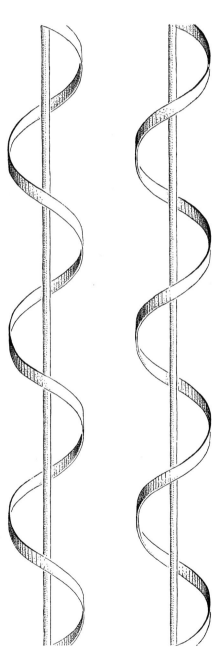

ASYMMETRY of a helix is either left-handed (left) or right-handed. Helix in proteins appears to be exclusively right-handed.

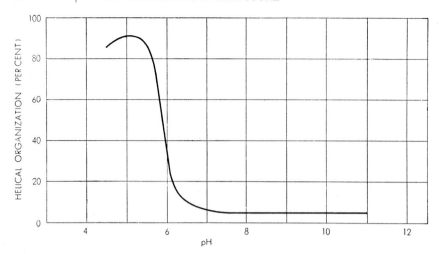

ALPHA HELIX BREAKDOWN is induced in solutions of some polypeptides when the pH (acidity or alkalinity) reaches a critical value at which hydrogen bonds are disrupted.

our laboratory has compared the breakdown temperatures of collagen molecules and tissues from various species and found that the tissue temperature is uniformly about 60 degrees higher. Thus we must conclude that the extra stability conferred by hydroxyproline goes directly to the molecule and not to the fibril.

The structure of collagen demonstrates three levels in the adaptation of polypeptide chains to fit the requirements of function. First there are the chains as found in gelatin, with their three amino acids lined up in just the right sequence. These randomly coiled and quite soluble molecules are transformed into relatively insoluble, girder-like building units when united into sets of three by hydrogen bonds. The subtly fashioned collagen molecules are still too fragile to withstand body temperatures. When arranged side by side, however, they form a crystalline structure which resists comparatively high temperatures and has fiber-like qualities with the vast range of strengths and textures required in the different types of tissues that are made of collagen.

The story of collagen, like that of other proteins, is still far from complete. But it now seems that it will rank among the first proteins whose molecular structure has been clearly discerned and related in detail to the functions it serves.

The Three-dimensional Structure of a Protein Molecule

by John C. Kendrew
December 1961

The way in which the chain of amino acid units in a protein molecule is coiled and folded in space has been worked out for the first time. The protein is myoglobin, the molecule of which contains 2,600 atoms

When the early explorers of America made their first landfall, they had the unforgettable experience of glimpsing a New World that no European had seen before them. Moments such as this—first visions of new worlds—are one of the main attractions of exploration. From time to time scientists are privileged to share excitements of the same kind. Such a moment arrived for my colleagues and me one Sunday morning in 1957, when we looked at something no one before us had seen: a three-dimensional picture of a protein molecule in all its complexity. This first picture was a crude one, and two years later we had an almost equally exciting experience, extending over many days that were spent feeding data to a fast computing machine, of building up by degrees a far sharper picture of this same molecule. The protein was myoglobin, and our new picture was sharp enough to enable us to deduce the actual arrangement in space of nearly all of its 2,600 atoms.

We had chosen myoglobin for our first attempt because, complex though it is, it is one of the smallest and presumably the simplest of protein molecules, some of which are 10 or even 100 times larger. The purpose of this article is to indicate some of the reasons why we thought it important to elucidate the three-dimensional architecture of a protein, to explain something of the methods we used and to describe our results.

In a real sense proteins are the "works" of living cells. Almost all chemical reactions that take place in cells are catalyzed by enzymes, and all known enzymes are proteins; an individual cell contains perhaps 1,000 different kinds of enzyme, each catalyzing a different and specific reaction. Proteins have many other important functions, being constituents of bone, muscle and tendon, of blood, of hair and skin and membranes. In addition to all this it is now evident that the hereditary information, transmitted from generation to generation in the nucleic acid of the chromosomes, finds its expression in the characteristic types of protein molecule synthesized by each cell. Clearly to understand the behavior of a living cell it is necessary first to find out how so wide a variety of functions can be assumed by molecules all made up for the most part of the same few basic units.

These units are amino acids, about 20 in number, joined together to form the long molecular chains known as polypeptides. Each link in a chain consists of the group —CO—CHR—NH—, where C, O, N and H represent atoms of carbon, oxygen, nitrogen and hydrogen respectively, and R represents any of the various groups of atoms in a side chain that differs for each of the 20 amino acids. All protein molecules contain polypeptide chains, and some of them contain no other constituents; in others there is an additional group of a different kind. For example, the hemoglobin in red blood corpuscles contains four polypeptide chains and four so-called heme groups: flat assemblages of atoms with an iron atom at the center. The function of the heme group is to combine reversibly with a molecule of oxygen, which is then carried by the blood from the lungs to the tissues. Myoglobin is, as it were, a junior relative of hemoglobin, being a quarter its size and consisting of a single polypeptide chain of about 150 amino acid units together with a single heme group. Myoglobin is contained within the cells of the tissues, and it acts as a temporary storehouse for the oxygen brought by the hemoglobin in the blood.

Following the classic researches on the insulin molecule by Frederick Sanger at the University of Cambridge, several groups of investigators have been able to discover the order in which the amino acids are arranged in the polypeptide chains of a number of proteins [see "The Chemical Structure of Proteins," by William H. Stein and Stanford Moore; SCIENTIFIC AMERICAN Offprint 80]. This laborious task does not, however, provide the whole story. A polypeptide chain of perhaps hundreds of links could be arranged in space in an almost infinite number of ways. Chemical methods give only the order of the links; equally important is their arrangement in space, the way in which particular side chains form crosslinks to bind the whole structure together into a nearly spherical object (as most proteins are known to be). Also of equal importance is the way in which certain key amino acid units, perhaps lying far apart in the sequence, are brought together by the three-dimensional folding to form a particular constellation of precise configuration—the so-called active site of the molecule—that enables the protein to perform its special functions. How is it possible to discover the three-dimensional arrangement of a molecule as complicated as a protein?

The key to the problem is that many proteins can be persuaded to crystallize, and often their crystals are as regular and as nearly perfect in shape as the crystals of simpler compounds. The fact that proteins crystallize is interesting in itself, for crystallization implies a regular three-dimensional array of identical molecules. If all the molecules did not have the same detailed shape, they could not form the repeating arrays that are necessary if the aggregate is to possess the regular external shape of a crystal. Therefore it appears that all protein molecules of a given type are identical—that is, they are not simply "colloidal" aggregates of indefinite shape. The existence of protein crystals means, in fact, that proteins do have a definite three-dimensional structure to solve. And the most powerful techniques for studying the structures of crystals are those of X-ray crystallography.

The X-Ray Approach

In 1912 Walter Friedrich, C. M. Paul Knipping and Max von Laue discovered that if a crystal is turned in various directions while a beam of X rays is sent through it, some of the X rays do not travel in a straight line. When the transmitted rays fall on a photographic plate, they produce not only a dark central spot but also a pattern of fainter spots around it. The reason for this diffraction pattern is that X rays are scattered, or reflected, by the electrons that form the outer part of each atom in the crystal.

The atoms are arranged in an orderly array, something like the trees in a regularly planted orchard. As one drives past an orchard in an automobile and looks into it along different directions, one sees one set after another of lines of trees coming into view end on. Similarly, if one could look at the atoms of a crystal, one would see different planes of atoms in different directions. The X-ray beam is reflected by these sets of planes much as light is reflected from the surface of a mirror; that is to say, the angle of reflection is equal to the angle of incidence. But it can be shown that because the reflection is a set of parallel planes rather than a single surface, as in a mirror, the reflected beam will "flash up" only at a particular angle of incidence between incident beam and planes, this angle becoming greater the closer together the planes of the set are. Thus each spot in the X-ray diffraction pattern corresponds to a particular set of planes; and the spots farthest out in the pattern (those made by X rays diffracted through the biggest angles) correspond to the most closely spaced sets of planes. In an X-ray camera the crystal is rotated in a predetermined manner so that one after another of the sets of planes comes into the correct reflecting position. As each set does so, the corresponding reflected beam flashes up and makes its imprint on the photographic plate.

Each type of crystal has its own characteristic arrangement of atoms and so will produce its own specific X-ray pattern, the features of which can be unambiguously, if tediously, predicted by calculation if the structure of the crystal is known. X-ray analysis involves the reverse calculation: Given the X-ray pattern, what is the crystal structure that must have produced it?

In analyzing complex crystals the calculation is carried out by applying a method known as Fourier synthesis to the repeating, three-dimensional configuration. To understand what this involves, consider first a one-dimensional analogy: a musical note. Physically, a steady musical note is a repeating sequence of rarefactions and condensations in the air between the listener and the instrument producing the note. If the den-

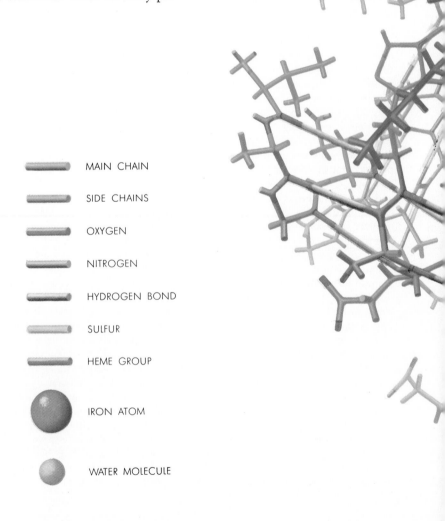

THREE-DIMENSIONAL MODEL of the myoglobin molecule is depicted in this painting by Irving Geis. The key to the model is at the left side of the painting. The molecule consists of some 150 amino acid units strung together in a single chain with a heme group attached to it. At the center of the heme group is a single atom of iron. Most of the amino acid units are arranged in helical sections such as the one running diagonally across the bottom of the painting. Each amino acid unit in the model is identified in the illustration on the following two pages. The model is the result of work by the author, R. E. Dickerson, B. E. Strandberg, R. G. Hart, D. R. Davies, D. C. Phillips, V. C. Shore and H. C. Watson.

MAIN CHAIN

SIDE CHAINS

OXYGEN

NITROGEN

HYDROGEN BOND

SULFUR

HEME GROUP

IRON ATOM

WATER MOLECULE

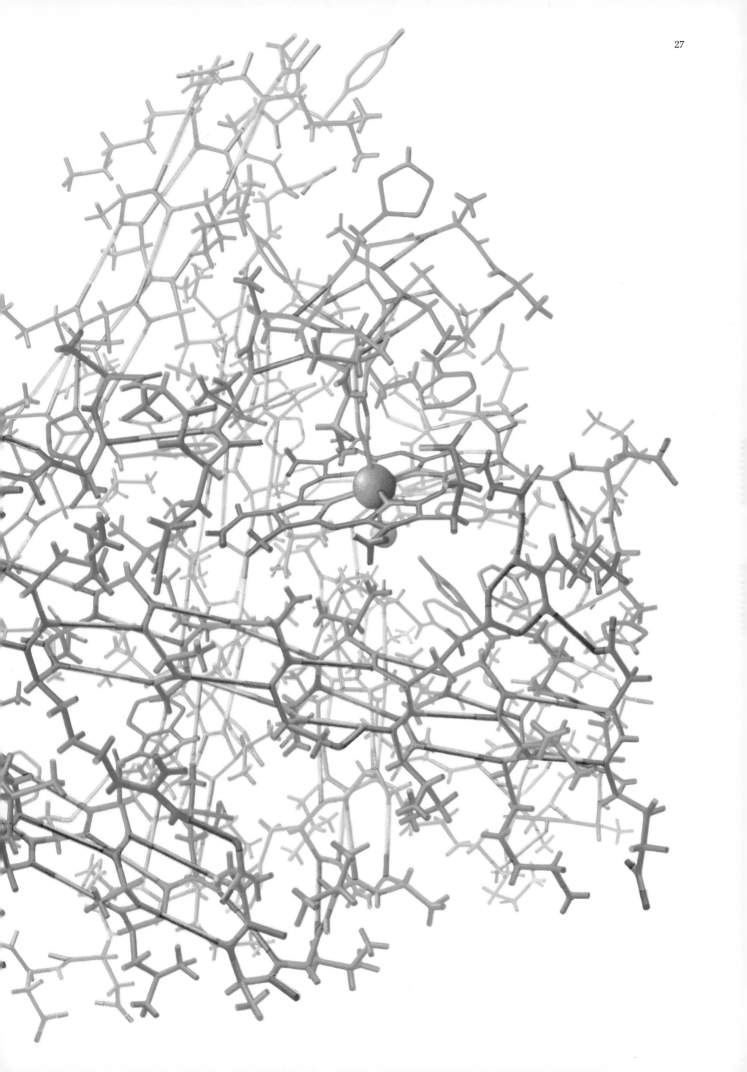

sity of the air is plotted along the path, the graph is a complex but perfectly repetitive wave form. More than 150 years ago the French physicist Jean Baptiste Fourier discovered that any such wave form can be decomposed, or analyzed, into a set of harmonics that are pure sine waves of shorter and shorter wavelength and thus of higher and higher frequency [see "The Reproduction of Sound," by Edward E. David, SCIENTIFIC AMERICAN, August, 1961]. The reverse process — Fourier synthesis — consists in combining a series of pure sine waves of the proper relative amplitude in the proper relative phases (that is, in or out of step with one another to the correct extent) so as to reproduce the original wave form, or note. In practice it is not necessary to use all the components to obtain a reasonably faithful reproduction. The greater the number of higher harmonics that are included, however, the more nearly perfect is the rendering of the note.

To an X-ray beam a crystal is an extended electron cloud, the density of

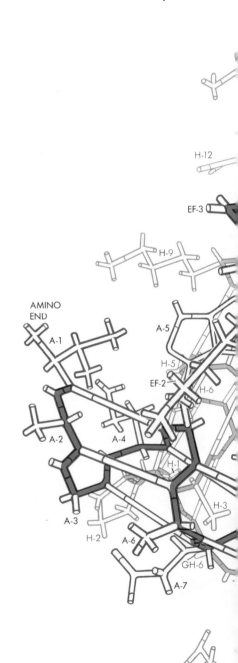

ALANINE	ALA	
ARGININE	ARG	
ASPARTIC ACID OR ASPARAGINE	ASP	
GLUTAMIC ACID OR GLUTAMINE	GLU	
GLUTAMIC ACID	GLU. C	
GLYCINE	GLY	
HISTIDINE	HIS	
ISOLEUCINE	ILEU	
LEUCINE	LEU	
LYSINE	LYS	
METHIONINE	MET	
PHENYLALANINE	PHE	
PROLINE	PRO	
SERINE	SER	
THREONINE	THR	
TYROSINE	TYR	
VALINE	VAL	

A- 1 VAL (AMINO END)
2 ALA
3 GLY
4 GLU
5 TYR
6 SER
7 GLU
8 ILEU
9 LEU
10 LYS
11 (NOT GLY)
12 TYR
13 (NOT GLY)
14 LEU
15 LEU
16 GLU
AB- 1 (NOT GLY)
B- 1 LEU
2 VAL OR THR
3 ALA
4 GLY
5 HIS
6 GLY
7 LYS
8 LEU
9 THR
10 LEU
11 ILEU
12 SER
13 LEU
14 PHE
15 LYS
16 SER

C- 1 HIS
2 PRO
3 GLU.C
4 THR
5 LEU
6 GLU
7 LYS
CD- 1 PHE
2 ASP
3 ARG
4 PHE
5 LYS
6 HIS
7 LEU
8 LYS
D- 1 THR
2 GLU.C
3 ALA
4 GLU.C
5 MET
6 LYS
7 ALA
E- 1 SER
2 GLU.C
3 ASP
4 LEU
5 LYS
6 VAL
7 HIS
8 GLY
9 ILEU
10 GLU
11 VAL
12 ASP
13 (NOT ALA, GLY)
14 ALA
15 LEU
16 GLY
17 ALA
18 ILEU
19 ASP
20 ARG
EF- 1 LYS
2 LYS
3 GLY
4 LEU
5 HIS
6 (NOT GLY)
7 (NOT GLY)
8 GLU
F- 1 GLU
2 ALA
3 PRO
4 THR
5 ALA
6 HIS
7 SER
8 HIS
9 ALA

FG- 1 (NOT GLY)
2 (NOT GLY)
3 PHE
4 (NOT ALA)
5 ILEU
G- 1 PRO
2 ILEU
3 LYS
4 TYR
5 (NOT ALA, GLY)
6 GLU
7 HIS
8 LEU
9 SER
10 (NOT GLY, ALA)
11 ALA
12 VAL OR THR
13 ILEU
14 HIS
15 VAL
16 ARG
17 ALA
18 THR
19 LYS
GH- 1 HIS
2 ASP
3 ASP
4 GLU
5 PHE
6 GLY
H- 1 ALA
2 PRO
3 ALA
4 ASP
5 GLY
6 ALA
7 MET
8 GLY
9 LYS
10 ALA
11 LEU
12 GLU.C
13 LEU
14 PHE
15 ARG
16 LYS
17 ASP.C
18 ILEU
19 ALA
20 ALA
21 LYS
22 TYR
23 LYS
24 GLU.C
HC- 1 LEU
2 GLY
3 TYR
4 GLY
5 GLU.C (CARBOXYL END)

SEQUENCE OF AMINO ACID UNITS in the model of myoglobin is indicated by the letters and numbers in the illustration on these two pages. The amino acid unit represented by each symbol is given in the table above; the key to the abbreviations is at top left in the table. The brackets in the table indicate those amino acid units which form a helical section. The direction of the main chain is traced in color in the illustration; the chain begins at far left (*amino end*) and ends near the top (*carboxyl end*). Here the heme group is indicated in gray. Not all the amino acid units in the model have been positively identified. In some cases it has only been determined that they cannot be certain units. The over-all configuration of the molecule, however, is known with a considerable degree of confidence.

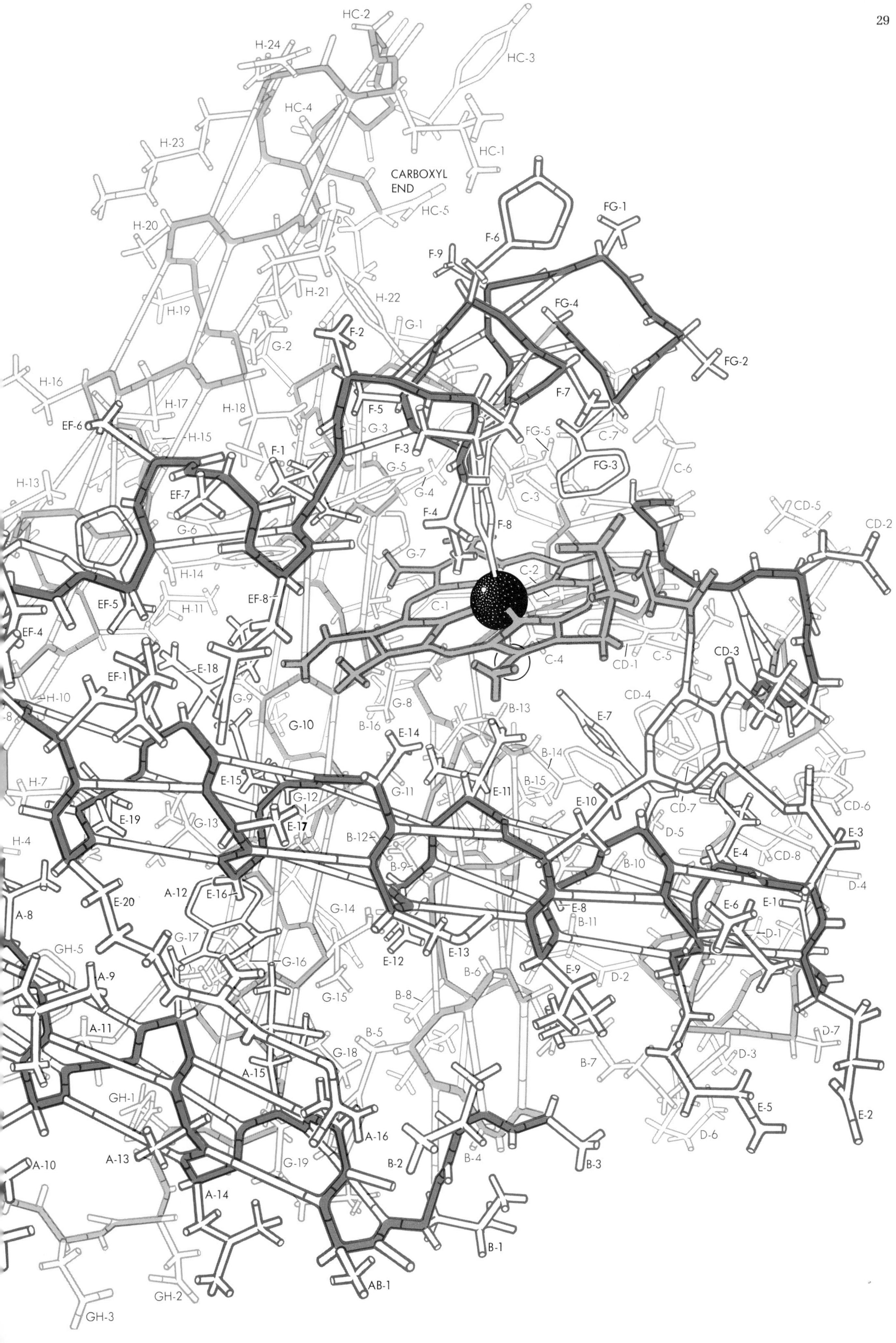
HC-2
H-24
HC-3
HC-4
H-23
HC-1
CARBOXYL
END
HC-5
FG-1
H-20
F-6
F-9
H-21
H-22
H-19
FG-4
G-1
F-2
G-2
FG-2
H-16
F-7
H-17
H-18
F-5
EF-6
G-3
H-15
FG-5
C-7
F-1
F-3
G-5
FG-3
C-6
H-13
EF-7
G-4
C-3
CD-5
F-4
CD-2
G-6
F-8
G-7
H-14
C-2
EF-5
EF-8
C-1
H-11
EF-4
CD-3
C-4
CD-1
C-5
E-18
EF-1
H-10
G-9
CD-4
B-13
E-7
G-8
G-10
B-16
E-14
B-14
B-15
H-7
E-15
G-11
E-11
E-10
G-12
CD-7
CD-6
E-19
G-13
E-17
D-5
E-3
H-4
B-12
E-4
CD-8
B-9
B-10
D-4
A-12
E-16
E-8
E-6
E-1
E-20
G-14
A-8
B-11
D-1
G-17
GH-5
G-16
E-12
E-13
E-9
D-2
A-9
B-6
G-15
B-8
D-7
A-11
B-5
G-18
D-3
B-7
A-15
GH-1
E-5
E-2
A-16
D-6
A-10
A-13
B-2
G-19
B-4
B-3
A-14
B-1
AB-1
GH-2
GH-3

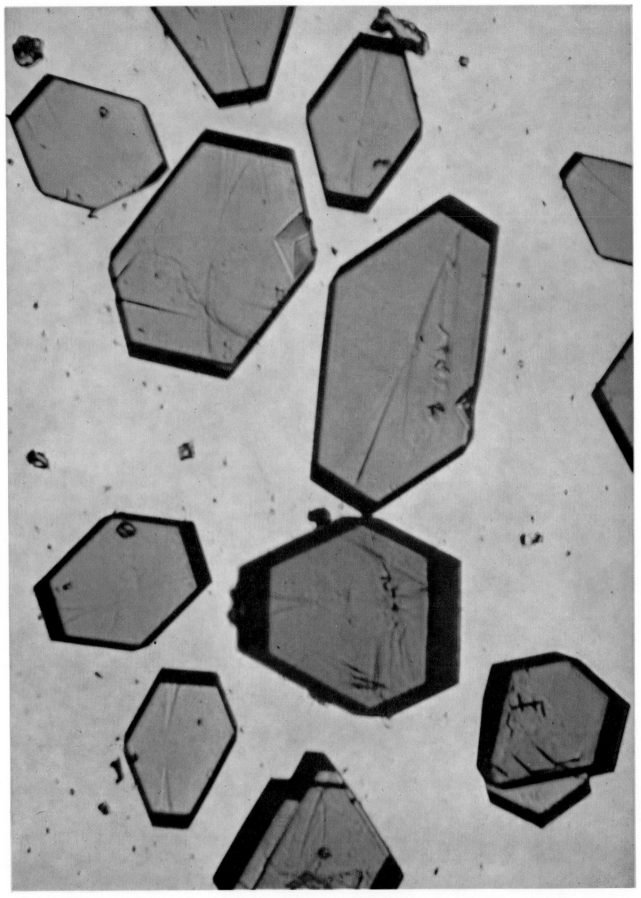

CRYSTALS OF MYOGLOBIN prepared from sperm-whale muscle are enlarged some 50 diameters. In them the molecules of myoglobin are stacked in regular array. By directing a beam of X rays at a single crystal and analyzing the pattern of the reflected rays, the author and his colleagues were able to plot the density of electrons in the molecule and thereby to locate the atoms in it.

SPECIAL X-RAY CAMERA is used to make X-ray diffraction photographs of the myoglobin crystal. The crystal is contained in the thin glass tube exactly at the center of the photograph. The beam of X rays comes out of metal tube just to the right of glass one.

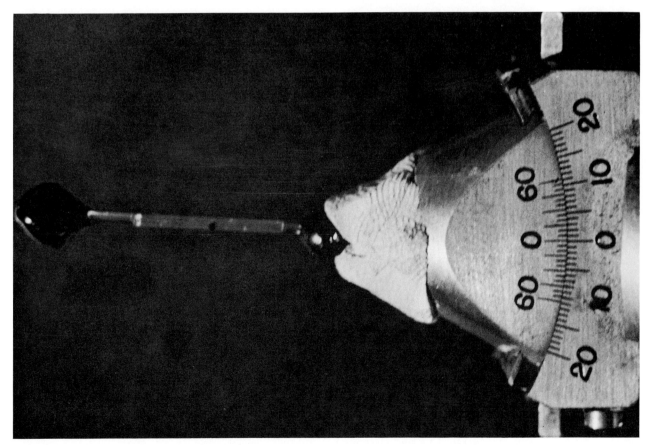

CRYSTAL OF MYOGLOBIN is the dark speck in the middle of the glass tube in this close-up. As the X-ray exposure is made the crystal is rotated so that the X rays reflected from planes of atoms in the crystal "flash up" to make spots on a photographic plate.

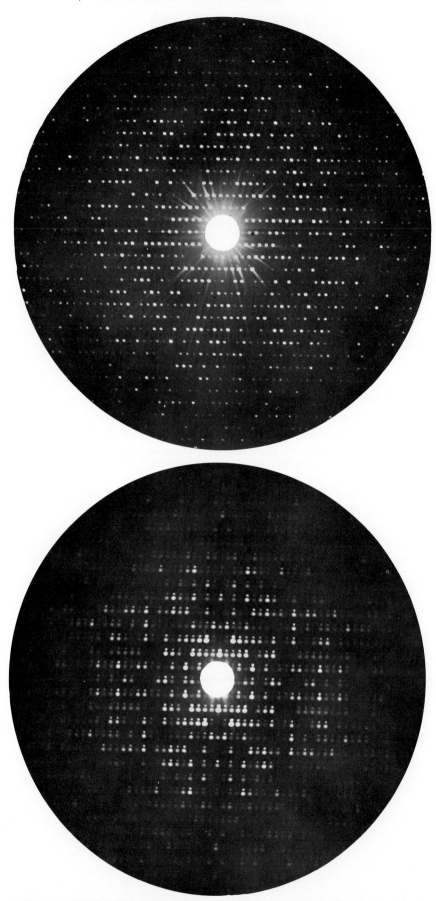

X-RAY PHOTOGRAPHS of myoglobin are patterns of spots. At top is a photograph of a normal crystal. At bottom is a photograph of a different type in which patterns of normal crystal and one labeled with heavy atoms are superimposed slightly out of register. Differences in density between two sets made it possible to determine phase of X-ray reflections.

which varies from place to place in three dimensions but in a regular, repeating way. (The density at any point depends on the types of atom in the neighborhood and their spatial arrangement.) The crystal can therefore be thought of as a kind of three-dimensional sound wave consisting of rarefactions and condensations of electrons rather than of air particles. This wave too can be decomposed into harmonics—that is, simple, repetitive patterns of density variation; along any single direction the density of each harmonic varies sinusoidally.

It turns out that each harmonic corresponds to one particular spot in the X-ray diffraction pattern. The reason is that each set of possible atomic planes in the crystal constitutes one element, so to speak, of the over-all periodic structure. That is just what a harmonic is: a component of the over-all periodic structure. From the position and darkness of a spot, the "wavelength" (the spacing between high-density peaks) and "amplitude" (the value of the density at the peaks) of the corresponding harmonic can be computed. So the problem is reduced to calculating the harmonics from the spot pattern and then adding them together to arrive at the total structure.

There is, however, a serious catch: the diffraction pattern provides information on the wavelength and amplitude of the harmonics but not on their relative phases. In the case of sound, phase is not particularly important in synthesizing a wave; the ear is rather insensitive to phase difference and hears very nearly the same note so long as the relative amplitudes of the harmonics are correct. On the other hand, the shape of the wave as seen by the eye varies greatly when the relative phases of the components are shifted.

In deriving crystal structure the correct shape of the three-dimensional "wave" is precisely what one is looking for. But the X-ray picture contains only half of the information required; it contains the amplitudes but not the phases. In simpler structures crystallographers get around the difficulty by a method of trial and error; from a plausible model structure they calculate the phases and use these in conjunction with the measured amplitudes to calculate a Fourier synthesis, that is to say, an enlarged picture of the distribution of the electrons (and hence of the atoms) in the structure. The result should be a good deal closer to the real structure than the original model, and from it crystallographers can calculate a new and improved set of phases. If the original model was good enough to put them on the right track, they gradually approach the true structure by a series of successive approximations, or refinements.

As in the case of the musical note, the greater the number of higher harmonics that are included, the sharper and more precise is the resulting picture. The higher harmonics of a musical note are, of course, the components of shortest wavelength (highest frequency); in a crystal structure the harmonics are correspondingly the reflections from the most closely spaced sets of planes. As has been mentioned, these reflections occur at the largest angles and show up as spots farthest out in the pattern. The resolution of the final picture—that is, the smallest scale of detail it can show—depends on the outer limit of the spots included in the analysis; the number of spots that have to be included goes up in proportion to the cube of the resolving power required.

The first X-ray photographs of protein crystals were made nearly 25 years ago, but for many years it was not possible even in principle to imagine how the structures of crystals so complex could be discovered. Their X-ray patterns contained many thousands of reflections, paralleling the complexity of the molecules themselves. There was no hope of proceeding by trial and error; the first model could never be good enough to provide a useful starting point. So although protein crystallographers discovered many interesting facts about protein crystals, they did not succeed in extracting much information bearing directly on the molecular structure.

In 1953 the whole prospect was transformed by a discovery of my University of Cambridge colleague Max F. Perutz, who had been studying hemoglobin crystals for many years. The hemoglobin molecule contains two free sulfhydryl groups (SH) of the amino acid cystine; by well-known reactions it is possible to attach atoms of mercury to these groups. Perutz found that if he made crystals of hemoglobin labeled with mercury, their X-ray pattern differed significantly from that of unlabeled crystals, even though the mass of a mercury atom is very small compared with that of a complete hemoglobin molecule. This made it possible to apply the so-called method of isomorphous replacement in the Fourier synthesis.

A full explanation of the method is beyond the scope of the present article. Suffice it to say that by comparing in detail the X-ray patterns of crystals with and without heavy atoms it is possible to deduce the phases of all the reflec-

tions, and this without any of the guesswork of the trial-and-error method. Thus Perutz' observation for the first time made it possible, in principle at least, to solve the complex X-ray pattern of a protein crystal and to produce a model of the structure of the molecule.

In our studies of myoglobin we could not follow Perutz' method for attaching mercury atoms, because myoglobin lacks free sulfhydryl groups. We were, however, successful in finding other ways to attach four or five kinds of heavy atom at different sites in the molecule, and we were then able to proceed to a study of the three-dimensional structure of the crystal. A complete solution would involve including all the reflections in the X-ray pattern in our calculation—some 25,000 in all. At the time this work began no computers in existence were fast enough or large enough to handle so great an amount of data; besides, we thought it better in the first instance to test the new method on a smaller scale.

The Six-Angstrom-Unit Picture

As has already been indicated, if we include only the central reflections of the pattern, we obtain a low-resolution, or crude, representation of the structure. The higher the resolution that is desired, the farther out in the pattern must the reflections be measured. We decided that in the first stage of the project we would aim for a resolution of six angstrom units (an angstrom unit is one hundred-millionth of a centimeter). This would be sufficient to reveal the general arrangement of the polypeptide chains in the molecule, but not the configuration of the atoms within the chains or that of the side chains surrounding them. To achieve a six-angstrom resolution we had to measure 400 reflections from the unlabeled protein and from each of five types of crystal containing heavy atoms. Our calculations, which were completed in the summer of 1957, gave us the density of electrons at a large number of points in the crystal, a high electron density being found where many atoms are concentrated. Crystallographers usually represent a three-dimensional density distribution, or Fourier synthesis, by cutting an imaginary series of parallel sections through the structure. The density distribution in each section is represented by a series of density contours drawn on a lucite sheet. When all the sheets are stacked together, they give a representation in space of the density throughout the molecule.

As soon as we had constructed our

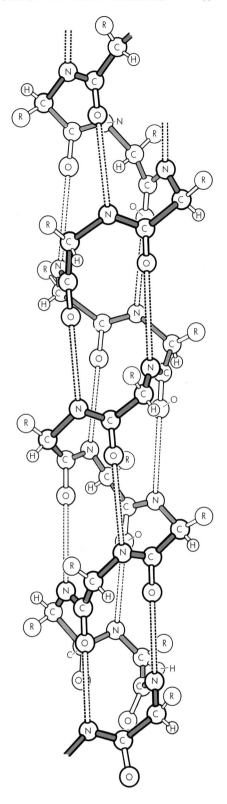

ALPHA HELIX of a protein molecule is a coiled chain of amino acid units. The backbones of the units form a repeating sequence of atoms of carbon (*C*), oxygen (*O*), hydrogen (*H*) and nitrogen (*N*). The *R* stands for the side chain that distinguishes one amino acid from another. The configuration of the helix is maintained by hydrogen bonds (*broken lines*). The hydrogen atom that participates in each of these bonds is not shown.

ELECTRON-DENSITY MAP of the myoglobin crystal is made up of lucite sheets, on each of which are traced the contours of elec- tron density at that depth. The dark band in the middle is the heme group. This map was made at a resolution of two angstrom units.

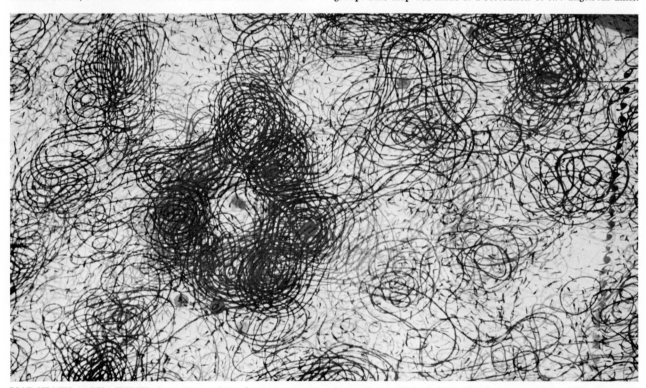

MAP SHOWS ALPHA HELIX when it is seen from the appropriate angle. Here the camera looks through the contours on a series of lucite sheets. Alpha helix is the dark ring at left center. Thus it is seen along its axis, as though it were a cylinder seen from the end.

first lucite density map of the myoglobin crystal, we could see at a glance that it contained the features we were looking for, namely a set of high-density rods of just the dimensions one would expect for a polypeptide chain. Closer examination showed that in fact it consisted of almost nothing but a complicated and intertwining set of these rods, sometimes going straight for a distance, then turning a corner and going off in a new direction. In addition to the rods we were able to see very dense peaks, which we took to be the heme groups themselves. The iron atom at the center of the heme group, being by far the heaviest atom in the molecule and therefore having the largest number of planetary electrons, would be expected to stand out as a prominent feature. It was not at all easy, however, to gain any impression of what the molecule was actually like, largely because the molecules are packed together in the crystal and it is hard to see where one begins and the next one ends.

Our next task was to dissect out a single molecule from the enlarged density map of the crystal so that we could look at it separately. Fortunately all protein crystals, including myoglobin, contain a good deal of liquid which fills up the interstices between neighboring molecules, and which at low resolution looks like a uniform sea of density, so that it can easily be distinguished from the irregular variations of density within the molecule itself. By looking for the liquid regions we were able to draw an outline surface around the molecule and so isolate it from its neighbors. Extracted in this way, the molecule stood forth as the complicated and asymmetrical object shown in the top illustration on page 38. The polypeptide chain winds irregularly around the structure, supporting the heme group in a kind of basket. For the most part the course of the polypeptide chain could be followed, but we could not be sure of its route everywhere, since its density became lower at the corners and it tended to fade into the background at those points. Our model did, however, give us a good general picture of the layout of the molecule, and it showed us that it was indeed much more complicated and irregular than most of the earlier theories of the structure of proteins had suggested.

The Two-Angstrom-Unit Picture

At a resolution of six angstroms we could not expect to see any details of the polypeptide chain or of the side chains attached to it. To see all the atoms

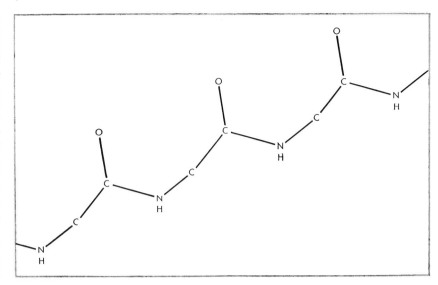

BACKBONE OF THE ALPHA HELIX is shown schematically in this diagram. The sequence of atoms in the helix is —CO—CHR—NH—. Here the HR attached to isolated C is omitted.

BACKBONE IS SUPERIMPOSED on contours made by plotting on a cylinder density along helix in crystal. Cylinder was then unrolled. Backbone thus appears to repeat.

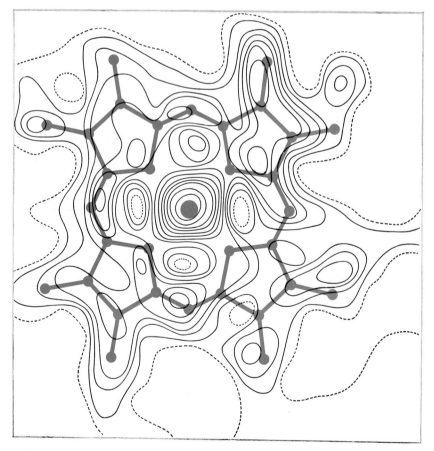

SKELETON OF THE HEME GROUP is outlined in this diagram. At the center of the group is the iron atom (*Fe*). There are four such groups in hemoglobin and one in myoglobin.

SKELETON IS SUPERIMPOSED on another section of the electron-density map of the myoglobin crystal. Here bonds to the iron atom are omitted to show contours around atom.

of a structure as separate blobs of density it would be necessary to work at a resolution higher than 1.5 angstroms, for neighboring atoms attached to one another by chemical bonds lie only from one to 1.5 angstroms apart. A Fourier synthesis of myoglobin at 1.5 angstroms resolution would involve 25,000 reflections; we decided in the second stage of the work to limit our ambitions to a resolution of two angstroms. Even this required that we include in our calculations nearly 10,000 reflections for the unsubstituted crystal and the same number for each of the heavy-atom derivatives. It was necessary to measure a formidable number of X-ray photographs, a task that took a team of six people many months to complete. At the end of this time the mass of data that we had accumulated was so great that it could be handled only by a truly fast computer. We were fortunate that a machine of this class—EDSAC Mark II —had recently come into service at the University of Cambridge, and we were able to use it for deducing the phases of all the 10,000 reflections and for the ensuing calculations of the Fourier synthesis itself. Fast though it is, we taxed the powers of EDSAC II to the utmost, and it was clear that any further improvement in resolution would demand the use of still more powerful machines.

Once more our results were plotted in the form of a three-dimensional contour map [*see top illustration on page 34*]. Since we were looking for finer details, it was necessary to cut sections through the structure at closer intervals than before; in fact, this time we had about 50 sections compared to 16 in the six-angstrom map. To construct the new map it was necessary to calculate the electron density at 100,000 points in the molecule. Indeed, the amount of information contained in the final synthesis was so great that drawing and building the density map was in itself a lengthy task, amounting to some six man-months of work. The result was a complicated set of dense and less dense regions that at first sight seemed completely irregular. Our first step was to see what we could learn from it about the configuration of the polypeptide chains, which in our earlier synthesis had appeared merely as solid rods.

Here I shall digress briefly to consider some earlier work on the fibrous protein of hair. In fibrous proteins the polypeptide chains probably run parallel to the axis of the fiber for considerable distances. Such protein fibers were among the earliest biological macromolecules to be examined by X-ray methods.

W. T. Astbury, in his classical work carried out at the University of Leeds in the early 1930's, showed that a human hair gave a characteristic X-ray pattern, which on stretching changed reversibly into quite a different pattern. He was able to show that the pattern of stretched hair—the so-called beta form—corresponded to the polypeptide chains being almost fully extended; it followed that in the unstretched, or alpha, form the chains must assume some kind of folded configuration. For many years different workers proposed a succession of more or less unsatisfactory models of the folded chain in unstretched hair, but finally in 1951 Linus Pauling and Robert B. Corey of the California Institute of Technology found the definitive solution, showing that the chain actually took up a helical or spiral shape, the now famous alpha helix [see illustration on page 33]. In this configuration the successive turns of the helix are held together by weak hydrogen bonds between NH groups on one turn and CO groups opposite them on the next turn. The alpha helix turned out to be present in several other fibrous proteins besides hair; and although there was no definite proof of the fact, indirect evidence indicated that the alpha helix, or something like it, could exist in the molecule of globular proteins too.

The first thing we wanted to do when we finished our Fourier synthesis at two angstroms resolution was to see whether or not there was anything to the idea of helical structures in a globular protein. Accordingly we looked through the stack of lucite sheets in a direction corresponding to the axis of one of the rods we had seen at low resolution. We were delighted to find that the dense rod now revealed itself as a hollow cylinder running dead straight through the structure [see bottom illustration on page 34]. Closer examination showed that the density followed a spiral course along the surface of the cylinder, indicating that the polypeptide chain indeed assumed a helical shape. Detailed measurement of the spiral density showed that it followed precisely the dimensions of the alpha helix deduced by Pauling and Corey 10 years earlier. In fact, it turned out that about three-quarters of the polypeptide chain in the molecule took the form of straight lengths of alpha helix, the helical segments being joined by irregular regions at the corners. In all there were eight such segments, varying in length from seven to 24 amino acid units. In each segment it was possible to fit the alpha helix exactly to the observed density in such a way that we could be reasonably sure of the placing of each atom, even though at this resolution we had not secured full separation between one atom and its neighbors.

The next object of interest was the heme group. Looking at the map from the appropriate angles, we now saw this group as a flat object with a region of high density at the iron atom in the center. A section through the map through the plane of the flat object shows a variation in density that closely follows the known chemical structure of the system of rings in the heme group [see illustrations on preceding page].

The Three-dimensional Model

When we came to study our structure in detail, we soon felt the need of a better way to represent the three-dimensional density distribution. We wanted some method that would enable us to fit actual atomic models to the features we could see. Our solution was to erect a forest of steel rods on which we placed colored clips to represent the distribution in space of points of high density, different colors representing different values of the density [see illustration on page 39]. The scale of this model was five centimeters per angstrom, so that the whole model would fit in a cube about six feet on a side. Each helical segment of polypeptide chain could be seen as a spiral of colored clips passing through the model, and we were then able to insert actual alpha helices made of skeleton-type models (similar to the familiar ball-and-spoke models but with the balls omitted) and to show that they precisely followed the dense trail of clips. In this way we were able to trace the polypeptide chain from beginning to end, right through the molecule, and to establish its configuration in each of the irregular corners joining neighboring helices. Once the course of the main chain had been delineated with atomic models, we were able to see the side chains emerging from it at appropriate intervals as dense branches of various sizes. At first we thought it unlikely that we would be able to identify many of the side chains, but after some practice we found that in fact we were able to do so surprisingly often. As mentioned earlier, side chains in proteins are of only 20 kinds (in myoglobin only 17), and they are of very different shapes and sizes, ranging from the one in glycine, which is only a single hydrogen atom (invisible to the crystallographer), to the chain in tryptophan, with a double-ring system of 10 carbon and nitrogen atoms. Our problem was reduced to deciding among 17 possible side chains in each case.

Some of our identifications were definite, others were tentative. Fortunately an independent check on our conclusions lay at hand. For several years A. B. Edmundson and C. H. W. Hirs, working in Stein and Moore's laboratory at the Rockefeller Institute, had been trying to establish the amino acid sequence of myoglobin by chemical methods. Their work is still incomplete, but they have broken down the molecule into a set of short pieces, or peptides, the compositions—and in some cases the internal sequences—of which they have determined. The order in which the peptides are arranged in the intact molecule has yet to be established chemically. We have been able, however, to place almost every one of the peptides with certainty in its correct position along the chain by comparing its composition with our X-ray identifications of the side chains. There are virtually no gaps left, nor are there peptides unplaced. Once assigned to their correct positions, the peptides often help to confirm doubtful X-ray identifications, and by putting the two types of evidence together we arrive at a nearly complete amino acid sequence for the whole molecule.

Simply to determine the amino acid sequence was not our main aim in undertaking the X-ray analysis of myoglobin. We were much more concerned with the three-dimensional arrangement of the side chains in the molecule and with the interactions between them that produce and maintain the molecule's characteristic configuration. To study these interactions we undertook to make a model of the whole molecule, with every side chain in place [see illustration on pages 26 and 27]. The result was an object still more complex than the low-resolution model, although of course all the features of the latter are still apparent in the former. We can now discern many of the types of interaction that protein chemists have postulated on the basis of physicochemical studies. For example, positively charged basic groups such as those of lysine and arginine are held by electrostatic attraction close to negatively charged acid groups such as aspartic or glutamic acid; several types of hydrogen-bond interaction can be seen, among them NH groups in the main chain bonded to the oxygen atom of serine or threonine; and everywhere we find a close interlocking of hydrocarbon groups such as CH_2 or CH_3, giving rise to the so-called van der Waals' attraction. The structure is not yet sufficiently complete in all details to allow a

38

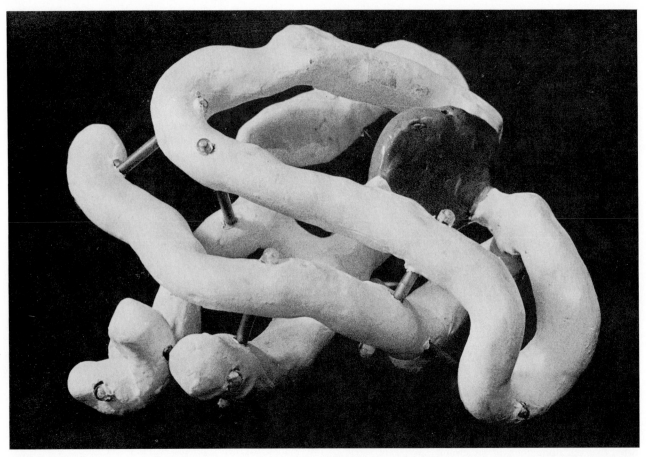

EARLY MODEL of the myoglobin molecule was made at a resolution of six angstrom units. This model has the same general configuration as that of the model depicted on pages 26 and 27, but it lacks detail. The heme group is the flat section at upper right.

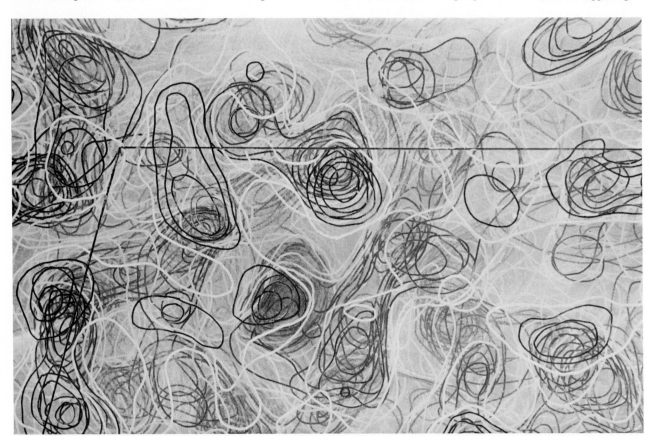

CLOSE-UP OF CONTOURS of map on which the six-angstrom model was based shows that contours are coarser than those in two-angstrom map. Early model was based on work of author, G. Bodo, H. M. Dintzis, R. G. Parrish, H. W. Wyckoff and D. C. Phillips.

full analysis of the interactions, but at least we can now see the general pattern of forces that maintains the integrity of the molecule.

We can also often see why helical segments end at a particular place; in many instances proline side chains are found at the ends of helices and, as was pointed out several years ago, proline is bound to interfere with helix formation because of its peculiar shape, unlike that of any other naturally occurring amino acid. Finally, we can examine the way in which the heme group itself is attached to the rest of the molecule; the iron atom is attached to a nitrogen in a histidine side chain (as had been suggested years ago on the basis of chemical evidence), and the flat ring system is stabilized by hydrocarbon side chains, especially ring side chains, lying parallel to it.

In similar studies of the larger hemoglobin molecule Perutz and his collaborators have shown that, at least to the resolution of 5.5 angstroms that they have so far achieved, there is an astonishing similarity between the three-dimensional structure of myoglobin and the structure of each of the four subunits formed by the individual polypeptide chains of hemoglobin. This is a most remarkable result considering that we are dealing with two distinct proteins, one found in muscle and the other in red blood cells, one derived from sperm whale and the other from horse. Furthermore, the amino acid compositions of the two proteins are known to differ substantially.

The amino acid sequences of the hemoglobin chains have been completely determined. We have found that when we lay the hemoglobin sequences alongside those of myoglobin, making appropriate allowances for slight differences in their length, there are many correspondences, often just at those points where a study of the myoglobin molecule indicates that a crucial stabilizing reaction takes place. We can even begin to find chemical explanations for some of the peculiarities of the congenitally abnormal hemoglobins present in individuals suffering from certain rare blood diseases.

Even in the present incomplete state of our studies on myoglobin we are beginning to think of a protein molecule in terms of its three-dimensional chemical structure and hence to find rational explanations for its chemical behavior and physiological function, to understand its affinities with related proteins and to glimpse the problems involved in explaining the synthesis of proteins in living organisms and the nature of the malfunctions resulting from errors in this process. It is evident that today students of the living organism do indeed stand on the threshold of a new world. Analyses of many other proteins, and at still higher resolutions (such as we hope soon to achieve with myoglobin), will be needed before this new world can be fully invaded, and the manifold interactions between the giant molecules of living cells must be comprehended in terms of well-understood concepts of chemistry. Nevertheless, the prospect of establishing a firm basis for an understanding of the enormous complexities of structure, of biogenesis and of function of living organisms in health and disease is now distinctly in view.

FOREST OF RODS was used to build up the two-angstrom model of the myoglobin molecule from electron-density map. Densities were indicated by clips on rods, and model was based on position of clips. Outline of heme group is visible at upper left center.

4

The Hemoglobin Molecule

M. F. Perutz
November 1964

*Its 10,000 atoms are assembled into four chains,
each a helix with several bends. The molecule has
one shape when ferrying oxygen molecules and a
slightly different shape when it is not*

In 1937, a year after I entered the University of Cambridge as a graduate student, I chose the X-ray analysis of hemoglobin, the oxygen-bearing protein of the blood, as the subject of my research. Fortunately the examiners of my doctoral thesis did not insist on a determination of the structure, otherwise I should have had to remain a graduate student for 23 years. In fact, the complete solution of the problem, down to the location of each atom in this giant molecule, is still outstanding, but the structure has now been mapped in enough detail to reveal the intricate three-dimensional folding of each of its four component chains of amino acid units, and the positions of the four pigment groups that carry the oxygen-combining sites.

The folding of the four chains in hemoglobin turns out to be closely similar to that of the single chain of myoglobin, an oxygen-bearing protein in muscle whose structure has been elucidated in atomic detail by my colleague John C. Kendrew and his collaborators. Correlation of the structure of the two proteins allows us to specify quite accurately, by purely physical methods, where each amino acid unit in hemoglobin lies with respect to the twists and turns of its chains.

Physical methods alone, however, do not yet permit us to decide which of the 20 different kinds of amino acid units occupies any particular site. This knowledge has been supplied by chemical analysis; workers in the U.S. and in Germany have determined the sequence of the 140-odd amino acid units along each of the hemoglobin chains. The combined results of the two different methods of approach now provide an accurate picture of many facets of the hemoglobin molecule.

In its behavior hemoglobin does not resemble an oxygen tank so much as a molecular lung. Two of its four chains shift back and forth, so that the gap between them becomes narrower when oxygen molecules are bound to the hemoglobin, and wider when the oxygen is released. Evidence that the chemical activities of hemoglobin and other proteins are accompanied by structural changes had been discovered before, but this is the first time that the nature of such a change has been directly demonstrated. Hemoglobin's change of shape makes me think of it as a breathing molecule, but paradoxically it expands, not when oxygen is taken up but when it is released.

When I began my postgraduate work in 1936 I was influenced by three inspiring teachers. Sir Frederick Gowland Hopkins, who had received a Nobel prize in 1929 for discovering the growth-stimulating effect of vitamins, drew our attention to the central role played by enzymes in catalyzing chemical reactions in the living cell. The few enzymes isolated at that time had all proved to be proteins. David Keilin, the discoverer of several of the enzymes that catalyze the processes of respiration, told us how the chemical affinities and catalytic properties of iron atoms were altered when the iron combined with different proteins. J. D. Bernal, the X-ray crystallographer, was my research supervisor. He and Dorothy Crowfoot Hodgkin had taken the first X-ray diffraction pictures of crystals of protein a year or two before I arrived, and they had discovered that protein molecules, in spite of their large size, have highly ordered structures. The wealth of sharp X-ray diffraction spots produced by a single crystal of an enzyme such as pepsin could be explained only if every one, or almost every one, of the 5,000 atoms in the pepsin molecule occupied a definite position that was repeated in every one of the myriad of pepsin molecules packed in the crystal. The notion is commonplace now, but it caused a sensation at a time when proteins were still widely regarded as "colloids" of indefinite structure.

In the late 1930's the importance of the nucleic acids had yet to be discovered; according to everything I had learned the "secret of life" appeared to be concealed in the structure of proteins. Of all the methods available in chemistry and physics, X-ray crystallography seemed to offer the only chance, albeit an extremely remote one, of determining that structure.

The number of crystalline proteins then available was probably not more than a dozen, and hemoglobin was an obvious candidate for study because of its supreme physiological importance, its ample supply and the ease with which it could be crystallized. All the same, when I chose the X-ray analysis of hemoglobin as the subject of my Ph.D. thesis, my fellow students regarded me with a pitying smile. The most complex organic substance whose structure had yet been determined by X-ray analysis was the molecule of the dye phthalocyanin, which contains 58 atoms. How could I hope to locate the thousands of atoms in the molecule of hemoglobin?

The Function of Hemoglobin

Hemoglobin is the main component of the red blood cells, which carry oxygen from the lungs through the arteries to the tissues and help to carry carbon dioxide through the veins back to the lungs. A single red blood cell contains about 280 million molecules of hemoglobin. Each molecule has 64,500 times the weight of a hydrogen atom and is made up of about 10,000 atoms of hydrogen, carbon, nitrogen, oxygen and sulfur, plus four atoms of iron, which are more important than all the rest. Each iron atom lies at the center of the group of atoms that form the pigment called heme, which gives blood its red color and its ability to combine with oxygen. Each heme group is enfolded in one of the four chains of amino acid units that collectively constitute the protein part of the molecule, which is called globin. The four chains of globin consist of two identical pairs. The members of one pair are known as alpha chains and those of the other as beta chains. Together the four chains contain a total of 574 amino acid units.

In the absence of an oxygen carrier a liter of arterial blood at body temperature could dissolve and transport no more than three milliliters of oxygen. The presence of hemoglobin increases this quantity 70 times. Without hemoglobin large animals could not get enough oxygen to exist. Similarly, hemoglobin is responsible for carrying more than 90 percent of the carbon dioxide transported by venous blood.

Each of the four atoms of iron in the hemoglobin molecule can take up one molecule (two atoms) of oxygen. The reaction is reversible in the sense that oxygen is taken up where it is plentiful, as in the lungs, and released where it is scarce, as in the tissues. The reaction is accompanied by a change in color: hemoglobin containing oxygen, known as oxyhemoglobin, makes arterial blood look scarlet; reduced, or oxygen-free, hemoglobin makes venous blood look purple. The term "reduced" for the oxygen-free form is really a misnomer because "reduced" means to the chemist that electrons have been added to an atom or a group of atoms. Actually, as James B. Conant of Harvard University demonstrated in 1923, the iron atoms in both reduced hemoglobin and oxyhemoglobin are in the same electronic condition: the divalent, or ferrous, state. They become oxidized to the trivalent, or ferric, state if hemoglobin is treated with a ferricyanide or removed from the red cells and exposed to the air for a considerable time; oxidation also occurs in certain blood diseases. Under these conditions hemoglobin turns brown and is known as methemoglobin, or ferrihemoglobin.

Ferrous iron acquires its capacity for binding molecular oxygen only through its combination with heme and globin. Heme alone will not bind oxygen, but the specific chemical environment of the globin makes the combina-

HEMOGLOBIN MOLECULE, as deduced from X-ray diffraction studies, is shown from above (*top*) and side (*bottom*). The drawings follow the representation scheme used in three-dimensional models built by the author and his co-workers. The irregular blocks represent electron-density patterns at various levels in the hemoglobin molecule. The molecule is built up from four subunits: two identical alpha chains (*light blocks*) and two identical beta chains (*dark blocks*). The letter "N" in the top view identifies the amino ends of the two alpha chains; the letter "C" identifies the carboxyl ends. Each chain enfolds a heme group (*colored disk*), the iron-containing structure that binds oxygen to the molecule.

X-RAY DIFFRACTION PATTERN was made from a single crystal of hemoglobin that was rotated during the photographic exposure. Electrons grouped around the centers of the atoms in the crystal scatter the incident X rays, producing a symmetrical array of spots. Spots that are equidistant from the center and opposite each other have the same density.

tion possible. In association with other proteins, such as those of the enzymes peroxidase and catalase, the same heme group can exhibit quite different chemical characteristics.

The function of the globin, however, goes further. It enables the four iron atoms within each molecule to interact in a physiologically advantageous manner. The combination of any three of the iron atoms with oxygen accelerates the combination with oxygen of the fourth; similarly, the release of oxygen by three of the iron atoms makes the fourth cast off its oxygen faster. By tending to make each hemoglobin molecule carry either four molecules of oxygen or none, this interaction ensures efficient oxygen transport.

I have mentioned that hemoglobin also plays an important part in bearing carbon dioxide from the tissues back to the lungs. This gas is not borne by the iron atoms, and only part of it is bound directly to the globin; most of it is taken up by the red cells and the noncellular fluid of the blood in the form of bicarbonate. The transport of bicarbonate is facilitated by the disappearance of

an acid group from hemoglobin for each molecule of oxygen discharged. The reappearance of the acid group when oxygen is taken up again in the lungs sets in motion a series of chemical reactions that leads to the discharge of carbon dioxide. Conversely, the presence of bicarbonate and lactic acid in the tissues accelerates the liberation of oxygen.

Breathing seems so simple, yet it appears as if this elementary manifestation of life owes its existence to the interplay of many kinds of atoms in a giant molecule of vast complexity. Elucidating the structure of the molecule should tell us not only what the molecule looks like but also how it works.

The Principles of X-Ray Analysis

The X-ray study of proteins is sometimes regarded as an abstruse subject comprehensible only to specialists, but the basic ideas underlying our work are so simple that some physicists find them boring. Crystals of hemoglobin and other proteins contain much water and, like living tissues, they tend to lose their regularly ordered structure on dry-

ing. To preserve this order during X-ray analysis crystals are mounted wet in small glass capillaries. A single crystal is then illuminated by a narrow beam of X rays that are essentially all of one wavelength. If the crystal is kept stationary, a photographic film placed behind it will often exhibit a pattern of spots lying on ellipses, but if the crystal is rotated in certain ways, the spots can be made to appear at the corners of a regular lattice that is related to the arrangement of the molecules in the crystal [*see illustration at left*]. Moreover, each spot has a characteristic intensity that is determined in part by the arrangement of atoms inside the molecules. The reason for the different intensities is best explained in the words of W. L. Bragg, who founded X-ray analysis in 1913—the year after Max von Laue had discovered that X rays are diffracted by crystals—and who later succeeded Lord Rutherford as Cavendish Professor of Physics at Cambridge:

"It is well known that the form of the lines ruled on a [diffraction] grating has an influence on the relative intensity of the spectra which it yields. Some spectra may be enhanced, or reduced, in intensity as compared with others. Indeed, gratings are sometimes ruled in such a way that most of the energy is thrown into those spectra which it is most desirable to examine. The form of the line on the grating does not influence the positions of the spectra, which depend on the number of lines to the centimetre, but the individual lines scatter more light in some directions than others, and this enhances the spectra which lie in those directions.

"The structure of the group of atoms which composes the unit of the crystal grating influences the strength of the various reflexions in exactly the same way. The rays are diffracted by the electrons grouped around the centre of each atom. In some directions the atoms conspire to give a strong scattered beam, in others their effects almost annul each other by interference. The exact arrangement of the atoms is to be deduced by comparing the strength of the reflexions from different faces and in different orders."

Thus there should be a way of reversing the process of diffraction, of proceeding backward from the diffraction pattern to an image of the arrangement of atoms in the crystal. Such an image can actually be produced, somewhat laboriously, as follows. It will be noted that spots on opposite sides of the center of an X-ray picture have the same

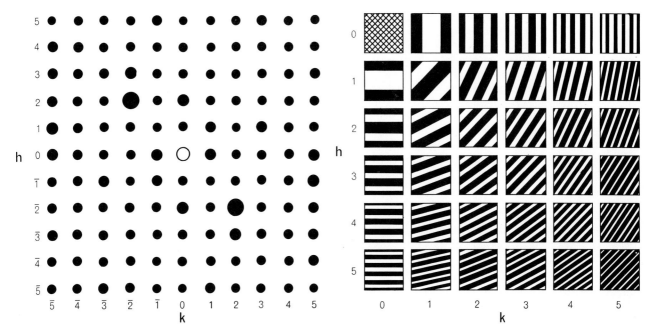

INTERPRETATION OF X-RAY IMAGE can be done with a special optical device to generate a set of diffraction fringes (*right*) from the spots in an X-ray image (*left*). Each pair of symmetrically related spots produces a unique set of fringes. Thus the spots indexed 2,$\bar{2}$ and $\bar{2}$,2 yield the fringes indexed 2,2. A two-dimensional image of the atomic structure of a crystal can be generated by printing each set of fringes on the same sheet of photographic paper. But the phase problem (*below*) must be solved first.

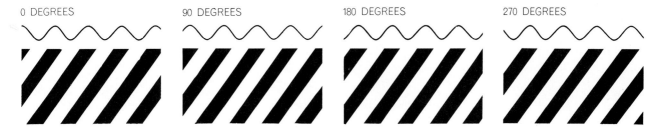

PHASE PROBLEM arises because the spots in an X-ray image do not indicate how the fringes are related in phase to an arbitrarily chosen common origin. Here four identical sets of fringes are related by different phases to the point of origin at the top left corner. The phase marks the distance of the wave crest from the origin, measured in degrees. One wavelength is 360 degrees.

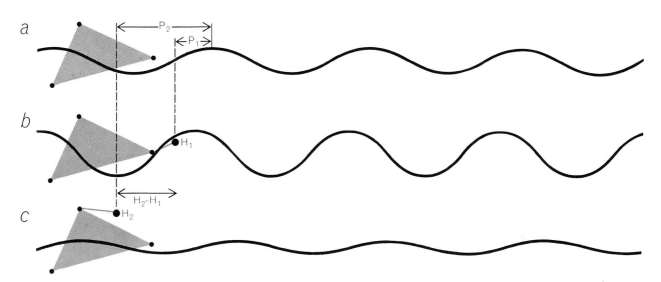

HEAVY-ATOM REPLACEMENT METHOD provides information about phases by changing the intensities of the X-ray diffraction pattern. In *a* a highly oversimplified protein (a triangle of three atoms) scatters a sinusoidal wave that represents the amplitude and phase of a single set of fringes. In *b* and *c*, after heavy atoms H_1 and H_2 are attached to the protein in different positions, the wave is changed in amplitude and phase. The heavy atoms can serve as points of common origin for measuring the magnitude of the phases (P_1 and P_2) of waves scattered by the unaltered protein. The distance between H_1 and H_2 must be accurately known.

degree of intensity. With the aid of a simple optical device each symmetrically related pair of spots can be made to generate a set of diffraction fringes, with an amplitude proportional to the square root of the intensity of the spots. The device, which was invented by Bragg and later developed by H. Lipson and C. A. Taylor at the Manchester College of Science and Technology, consists of a point source of monochromatic light, a pair of plane-convex lenses

and a microscope. The pair of spots in the diffraction pattern is represented by a pair of holes in a black mask that is placed between the two lenses. If the point source is placed at the focus of one of the lenses, the waves of parallel light emerging from the two holes will interfere with one another at the focus of the second lens, and their interference pattern, or diffraction pattern, can be observed or photographed through the microscope.

Imagine that each pair of symmetrically related spots in the X-ray picture is in turn represented by a pair of holes in a mask, and that its diffraction fringes are photographed. Each set of fringes will then be at right angles to the line joining the two holes, and the distance between the fringes will be inversely proportional to the distance between the holes. If the spots are numbered from the center along two mutually perpendicular lines by the indices h and k, the relation between any pair of spots and its corresponding set of fringes would be as shown in the top illustration on the opposite page.

The Phase Problem

An image of the atomic structure of the crystal can be generated by printing each set of fringes in turn on the same sheet of photographic paper, or by superposing all the fringes and making a print of the light transmitted through them. At this point, however, a fatal complication arises. In order to obtain the right image one would have to place each set of fringes correctly with respect to some arbitrarily chosen common origin [see middle illustration on preceding page]. At this origin the amplitude of any particular set of fringes may show a crest or trough or some intermediate value. The distance of the wave crest from the origin is called the phase. It is almost true to say that by superposing sets of fringes of given amplitude one can generate an infinite number of different images, depending on the choice of phase for each set of fringes. By itself the X-ray picture tells us only about the amplitudes and nothing about the phases of the fringes to be generated by each pair of spots, which means that half the information needed for the production of the image is missing.

The missing information makes the diffraction pattern of a crystal like a hieroglyphic without a key. Having spent years hopefully measuring the intensities of several thousand spots in the diffraction pattern of hemoglobin, I found myself in the tantalizing position of an explorer with a collection of tablets engraved in an unknown script. For some time Bragg and I tried to develop methods for deciphering the phases, but with only limited success. The solution finally came in 1953, when I discovered that a method that had been developed by crystallographers for solving the phase problem in simpler structures could also be applied to proteins.

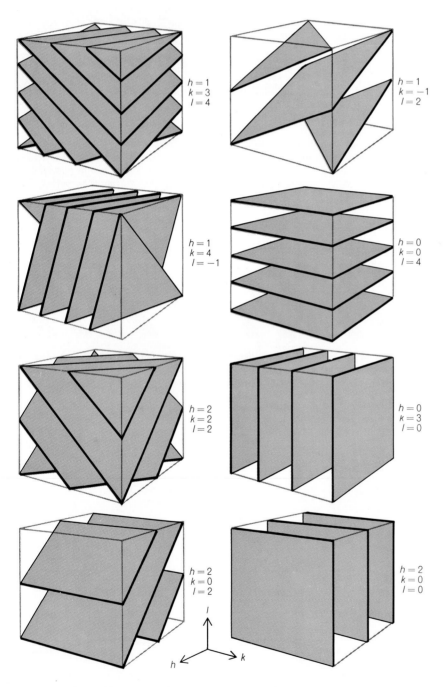

THREE-DIMENSIONAL FRINGES are needed to build up an image of protein molecules. For this purpose many different X-ray diffraction images are prepared and symmetrically related pairs of spots are indexed in three dimensions: h, k and l and $\bar{h}$, $\bar{k}$ and $\bar{l}$. Each pair of spots yields a three-dimensional fringe like those shown here. Fringes from thousands of spots must be superposed in proper phase to build up an image of the molecule.

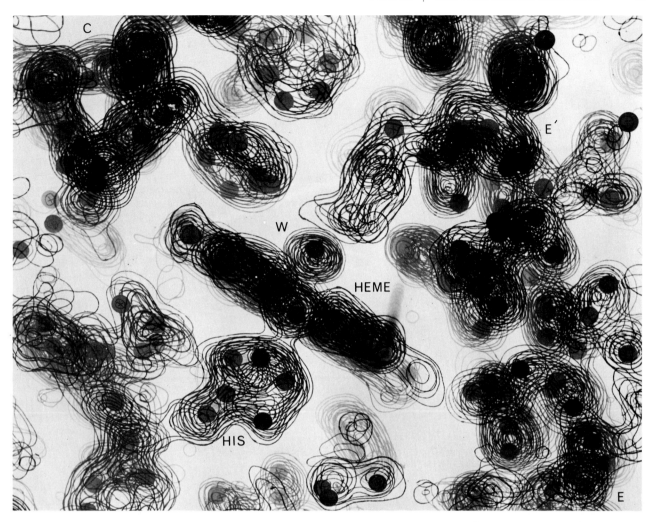

CONTOUR MAPS, drawn on stacked sheets of clear plastic, show a portion of the myoglobin molecule as revealed by superposition of three-dimensional fringe patterns. The maps were made by John C. Kendrew and his associates at the University of Cambridge. Myoglobin is very similar to the beta chain of hemoglobin. The heme group is seen edge on. *His* is an amino acid subunit of histidine that is attached to the iron atom of the heme group. *W* is a water molecule linked to the iron atom. The region between *E* and *E'* represents amino acid subunits arranged in an alpha helix. *C* is an alpha helix seen end on. The black dots mark atomic positions.

In this method the molecule of the compound under study is modified slightly by attaching heavy atoms such as those of mercury to definite positions in its structure. The presence of a heavy atom produces marked changes in the intensities of the diffraction pattern, and this makes it possible to gather information about the phases. From the difference in amplitude in the absence or presence of a heavy atom, the distance of the wave crest from the heavy atom can be determined for each set of fringes. Thus with the heavy atom serving as a common origin the magnitude of the phase can be measured. The bottom illustration on page 108 shows how the phase of a single set of fringes, represented by a sinusoidal wave that is supposedly scattered by the oversimplified protein molecule, can be measured from the increase in amplitude produced by the heavy atom H_1.

Unfortunately this still leaves an am-

biguity of sign; the experiment does not tell us whether the phase is to be measured from the heavy atom in the forward or the backward direction. If n is the number of diffracted spots, an ambiguity of sign in each set of fringes would lead to 2^n alternative images of the structure. The Dutch crystallographer J. M. Bijvoet had pointed out some years earlier in another context that the ambiguity could be resolved by examining the diffraction pattern from a second heavy-atom compound.

The bottom illustration on page 108 shows that the heavy atom H_2, which is attached to the protein in a position different from that of H_1, diminishes the amplitude of the wave scattered by the protein. The degree of attenuation allows us to measure the distance of the wave crest from H_2. It can now be seen that the wave crest must be in front of H_1; otherwise its distance from H_1 could not be reconciled with its distance from

H_2. The final answer depends on knowing the length and direction of the line joining H_2 to H_1. These quantities are best calculated by a method that does not easily lend itself to exposition in nonmathematical language. It was devised by my colleague Michael G. Rossmann.

The heavy-atom method can be applied to hemoglobin by attaching mercury atoms to the sulfur atoms of the amino acid cysteine. The method works, however, only if this attachment leaves the structure of the hemoglobin molecules and their arrangement in the crystal unaltered. When I first tried it, I was not at all sure that these stringent demands would be fulfilled, and as I developed my first X-ray photograph of mercury hemoglobin my mood alternated between sanguine hopes of immediate success and desperate forebodings of all the possible causes of failure. When the diffraction spots ap-

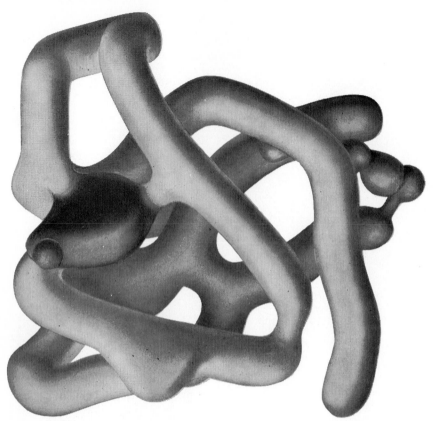

MYOGLOBIN MOLECULE, as first reconstructed at low resolution by Kendrew and his co-workers in 1957, had this rather repulsive visceral appearance. The sausage-like knot marks the path of the amino acid chain of the molecule. The dark disklike shape (here placed at an incorrect angle) is the heme group. A more detailed and more correct view of myoglobin, as seen from the other side, appears at bottom right on the opposite page.

compounds of the protein, each with heavy atoms attached to different positions in the molecule. Then the results have to be corrected by various geometric factors before they are finally used to build up an image through the superposition of tens of thousands of fringes. In the final calculation tens of millions of numbers may have to be added or subtracted. Such a task would have been quite impossible before the advent of high-speed computers, and we have been fortunate in that the development of computers has kept pace with the expanding needs of our X-ray analyses.

While I battled with technical difficulties of various sorts, my colleague John Kendrew successfully applied the heavy-atom method to myoglobin, a protein closely related to hemoglobin [see the article "The Three-dimensional Structure of a Protein Molecule," by John C. Kendrew, beginning on page 25]. Myoglobin is simpler than hemoglobin because it consists of only one chain of amino acid units and one heme group, which binds a single molecule of oxygen. The complex interaction phenomena involved in hemoglobin's dual function as a carrier of oxygen and of carbon dioxide do not occur in myoglobin, which acts simply as an oxygen store.

Together with Howard M. Dintzis and G. Bodo, Kendrew was brilliantly successful in managing to prepare as many as five different crystalline heavy-atom compounds of myoglobin, which meant that the phases of the diffraction spots could be established very accurately. He also pioneered the use of high-speed computers in X-ray analysis. In 1957 he and his colleagues obtained the first three-dimensional representation of myoglobin [see illustration on this page].

It was a triumph, and yet it brought a tinge of disappointment. Could the search for ultimate truth really have revealed so hideous and visceral-looking an object? Was the nugget of gold a lump of lead? Fortunately, like many other things in nature, myoglobin gains in beauty the closer you look at it. As Kendrew and his colleagues increased the resolution of their X-ray analysis in the years that followed, some of the intrinsic reasons for the molecule's strange shape began to reveal themselves. This shape was found to be not a freak but a fundamental pattern of nature, probably common to myoglobins and hemoglobins throughout the vertebrate kingdom.

In the summer of 1959, nearly 22 years after I had taken the first X-ray

peared in exactly the same position as in the mercury-free protein but with slightly altered intensities, just as I had hoped, I rushed off to Bragg's room in jubilant excitement, expecting that the structure of hemoglobin and of many other proteins would soon be determined. Bragg shared my excitement, and luckily neither of us anticipated the formidable technical difficulties that were to hold us up for another five years.

Resolution of the Image

Having solved the phase problem, at least in principle, we were confronted with the task of building up a structural image from our X-ray data. In simpler structures atomic positions can often be found from representations of the structure projected on two mutually perpendicular planes, but in proteins a three-dimensional image is essential. This can be attained by making use of the three-dimensional nature of the diffraction pattern. The X-ray diffraction pattern on page 107 can be regarded as a section through a sphere that is filled with layer after layer of diffraction

spots. Each pair of spots can be made to generate a set of three-dimensional fringes like the ones shown on page 6. When their phases have been measured, they can be superposed by calculation to build up a three-dimensional image of the protein. The final image is represented by a series of sections through the molecule, rather like a set of microtome sections through a piece of tissue, only on a scale 1,000 times smaller [see illustration on preceding page].

The resolution of the image is roughly equal to the shortest wavelength of the fringes used in building it up. This means that the resolution increases with the number of diffracted spots included in the calculation. If the image is built up from part of the diffraction pattern only, the resolution is impaired.

In the X-ray diffraction patterns of protein crystals the number of spots runs into tens of thousands. In order to determine the phase of each spot accurately, its intensity (or blackness) must be measured accurately several times over: in the diffraction pattern from a crystal of the pure protein and in the patterns from crystals of several

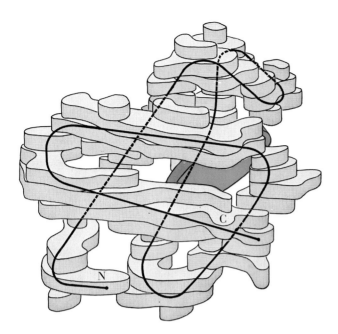

HEMOGLOBIN CHAINS, alpha at left and beta at right, are re-drawn from models built by the author and his colleagues. The superposed lines show the course of the central chain. A heme group (*color*) is partly visible, tucked in the back of each model.

pictures of hemoglobin, its structure emerged at last. Michael Rossmann, Ann F. Cullis, Hilary Muirhead, Tony C. T. North and I were able to prepare a three-dimensional electron-density map of hemoglobin at a resolution of 5.5 angstrom units, about the same as that obtained for the first structure of myoglobin two years earlier. This resolution is sufficient to reveal the shape of the chain forming the backbone of a protein molecule but not to show the position of individual amino acids.

As soon as the numbers printed by the computer had been plotted on contour maps we realized that each of the four chains of hemoglobin had a shape closely resembling that of the single chain of myoglobin. The beta chain and myoglobin look like identical twins, and the alpha chains differ from them merely by a shortcut across one small loop [*see illustration below*].

Kendrew's myoglobin had been extracted from the muscle of the sperm whale; the hemoglobin we used came from the blood of horses. More recent observations indicate that the myoglobins of the seal and the horse, and the hemoglobins of man and cattle, all have the same structure. It seems as though the apparently haphazard and irregular folding of the chain is a pattern specifically devised for holding a heme group in place and for enabling it to carry oxygen.

What is it that makes the chain take up this strange configuration? The extension of Kendrew's analysis to a high-

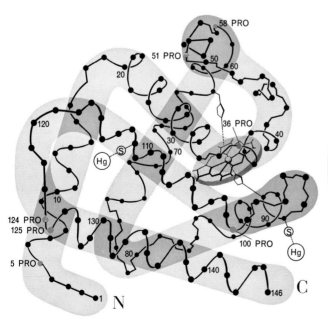

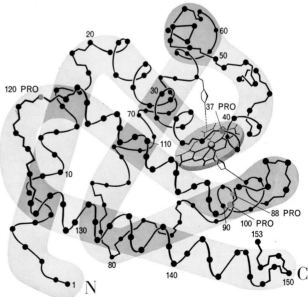

BETA CHAIN AND MYOGLOBIN appear at left and right. Every 10th amino acid subunit is marked, as are proline subunits (*color*), which often coincide with turns in the chain. Balls marked "Hg" show where mercury atoms can be attached to sulfur atoms (S).

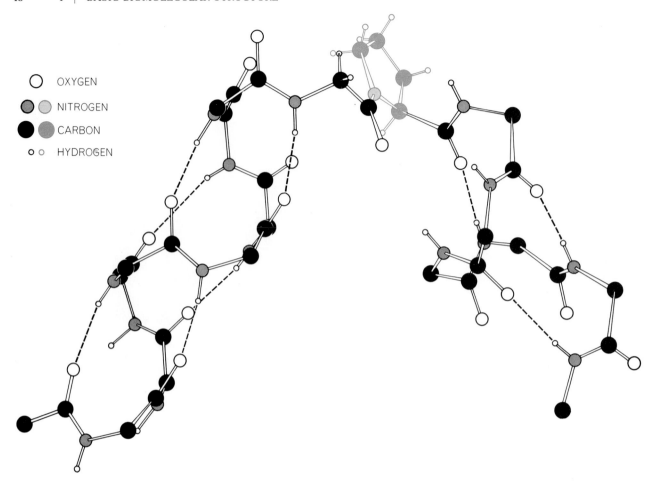

OXYGEN

NITROGEN

CARBON

HYDROGEN

CORNER IN HEMOGLOBIN MOLECULE occurs where a subunit of the amino acid proline (*color*) falls between two helical regions in the beta chain. The chain is shown bare; all hydrogen atoms and amino acid side branches, except for proline, are removed.

er resolution shows that the chain of myoglobin consists of a succession of helical segments interrupted by corners and irregular regions. The helical segments have the geometry of the alpha helix predicted in 1951 by Linus Pauling and Robert B. Corey of the California Institute of Technology. The heme group lies embedded in a fold of the chain, so that only its two acid groups protrude at the surface and are in contact with the surrounding water. Its iron atom is linked to a nitrogen atom of the amino acid histidine.

I have recently built models of the alpha and beta chains of hemoglobin and found that they follow an atomic pattern very similar to that of myoglobin. If two protein chains look the same, one would expect them to have much the same composition. In the language of protein chemistry this implies that in the myoglobins and hemoglobins of all vertebrates the 20 different kinds of amino acid should be present in about the same proportion and arranged in similar sequence.

Enough chemical analyses have been done by now to test whether or not this

is true. Starting at the Rockefeller Institute and continuing in our laboratory, Allen B. Edmundson has determined the sequence of amino acid units in the molecule of sperm-whale myoglobin. The sequences of the alpha and beta chains of adult human hemoglobin have been analyzed independently by Gerhardt Braunitzer and his colleagues at the Max Planck Institute for Biochemistry in Munich, and by William H. Konigsberg, Robert J. Hill and their associates at the Rockefeller Institute. Fetal hemoglobin, a variant of the human adult form, contains a chain known as gamma, which is closely related to the beta chain. Its complete sequence has been analyzed by Walter A. Schroeder and his colleagues at the California Institute of Technology. The sequences of several other species of hemoglobin and that of human myoglobin have been partially elucidated.

The sequence of amino acid units in proteins is genetically determined, and changes arise as a result of mutation. Sickle-cell anemia, for instance, is an inherited disease due to a mutation in one of the hemoglobin genes. The mu-

tation causes the replacement of a single amino acid unit in each of the beta chains. (The glutamic acid unit normally present at position No. 6 is replaced by a valine unit.) On the molecular scale evolution is thought to involve a succession of such mutations, altering the structure of protein molecules one amino acid unit at a time. Consequently when the hemoglobins of different species are compared, we should expect the sequences in man and apes, which are close together on the evolutionary scale, to be very similar, and those of mammals and fishes, say, to differ more widely. Broadly speaking, this is what is found. What was quite unexpected was the degree of chemical diversity among the amino acid sequences of proteins of similar three-dimensional structure and closely related function. Comparison of the known hemoglobin and myoglobin sequences shows only 15 positions—no more than one in 10—where the same amino acid unit is present in all species. In all the other positions one or more replacements have occurred in the course of evolution.

What mechanism makes these diverse

chains fold up in exactly the same way? Does a template force them to take up this configuration, like a mold that forces a car body into shape? Apart from the topological improbability of such a template, all the genetic and physicochemical evidence speaks against it, suggesting instead that the chain folds up spontaneously to assume one specific structure as the most stable of all possible alternatives.

Possible Folding Mechanisms

What is it, then, that makes one particular configuration more stable than all others? The only generalization to emerge so far, mainly from the work of Kendrew, Herman C. Watson and myself, concerns the distribution of the so-called polar and nonpolar amino acid units between the surface and the interior of the molecule.

Some of the amino acids, such as glutamic acid and lysine, have side groups of atoms with positive or negative electric charge, which strongly attract the surrounding water. Amino acid side groups such as glutamine or tyrosine, although electrically neutral as a whole, contain atoms of nitrogen or oxygen in which positive and negative charges are sufficiently separated to form dipoles; these also attract water, but not so strongly as the charged groups do. The attraction is due to a separation of charges in the water molecule itself, making it dipolar. By attaching themselves to electrically charged groups, or to other dipolar groups, the water molecules minimize the strength of the electric fields surrounding these groups and stabilize the entire structure by lowering the quantity known as free energy.

The side groups of amino acids such as leucine and phenylalanine, on the other hand, consist only of carbon and hydrogen atoms. Being electrically neutral and only very weakly dipolar, these groups repel water as wax does. The reason for the repulsion is strange and intriguing. Such hydrocarbon groups, as they are called, tend to disturb the haphazard arrangement of the liquid water molecules around them, making it ordered as it is in ice. The increase in order makes the system less stable; in physical terms it leads to a reduction of the quantity known as entropy, which is the measure of the disorder in a system. Thus it is the water molecules' anarchic distaste for the orderly regimentation imposed on them by the hydrocarbon side groups that forces these side groups to turn away from water and to stick to one another.

Our models have taught us that most electrically charged or dipolar side groups lie at the surface of the protein molecule, in contact with water. Nonpolar side groups, in general, are either confined to the interior of the molecule or so wedged into crevices on its surface as to have the least contact with water. In the language of physics, the distribution of side groups is of the kind leading to the lowest free energy and the highest entropy of the protein molecules and the water around them. (There is a reduction of entropy due to the orderly folding of the protein chain itself, which makes the system less stable, but this is balanced, at moderate temperatures, by the stabilizing contributions of the other effects just described.) It is too early to say whether these are the only generalizations to be made about the forces that stabilize one particular configuration of the protein chain in preference to all others.

At least one amino acid is known to be a misfit in an alpha helix, forcing the chain to turn a corner wherever the unit occurs. This is proline [see illustration on opposite page]. There is, however, only one corner in all the hemoglobins and myoglobins where a proline is always found in the same position: position No. 36 in the beta chain and No. 37 in the myoglobin chain [see bottom illustration on page 47]. At other corners the appearance of prolines is haphazard and changes from species to species. Elkan R. Blout of the Harvard Medical School finds that certain amino acids such as valine or threonine, if present in large numbers, inhibit the formation of alpha helices, but these do not seem to have a decisive influence in myoglobin and hemoglobin.

Since it is easier to determine the sequence of amino acid units in proteins than to unravel their three-dimensional structure by X rays, it would be useful to be able to predict the structure from the sequence. In principle enough is probably known about the forces between atoms and about the way they tend to arrange themselves to make such predictions feasible. In practice the enormous number of different ways in which a long chain can be twisted still makes the problem one of baffling complexity.

Assembling the Four Chains

If hemoglobin consisted of four identical chains, a crystallographer would expect them to lie at the corners of a regular tetrahedron. In such an arrangement each chain can be brought into congruence with any of its three neighbors by a rotation of 180 degrees about one of three mutually perpendicular

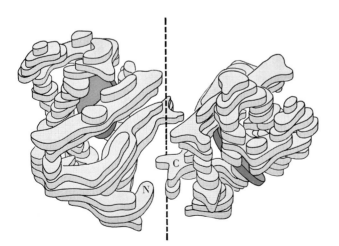

FOUR CHAINS OF HEMOGLOBIN are arranged in symmetrical fashion. Two alpha chains (*left*) and two beta chains (*right*) face each other across an axis of symmetry (*broken vertical lines*). In the assembled molecule the two alpha chains are inverted over the two beta chains and nested down between them. When arranged in this manner, the four chains lie at the corners of a tetrahedron.

axes of symmetry. Since the alpha and beta chains are chemically different, such perfect symmetry is unattainable, but the actual arrangement comes very close to it. As a first step in the assembly of the molecule two alpha chains are placed near a twofold symmetry axis, so that a rotation of 180 degrees brings one chain into congruence with its partner [see illustration on preceding page].

Next the same is done with the two beta chains. One pair, say the alpha chains, is then inverted and placed over the top of the other pair so that the four chains lie at the corners of a tetrahedron. A true twofold symmetry axis now passes vertically through the molecule, and "pseudo-axes" in two directions perpendicular to the first relate the alpha to the beta chains. Thus the arrangement is tetrahedral, but because of the chemical differences between the alpha and beta chains the tetrahedron is not quite regular.

The result is an almost spherical molecule whose exact dimensions are $64 \times 55 \times 50$ angstrom units. It is astonishing to find that four objects as irregular as the alpha and beta chains can fit together so neatly. On formal grounds one would expect a hole to pass through the center of the molecule because chains of amino acid units, being asymmetrical, cannot cross any symmetry axis. Such a hole is in fact found [see top illustration on page 41].

The most unexpected feature of the oxyhemoglobin molecule is the way the four heme groups are arranged. On the basis of their chemical interaction one would have expected them to lie close together. Instead each heme group lies in a separate pocket on the surface of the molecule, apparently unaware of the existence of its partners. Seen at the present resolution, therefore, the structure fails to explain one of the most important physiological properties of hemoglobin.

In 1937 Felix Haurowitz, then at the German University of Prague, discov-

ered an important clue to the molecular explanation of hemoglobin's physiological action. He put a suspension of needle-shaped oxyhemoglobin crystals away in the refrigerator. When he took the suspension out some weeks later, the oxygen had been used up by bacterial infection and the scarlet needles had been replaced by hexagonal plates of purple reduced hemoglobin. While Haurowitz observed the crystals under the microscope, oxygen penetrated between the slide and the cover slip, causing the purple plates to dissolve and the scarlet needles of hemoglobin to re-form. This transformation convinced Haurowitz that the reaction of hemoglobin with oxygen must be accompanied by a change in the structure of the hemoglobin molecule. In myoglobin, on the other hand, no evidence for such a change has been detected.

Haurowitz' observation and the enigma posed by the structure of oxyhemoglobin caused me to persuade a graduate student, Hilary Muirhead, to attempt an X-ray analysis at low resolution of the reduced form. For technical reasons human rather than horse hemoglobin was used at first, but we have now found that the reduced hemoglobins of man and the horse have very similar structures, so that the species does not matter here.

Unlike me, Miss Muirhead succeeded in solving the structure of her protein in time for her Ph.D. thesis. When we examined her first electron-density maps, we looked for two kinds of structural change: alterations in the folding of the individual chains and displacements of the chains with respect to each other. We could detect no changes in folding large enough to be sure that they were not due to experimental error. We did discover, however, that a striking displacement of the beta chains had taken place. The gap between them had widened and they had been shifted sideways, increasing the distance between their respective iron atoms from 33.4 to 40.3 angstrom units [see illustration on opposite page]. The arrangement of the two alpha chains had remained unaltered, as far as we could judge, and the distance between the iron atoms in the beta chains and their nearest neighbors in the alpha chains had also remained the same. It looked as though the two beta chains had slid apart, losing contact with each other and somewhat changing their points of contact with the alpha chains.

F. J. W. Roughton and others at the University of Cambridge suggest that the change to the oxygenated form of

RESIDUE NUMBER	HEMOGLOBIN			MYOGLOBIN
	ALPHA	BETA	GAMMA	
81	MET	LEU	LEU	HIS
82	PRO	LYS	LYS	GLU
83	ASN	GLY	GLY	ALA
84	ALA	THR	THR	GLU
85	LEU	PHE	PHE	LEU
86	SER	ALA	ALA	LYS
87	ALA	THR	GLN	PRO
88	LEU	LEU	LEU	LEU
89	SER	SER	SER	ALA
90	ASP	GLU	GLU	GLN
91	LEU	LEU	LEU	SER
92	HIS	HIS	HIS	HIS
93	ALA	CYS	CYS	ALA
94	HIS	ASP	ASN	THR
95	LYS	LYS	LYS	LYS
96	LEU	LEU	LEU	HIS
97	ARG	HIS	HIS	LYS
98	VAL	VAL	VAL	ILEU
99	ASP	ASP	ASP	PRO
100	PRO	PRO	PRO	ILEU
101	VAL	GLU	GLU	LYS
102	ASP	ASN	ASN	TYR

ALA ALANINE	GLY GLYCINE	PRO PROLINE
ARG ARGININE	HIS HISTIDINE	SER SERINE
ASN ASPARAGINE	ILEU ISOLEUCINE	THR THREONINE
ASP ASPARTIC ACID	LEU LEUCINE	TYR TYROSINE
CYS CYSTEINE	LYS LYSINE	VAL VALINE
GLN GLUTAMINE	MET METHIONINE	
GLU GLUTAMIC ACID	PHE PHENYLALANINE	

AMINO ACID SEQUENCES are shown for corresponding stretches of the alpha and beta chains of hemoglobin from human adults, the gamma chain that replaces the beta chain in fetal human hemoglobin and sperm-whale myoglobin. Colored bars show where the same amino acid units are found either in all four chains or in the first three. Site numbers for the alpha chain and myoglobin are adjusted slightly because they contain a different number of amino acid subunits overall than do the beta and gamma chains. Over their full length of more than 140 subunits the four chains have only 20 amino acid subunits in common.

hemoglobin takes place after three of the four iron atoms have combined with oxygen. When the change has occurred, the rate of combination of the fourth iron atom with oxygen is speeded up several hundred times. Nothing is known as yet about the atomic mechanism that sets off the displacement of the beta chains, but there is one interesting observation that allows us at least to be sure that the interaction of the iron atoms and the change of structure do not take place unless alpha and beta chains are both present.

Certain anemia patients suffer from a shortage of alpha chains; the beta chains, robbed of their usual partners, group themselves into independent assemblages of four chains. These are known as hemoglobin H and resemble normal hemoglobin in many of their properties. Reinhold Benesch and Ruth E. Benesch of the Columbia University College of Physicians and Surgeons have discovered, however, that the four iron atoms in hemoglobin H do not interact, which led them to predict that the combination of hemoglobin H with oxygen should not be accompanied by a change of structure. Using crystals grown by Helen M. Ranney of the Albert Einstein College of Medicine, Lelio Mazzarella and I verified this prediction. Oxygenated and reduced hemoglobin H both resemble normal human reduced hemoglobin in the arrangement of the four chains.

The rearrangement of the beta chains must be set in motion by a series of atomic displacements starting at or near the iron atoms when they combine with oxygen. Our X-ray analysis has not yet reached the resolution needed to discern these, and it seems that a deeper understanding of this intriguing phenomenon may have to wait until we succeed in working out the structures of reduced hemoglobin and oxyhemoglobin at atomic resolution.

Allosteric Enzymes

There are many analogies between the chemical activities of hemoglobin and those of enzymes catalyzing chemical reactions in living cells. These analogies lead one to expect that some enzymes may undergo changes of structure on coming into contact with the substances whose reactions they catalyze. One can imagine that the active sites of these enzymes are moving mechanisms rather than static surfaces magically endowed with catalytic properties.

Indirect and tentative evidence suggests that changes of structure involv-

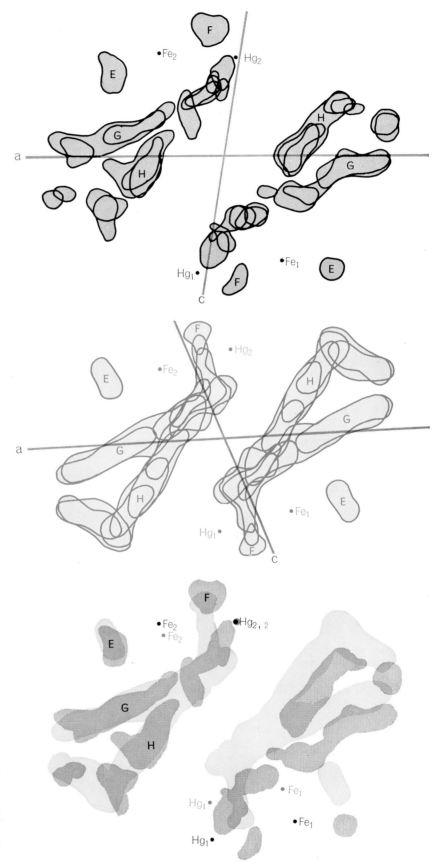

MOVEMENT OF HEMOGLOBIN CHAINS was discovered by comparing portions of the two beta chains in "reduced" (oxygen-free) human hemoglobin (top) with the same portions of horse hemoglobin containing oxygen (middle). The bottom illustration shows the outlines of the top and middle pictures superposed so that the mercury atoms (Hg_2) and helical regions (E, F, G, H) of the two chains at left coincide. The iron atoms (Fe_2) do not quite match. The chains at right are now seen to be shifted with respect to each other.

ing a rearrangement of subunits like that of the alpha and beta chains of hemoglobin do indeed occur and that they may form the basis of a control mechanism known as feedback inhibition. This is a piece of jargon that biochemistry has borrowed from electrical engineering, meaning nothing more complicated than that you stop being hungry when you have had enough to eat.

Constituents of living matter such as amino acids are built up from simpler substances in a series of small steps, each step being catalyzed by an enzyme that exists specifically for that purpose. Thus a whole series of different enzymes may be needed to make one amino acid. Such a series of enzymes appears to have built-in devices for ensuring the right balance of supply and demand. For example, in the colon bacillus the amino acid isoleucine is made from the amino acid threonine in several steps. The first enzyme in the series has an affinity for threonine: it catalyzes the removal of an amino group from it. H. Edwin Umbarger of the Long Island Biological Association in Cold Spring Harbor, N.Y., discovered that the action of the enzyme is inhibited by isoleucine, the end product of the last enzyme in the series. Jean-Pierre Changeux of the Pasteur Institute later showed that isoleucine acts not, as one might have expected, by blocking the site on the enzyme molecule that would otherwise combine with threonine but probably by combining with a different site on the molecule.

The two sites on the molecule must therefore interact, and Jacques Monod, Changeux and François Jacob have suggested that this is brought about by a rearrangement of subunits similar to that which accompanies the reaction of hemoglobin with oxygen. The enzyme is thought to exist in two alternative structural states: a reactive one when the supply of isoleucine has run out and an unreactive one when the supply exceeds demand. The discoverers have coined the name "allosteric" for enzymes of this kind.

The molecules of the enzymes suspected of having allosteric properties are all large ones, as one would expect them to be if they are made up of several subunits. This makes their X-ray analysis difficult. It may not be too hard to find out, however, whether or not a change of structure occurs, even if it takes a long time to unravel it in detail. In the meantime hemoglobin will serve as a useful model for the behavior of more complex enzyme systems.

The Structure of the Hereditary Material

5

F. H. C. Crick
October 1954

An account of the investigations which have led to the formulation of an understandable structure for DNA. The chemical reactions of this material within the nucleus govern the process of reproduction

Viewed under a microscope, the process of mitosis, by which one cell divides and becomes two, is one of the most fascinating spectacles in the whole of biology. No one who watches the event unfold in speeded-up motion pictures can fail to be excited and awed. As a demonstration of the powers of dynamic organization possessed by living matter, the act of division is impressive enough, but even more stirring is the appearance of two identical sets of chromosomes where only one existed before. Here lies biology's greatest challenge: How are these fundamental bodies duplicated? Unhappily the copying process is beyond the resolving power of microscopes, but much is being learned about it in other ways.

One approach is the study of the nature and behavior of whole living cells; another is the investigation of substances extracted from them. This article will discuss only the second approach, but both are indispensable if we are ever to solve the problem; indeed some of the most exciting results are being obtained by what might loosely be described as a combination of the two methods.

Chromosomes consist mainly of three kinds of chemical: protein, desoxyribonucleic acid (DNA) and ribonucleic acid (RNA). (Since RNA is only a minor component, we shall not consider it in detail here.) The nucleic acids and the proteins have several features in common. They are all giant molecules, and each type has the general structure of a main backbone with side groups attached. The proteins have about 20 different kinds of side groups; the nucleic acids usually only four (and of a different type). The smallness of these numbers itself is striking, for there is no obvious chemical reason why many more types of side groups should not occur. Another interesting feature is that no protein or nucleic acid occurs in more than one optical form; there is never an optical isomer, or mirror-image molecule. This shows that the shape of the molecules must be important.

These generalizations (with minor exceptions) hold over the entire range of living organisms, from viruses and bacteria to plants and animals. The impression is inescapable that we are dealing with a very basic aspect of living matter, and one having far more simplicity than we would have dared to hope. It encourages us to look for simple explanations for the formation of these giant molecules.

The most important role of proteins is that of the enzymes—the machine tools of the living cell. An enzyme is specific, often highly specific, for the reaction which it catalyzes. Moreover, chemical and X-ray studies suggest that the structure of each enzyme is itself rigidly determined. The side groups of a given enzyme are probably arranged in a fixed order along the polypeptide backbone. If we could discover how a cell produces the appropriate enzymes, in particular how it assembles the side groups of each enzyme in the correct order, we should have gone a long way toward explaining the simpler forms of life in terms of physics and chemistry.

We believe that this order is controlled by the chromosomes. In recent years suspicion has been growing that the key to the specificity of the chromosomes lies not in their protein but in their DNA. DNA is found in all chromosomes —and only in the chromosomes (with minor exceptions). The amount of DNA per chromosome set is in many cases a fixed quantity for a given species. The sperm, having half the chromosomes of the normal cell, has about half the amount of DNA, and tetraploid cells in the liver, having twice the normal chromosome complement, seem to have twice the amount of DNA. This constancy of the amount of DNA is what one might expect if it is truly the material that determines the hereditary pattern.

Then there is suggestive evidence in two cases that DNA alone, free of protein, may be able to carry genetic information. The first of these is the discovery that the "transforming principles" of bacteria, which can produce an inherited change when added to the cell, appear to consist only of DNA. The second is the fact that during the infection of a bacterium by a bacteriophage the DNA of the phage penetrates into the bacterial cell while most of the protein, perhaps all of it, is left outside.

The Chemical Formula

DNA can be extracted from cells by mild chemical methods, and much experimental work has been carried out to discover its chemical nature. This work has been conspicuously successful. It is now known that DNA consists of a very long chain made up of alternate sugar and phosphate groups [see diagram on below]. The sugar is always the same sugar, known as desoxyribose. And it is always joined onto the phosphate in the same way, so that the long chain is perfectly regular, repeating the same phosphate-sugar sequence over and over again.

But while the phosphate-sugar chain is perfectly regular, the molecule as a whole is not, because each sugar has a "base" attached to it and the base is not always the same. Four different types of base are commonly found: two of them are purines, called adenine and guanine, and two are pyrimidines, known as thymine and cytosine. So far as is known the order in which they follow one another along the chain is irregular, and probably varies from one piece of DNA to another. In fact, we suspect that the order of the bases is what confers specificity on a given DNA. Because the sequence of the bases is not known, one can only say that the general formula for DNA is established. Nevertheless this formula should be reckoned one of the major achievements of biochemistry, and it is the foundation for all the ideas described in the rest of this article.

At one time it was thought that the four bases occurred in equal amounts, but in recent years this idea has been shown to be incorrect. E. Chargaff and his colleagues at Columbia University, A. E. Mirsky and his group at the Rockefeller Institute for Medical Research and G. R. Wyatt of Canada have accurately measured the amounts of the bases in many instances and have shown that the relative amounts appear to be fixed for any given species, irrespective of the individual or the organ from which the DNA was taken. The proportions usually differ for DNA from different species, but species related to one another may not differ very much.

Although we know from the chemical formula of DNA that it is a chain, this does not in itself tell us the shape of the molecule, for the chain, having many single bonds around which it may rotate, might coil up in all sorts of shapes. However, we know from physical-chemical measurements and electron-microscope pictures that the molecule usually is long, thin and fairly straight, rather like a stiff bit of cord. It is only about 20 Angstroms thick (one Angstrom = one 100-millionth of a centimeter). This is very small indeed, in fact not much more than a dozen atoms thick.

The length of the DNA seems to depend somewhat on the method of preparation. A good sample may reach a length of 30,000 Angstroms, so that the structure is more than 1,000 times as long as it is thick. The length inside the cell may be much greater than this, because there is always the chance that the extraction process may break it up somewhat.

Pictures of the Molecule

None of these methods tells us anything about the detailed arrangement in space of the atoms inside the molecule. For this it is necessary to use X-ray diffraction. The average distance between bonded atoms in an organic molecule is about 1½ Angstroms; between unbonded atoms, three to four Angstroms. X-rays have a small enough wavelength (1½ Angstroms) to resolve the atoms, but unfortunately an X-ray diffraction photograph is not a picture in the ordinary sense of the word. We cannot focus X-rays as we can ordinary light; hence a picture can be obtained only by roundabout methods. Moreover, it can show clearly only the periodic, or regularly repeated, parts of the structure.

With patience and skill several English workers have obtained good diffraction pictures of DNA extracted from cells and drawn into long fibers. The first studies, even before details emerged, produced two surprises. First, they revealed that the DNA structure could take two forms. In relatively low hu-

midity, when the water content of the fibers was about 40 per cent, the DNA molecules gave a crystalline pattern, showing that they were aligned regularly in all three dimensions. When the humidity was raised and the fibers took up more water, they increased in length by about 30 per cent and the pattern tended to become "paracrystalline," which means that the molecules were packed side by side in a less regular manner, as if the long molecules could slide over one another somewhat. The second surprising result was that DNA from different species appeared to give identical X-ray patterns, despite the fact that the amounts of the four bases present varied. This was particularly odd because of the existence of the crystalline form just mentioned. How could the structure appear so regular when the bases varied? It seemed that the broad arrangement of the molecule must be independent of the exact sequence of the bases, and it was therefore thought that the bases play no part in holding the structure together. As we shall see, this turned out to be wrong.

The early X-ray pictures showed a third intriguing fact: namely, that the repeats in the crystallographic pattern came at much longer intervals than the chemical repeat units in the molecule. The distance from one phosphate to the next cannot be more than about seven Angstroms, yet the crystallographic repeat came at intervals of 28 Angstroms in the crystalline form and 34 Angstroms

FRAGMENT OF CHAIN of deoxyribonucleic acid shows the three basic units that make up the molecule. Repeated over and over in a long chain, they make it 1,000 times as long

in the paracrystalline form; that is, the chemical unit repeated several times before the structure repeated crystallographically.

J. D. Watson and I, working in the Medical Research Council Unit in the Cavendish Laboratory at Cambridge, were convinced that we could get somewhere near the DNA structure by building scale models based on the X-ray patterns obtained by M. H. F. Wilkins, Rosalind Franklin and their co-workers at Kings' College, London. A great deal is known about the exact distances between bonded atoms in molecules, about the angles between the bonds and about the size of atoms—the so-called van der Waals' distance between adjacent nonbonded atoms. This information is easy to embody in scale models. The problem is rather like a three-dimensional jig saw puzzle with curious pieces joined together by rotatable joints (single bonds between atoms).

The Helix

To get anywhere at all we had to make some assumptions. The most important one had to do with the fact that the crystallographic repeat did not coincide with the repetition of chemical units in the chain but came at much longer intervals. A possible explanation was that all the links in the chain were the same but the X-rays were seeing every tenth link, say, from the same angle and the others from different angles. What sort of chain might produce this pattern? The answer was easy: the chain might be coiled in a helix. (A helix is often loosely called a spiral; the distinction is that a helix winds not around a cone but around a cylinder, as a winding staircase usually does.) The distance between crystallographic repeats would then correspond to the distance in the chain between one turn of the helix and the next.

We had some difficulty at first because we ignored the bases and tried to work only with the phosphate-sugar backbone. Eventually we realized that we had to take the bases into account, and this led us quickly to a structure which we now believe to be correct in its broad outlines.

This particular model contains a pair of DNA chains wound around a common axis. The two chains are linked together by their bases. A base on one chain is joined by very weak bonds to a base at the same level on the other chain, and all the bases are paired off in this way right along the structure. In the diagram on page 56, the two ribbons represent the phosphate-sugar chains, and the pairs of bases holding them together are symbolized as horizontal rods. Paradoxically, in order to make the structure as symmetrical as possible we had to have the two chains run in opposite directions; that is, the sequence of the atoms goes one way in one chain and the opposite way in the other. Thus the figure looks exactly the same whichever end is turned up.

Now we found that we could not arrange the bases any way we pleased; the four bases would fit into the structure only in certain pairs. In any pair there must always be one big one (purine) and one little one (pyrimidine). A pair of pyrimidines is too short to bridge the gap between the two chains, and a pair of purines is too big to fit into the space.

At this point we made an additional assumption. The bases can theoretically exist in a number of forms depending upon where the hydrogen atoms are attached. We assumed that for each base one form was much more probable than all the others. The hydrogen atoms can be thought of as little knobs attached to the bases, and the way the bases fit together depends crucially upon where these knobs are. With this assumption the only possible pairs that will fit in are: adenine with thymine and guanine with cytosine.

The way these pairs are formed is shown in the diagrams on page 58. The dotted lines show the hydrogen bonds, which hold the two bases of a pair together. They are very weak bonds; their energy is not many times greater than the energy of thermal vibration at room temperature. (Hydrogen bonds are the main forces holding different water molecules together, and it is because of them that water is a liquid at room temperatures and not a gas.)

Adenine must always be paired with

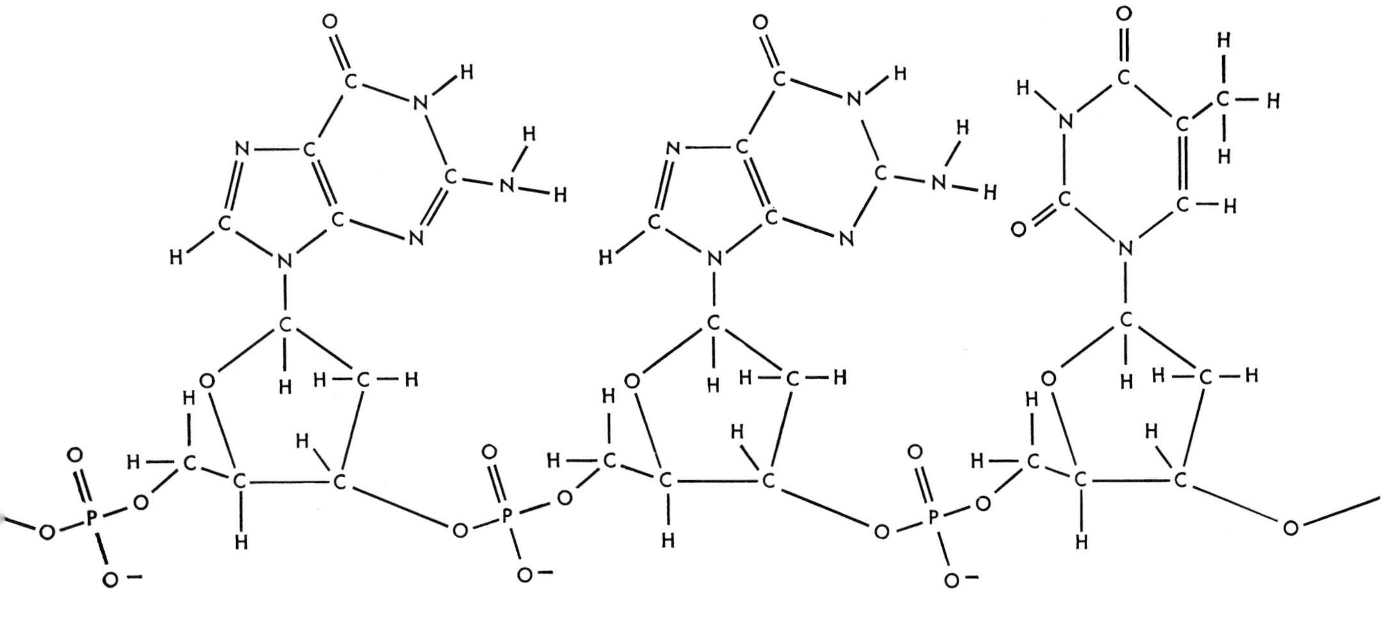

as it is thick. The backbone is made up of pentose sugar molecules (marked by the middle colored square), linked by phosphate groups (bottom square). The bases (top square), adenine, cytosine, guanine and thymine protrude off each sugar in irregular order.

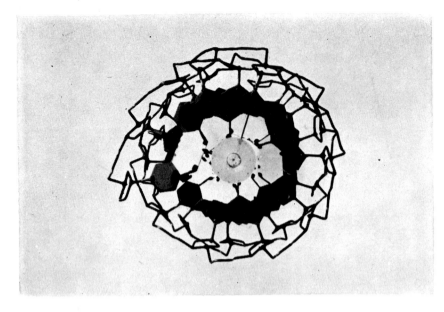

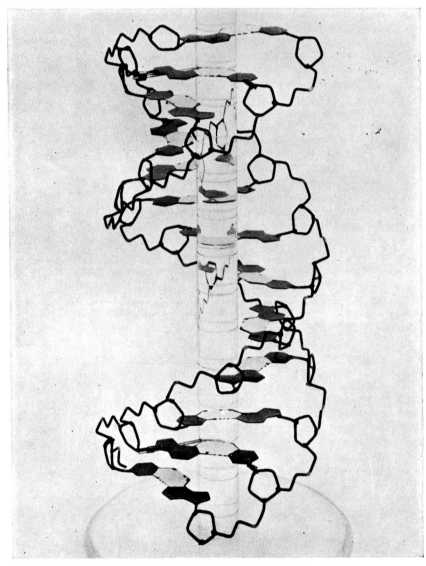

STRUCTURAL MODEL shows a pair of DNA chains wound as a helix about the fiber axis. The pentose sugars can be plainly seen. From every one on each chain protrudes a base, linked to an opposing one at the same level by a hydrogen bond. These base-to-base links act as horizontal supports, holding the chains together. Upper photograph is a top view.

thymine, and guanine with cytosine; it is impossible to fit the bases together in any other combination in our model. (This pairing is likely to be so fundamental for biology that I cannot help wondering whether some day an enthusiastic scientist will christen his new-born twins Adenine and Thymine!) The model places no restriction, however, on the sequence of pairs along the structure. Any specified pair can follow any other. This is because a pair of bases is flat, and since in this model they are stacked roughly like a pile of coins, it does not matter which pair goes above which.

It is important to realize that the specific pairing of the bases is the direct result of the assumption that both phosphate-sugar chains are helical. This regularity implies that the distance from a sugar group on one chain to that on the other at the same level is always the same, no matter where one is along the chain. It follows that the bases linked to the sugars always have the same amount of space in which to fit. It is the regularity of the phosphate-sugar chains, therefore, that is at the root of the specific pairing.

The Picture Clears

At the moment of writing, detailed interpretation of the X-ray photographs by Wilkins' group at Kings' College has not been completed, and until this has been done no structure can be considered proved. Nevertheless there are certain features of the model which are so strongly supported by the experimental evidence that it is very likely they will be embodied in the final correct structure. For instance, measurements of the density and water content of the DNA fibers, taken with evidence showing that the fibers can be extended in length, strongly suggest that there are two chains in the structural unit of DNA. Again, recent X-ray pictures have shown clearly a most striking general pattern which we can now recognize as the characteristic signature of a helical structure. In particular there are a large number of places where the diffracted intensity is zero or very small, and these occur exactly where one expects from a helix of this sort. Another feature one would expect is that the X-ray intensities should approach cylindrical symmetry, and it is now known that they do this. Recently Wilkins and his co-workers have given a brilliant analysis of the details of the X-ray pattern of the crystalline form, and have shown that they

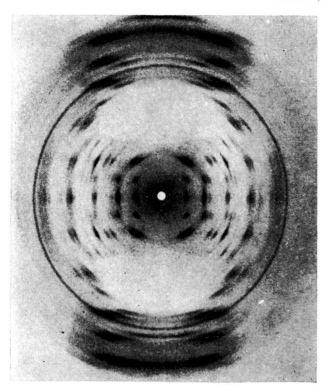

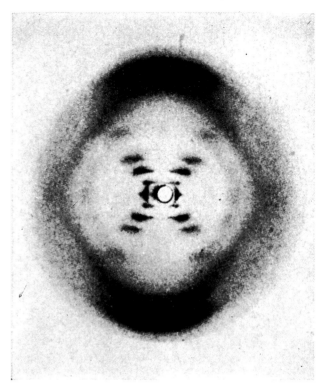

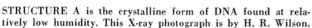

STRUCTURE A is the crystalline form of DNA found at relatively low humidity. This X-ray photograph is by H. R. Wilson.

STRUCTURE B is the paracrystalline form of DNA. The molecules are less regularly arranged. Picture is by R. E. Franklin.

are consistent with a structure of this type, though in the crystalline form the bases are tilted away from the fiber axis instead of perpendicular, as in our model. Our construction was based on the paracrystalline form.

Many of the physical and chemical properties of DNA can now be understood in terms of this model. For example, the comparative stiffness of the structure explains rather naturally why DNA keeps a long, fiber-like shape in solution. The hydrogen bonds of the bases account for the behavior of DNA in response to changes in pH. Most striking of all is the fact that in every kind of DNA so far examined—and over 40 have been analyzed—the amount of adenine is about equal to the amount of thymine and the guanine equal to the cytosine, while the cross-ratios (between, say, adenine and guanine) can vary considerably from species to species. This remarkable fact, first pointed out by Chargaff, is exactly what one would expect according to our model, which requires that every adenine be paired with a thymine and every guanine with a cytosine.

It may legitimately be asked whether the artificially prepared fibers of extracted DNA, on which our model is based, are really representative of intact DNA in the cell. There is every indication that they are. It is difficult to see

how the very characteristic features of the model could be produced as artefacts by the extraction process. Moreover, Wilkins has shown that intact biological material, such as sperm heads and bacteriophage, gives X-ray patterns very similar to those of the extracted fibers.

The present position, therefore, is that in all likelihood this statement about DNA can safely be made: its structure consists of two helical chains wound around a common axis and held together by hydrogen bonds between specific pairs of bases.

The Mold

Now the exciting thing about a model of this type is that it immediately suggests how the DNA might produce an exact copy of itself. The model consists of two parts, each of which is the complement of the other. Thus either chain may act as a sort of mold on which a complementary chain can be synthesized. The two chains of a DNA, let us say, unwind and separate. Each begins to build a new complement onto itself. When the process is completed, there are two pairs of chains where we had only one. Moreover, because of the specific pairing of the bases the sequence of the pairs of bases will have been duplicated exactly; in other words, the mold has not only assembled the build-

ing blocks but has put them together in just the right order.

Let us imagine that we have a single helical chain of DNA, and that floating around it inside the cell is a supply of precursors of the four sorts of building blocks needed to make a new chain. Unfortunately we do not know the makeup of these precursor units; they may be, but probably are not, nucleotides, consisting of one phosphate, one sugar and one base. In any case, from time to time a loose unit will attach itself by its base to one of the bases of the single DNA chain. Another loose unit may attach itself to an adjoining base on the chain. Now if one or both of the two newly attached units is not the correct mate for the one it has joined on the chain, the two newcomers will be unable to link together, because they are not the right distance apart. One or both will soon drift away, to be replaced by other units. When, however, two adjacent newcomers are the correct partners for their opposite numbers on the chain, they will be in just the right position to be linked together and begin to form a new chain. Thus only the unit with the proper base will gain a permanent hold at any given position, and eventually the right partners will fill in the vacancies all along the forming chain. While this is going on, the other single chain of the original pair also will

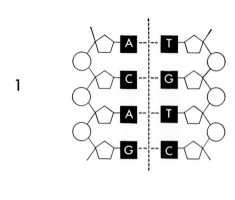

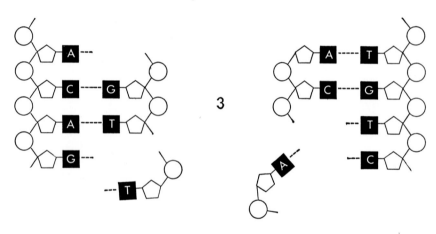

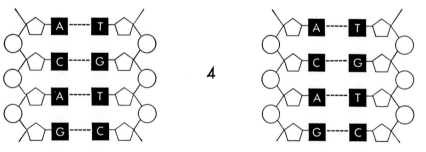

REPLICATION mechanism by which DNA might duplicate itself is shown in diagram. A helix of two DNA chains unwinds and separates (1). Two complementary chains of DNA (2) within the cell begin to attach DNA precursor units floating loosely (3). When the proper bases are joined, two new helixes will build up (4). Letters represent the bases.

be forming a new chain complementary to itself.

At the moment this idea must be regarded simply as a working hypothesis. Not only is there little direct evidence for it, but there are a number of obvious difficulties. For example, certain organisms contain small amounts of a fifth base, 5-methyl cytosine. So far as the model is concerned, 5-methyl cytosine fits just as well as cytosine and it may turn out that it does not matter to the organism which is used, but this has yet to be shown.

A more fundamental difficulty is to explain how the two chains of DNA are unwound in the first place. There would have to be a lot of untwisting, for the total length of all the DNA in a single chromosome is something like four centimeters (400 million Angstroms). This means that there must be more than 10 million turns in all, though the DNA may not be all in one piece.

The duplicating process can be made to appear more plausible by assuming that the synthesis of the two new chains begins as soon as the two original chains start to unwind, so that only a short stretch of the chain is ever really single. In fact, we may postulate that it is the growth of the two new chains that unwinds the original pair. This is likely in terms of energy because, for every hydrogen bond that has to be broken, two new ones will be forming. Moreover, plausibility is added to the idea by the fact that the paired chain forms a rather stiff structure, so that the growing chain would tend to unwind the old pair.

The difficulty of untwisting the two chains is a topological one, and is due to the fact that they are intertwined. There would be no difficulty in "unwinding" a single helical chain, because there are so many single bonds in the chain about which rotation is possible. If in the twin structure one chain should break, the other one could easily spin around. This might relieve accumulated strain, and then the two ends of the broken chain, still being in close proximity, might be joined together again. There is even some evidence suggesting that in the process of extraction the chains of DNA may be broken in quite a number of places and that the structure nevertheless holds together by means of the hydrogen bonding, because there is never a break in both chains at the same level. Nevertheless, in spite of these tentative suggestions, the difficulty of untwisting remains a formidable one.

There remains the fundamental puzzle as to how DNA exerts its hereditary

ONE LINKAGE of base to base across the pair of DNA chains is between adenine and thymine. For the structure proposed, the link of a large base with a small one is required to fit chains together.

ANOTHER LINKAGE is comprised of guanine with cytosine. Assuming the existence of hydrogen bonds between the bases, these two pairings, and only these, will explain the actual configuration.

influence. A genetic material must carry out two jobs: duplicate itself and control the development of the rest of the cell in a specific way. We have seen how it might do the first of these, but the structure gives no obvious clue concerning how it may carry out the second. We suspect that the sequence of the bases acts as a kind of genetic code. Such an arrangement can carry an enormous amount of information. If we imagine that the pairs of bases correspond to the dots and dashes of the Morse code, there is enough DNA in a single cell of the human body to encode about 1,000 large textbooks. What we want to know, however, is just how this is done in terms of atoms and molecules. In particular, what precisely is it a code for? As we have seen, the three key components of living matter—protein, RNA and DNA—are probably all based on the same general plan. Their backbones are regular, and the variety comes from the sequence of the side groups. It is therefore very natural to suggest that the sequence of the bases of the DNA is in some way a code for the sequence of the

amino acids in the polypeptide chains of the proteins which the cell must produce. The physicist George Gamow has recently suggested in a rather abstract way how this information might be transmitted, but there are some difficulties with the actual scheme he has proposed, and so far he has not shown how the idea can be translated into precise molecular configurations.

What then, one may reasonably ask, are the virtues of the proposed model, if any? The prime virtue is that the configuration suggested is not vague but can be described in terms acceptable to a chemist. The pairing of the bases can be described rather exactly. The precise positions of the atoms of the backbone is less certain, but they can be fixed within limits, and detailed studies of the X-ray data, now in progress at Kings' College, may narrow these limits considerably. Then the structure brings together two striking pieces of evidence which at first sight seem to be unrelated—the analytical data, showing the one-to-one ratios for adenine-thymine and guanine-cytosine, and the helical

nature of the X-ray pattern. These can now be seen to be two facets of the same thing. Finally, is it not perhaps a remarkable coincidence, to say the least, to find in this key material a structure of exactly the type one would need to carry out a specific replication process; namely, one showing both variety and complementarity?

The model is also attractive in its simplicity. While it is obvious that whole chromosomes have a fairly complicated structure, it is not unreasonable to hope that the molecular basis underlying them may be rather simple. If this is so, it may not prove too difficult to devise experiments to unravel it. It would, of course, help enormously if biochemists could discover the immediate precursors of DNA. If we knew the monomers from which nature makes DNA, RNA and protein, we might be able to carry out very spectacular experiments in the test tube. Be that as it may, we now have for the first time a well-defined model for DNA and for a possible replication process, and this in itself should make it easier to devise crucial experiments.

6 Hybrid Nucleic Acids

by S. Spiegelman
May 1964

One strand of nucleic acid will combine with another wherever the subunits of the two strands are complementary. Artificial combinations clarify the flow of information in the living cell

One of the most useful techniques for studying how genes work depends on the remarkable fact that certain chainlike molecules found in the living cell can "recognize" other chains whose molecular composition is complementary to their own. If one molecule is composed of subunits that can be symbolized by the sequence CATCATCAT..., it will recognize the complementary sequence GTAGTA-GTA... in a second molecule. As we shall see, these particular letters represent the chemical subunits that transmit the genetic information. When two such complementary chains are brought together under suitable conditions, they will "hybridize," or combine, to form a double-strand molecule in which the subunits C and G and A and T are linked by the weak chemical bond known as the hydrogen bond. This article will describe how hybridization has been exploited to study the cell's mechanism for manufacturing proteins.

A typical living cell synthesizes hundreds of different proteins, most of which serve as the enzymes, or biological catalysts, that mediate the myriad chemical reactions involved in growth and reproduction. Proteins are large chainlike molecules made out of some 20 different kinds of amino acids. According to current theory the sequence of amino acid units in a protein is specified by a single gene, and the genes are strung together in the chainlike molecules of deoxyribonucleic acid (DNA). The subunits of DNA that constitute the genetic code are four "bases": adenine (A), thymine (T), guanine (G) and cytosine (C). Normally DNA consists of two complementary chains linked by hydrogen bonds to form a double helix. Wherever A occurs in one chain, T occurs in the other; similarly,

G pairs with C. It is evident that each chain contains all the information needed to specify the complementary chain.

The flow of information in a cell begins with the base-pairing found in the double helix of DNA. Three principal modes of information transfer are distinguished by the end purposes they serve [see illustrations on opposite page]. The first is a duplication, which provides exact copies of the DNA molecule for transmission from one generation of cells to the next. The copying process utilizes the same "language" and the same "alphabet" that are present in the original material.

The second mode of transfer is a "transcription," which uses the same language but a slightly different alphabet. In this step DNA is transcribed into ribonucleic acid (RNA), a chainlike molecule that, like DNA, has four code units. Three are the same as those found in DNA: A, G and C. The fourth is uracil (U), which takes the place of thymine (T). One particular variety of RNA carries the actual program for protein synthesis. Although this variety of RNA is frequently called "messenger RNA," I prefer to speak of "translatable RNA" or "RNA messages." A "messenger" cannot be translated, but a message can.

The third mode of information transfer converts the information from the four-element language of translatable RNA to the 20-element language of the proteins. This step is properly regarded as a translation. Since every translation calls for a dictionary, it is not surprising that the cell uses one also. The cellular dictionary is made up of a collection of comparatively small RNA molecules known as transfer RNA (or soluble

RNA), which have the task of delivering specific amino acids to the site of protein synthesis. Each amino acid is attached to a transfer-RNA molecule by a specific activating enzyme.

The actual synthesis of protein molecules is accomplished with the help of ribosomes, which evidently serve to hold the translatable RNA "tape" in position while the message is being "read." Ribosomes are small spherical particles composed of protein and two kinds of RNA. One kind is about a million times heavier than a hydrogen atom; the other is about 600,000 times heavier. They are respectively called 23S RNA and 16S RNA, designations that refer to how fast they settle out of solution when they are spun at high speed in an ultracentrifuge.

Thus we see that cellular RNA is divided into two major categories: translatable and nontranslatable. The translatable variety (messenger RNA) constitutes only about 5 per cent of all the RNA in a cell; it is usually unstable and must be continuously resynthesized. The nontranslatable varieties of RNA (transfer RNA and the two kinds of ribosomal RNA) make up about 95 per cent of the RNA found in a cell and are extremely stable.

This picture of the genetic mechanism has arisen from the contributions of a large number of investigators using a wide variety of methods of analyzing gene function. I shall focus attention on some of the things that have been learned about the translatable and nontranslatable forms of RNA by exploiting the ability of RNA to hybridize with DNA of complementary composition. In effect this technique enables one to return an RNA molecule to the site of its synthesis on a particular stretch of DNA.

Early in 1958 my colleagues Masayasu Nomura and Benjamin D. Hall and I at the University of Illinois undertook to re-examine a remarkable experiment described in 1955 by Elliot Volkin and Lazarus Astrachan of the Oak Ridge National Laboratory. These workers had used radioactive isotopes to identify and study the RNA produced when the colon bacillus is infected with the bacterial virus designated T2. Infection occurs when T2 injects into the cell of the bacterium a double helix of DNA bearing all the information needed for the synthesis of new virus particles. Volkin and Astrachan had concluded that the RNA synthesized in the infected cells mimicked the composition of the T2 DNA.

At the time neither the experimenters nor anyone else thought that the RNA might represent a genetic message formed on a DNA template. It was suggested, rather, that this new kind of

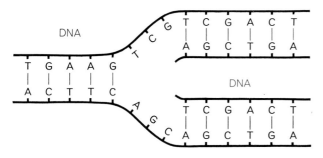 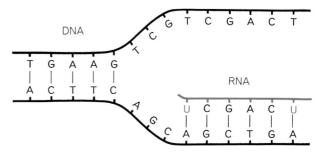

FLOW OF GENETIC INFORMATION involves duplication (*left*), transcription (*right*) and translation (*below*). Genetic information resides in giant chainlike molecules of deoxyribonucleic acid (DNA), in which the code "letters" are four bases: adenine (A), thymine (T), guanine (G) and cytosine (C). DNA normally consists of two complementary strands in which A pairs with T and G with C. During duplication, by an unknown mechanism, a new complementary strand is synthesized on each of the parent strands. In transcription only one strand of the DNA serves as a template and the new molecule formed is ribonucleic acid (RNA). In RNA the base uracil (U) takes the place of thymine as the partner of adenine. RNA molecules can be translatable or nontranslatable.

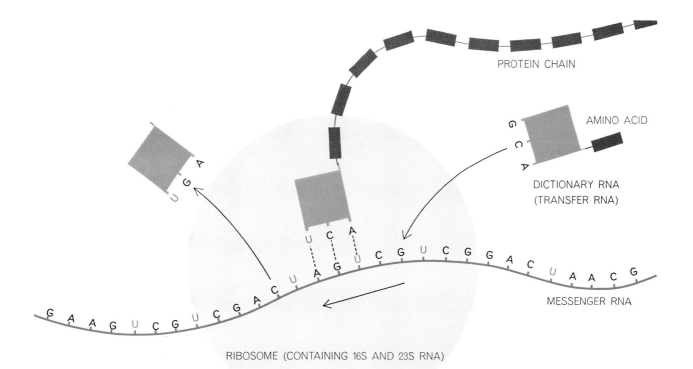

TRANSLATION PROCESS converts genetic information from the four-letter "language" of nucleic acids (DNA and RNA) into the 20-letter language of proteins. The letters of the protein language are the 20 amino acids that link together to form protein chains. If the DNA code is transcribed into translatable, or messenger, RNA, the RNA message becomes associated with one or more particles called ribosomes, which mediate the actual synthesis of protein. Ribosomes are made up of protein and two kinds of nontranslatable RNA, identified as 16S and 23S. Still another form of RNA called dictionary, or transfer, RNA delivers amino acids to the site of protein synthesis. It appears that a group of three bases in messenger RNA identifies each particular amino acid. According to one hypothesis the code group is "recognized" by a complementary set of bases in dictionary RNA. Evidently the ribosome serves as a "jig" for positioning amino acid subunits on the growing protein chain as the messenger RNA "tape" travels by.

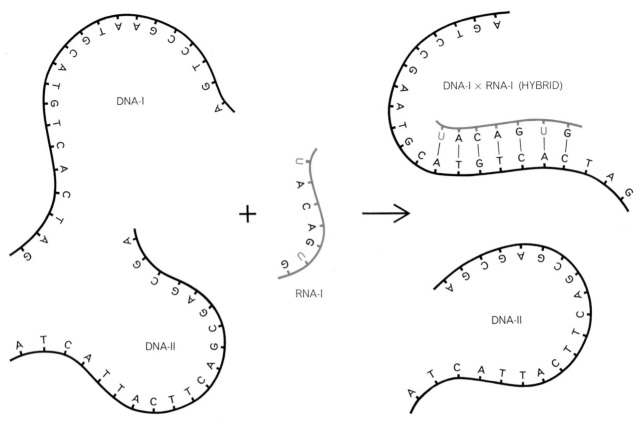

HYBRIDIZATION can occur when the base sequence in a strand of RNA matches up with that in single-strand ("denatured") DNA.

Here RNA-I is "challenged" with genetically related DNA-I and unrelated DNA-II. Only the genetically related strands hybridize.

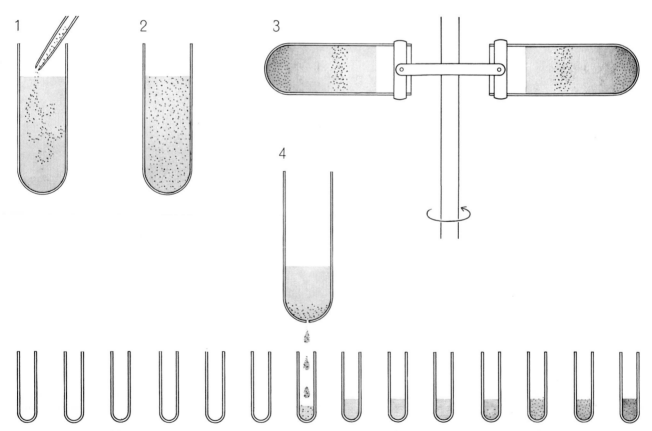

DENSITY-GRADIENT TECHNIQUE reveals if hybridization has taken place between RNA and DNA. The sample in question is added to a solution of cesium chloride (1 and 2). After centrifuga-

tion (3) the salt solution attains a smooth gradation in density. RNA (color), DNA (black) and RNA-DNA hybrids form layers according to their density. Fractions (4) can then be analyzed.

RNA was a precursor of the DNA needed to complete new virus particles. No doubt the experiment was misinterpreted and then neglected because it came so early in the modern history of DNA investigation. The helical model of DNA had been proposed only two years before by James D. Watson and F. H. C. Crick. Moreover, the experiment involved rather complex calculations and assumptions to support the view that the infected cells contained a distinctive new kind of RNA. It is clear in retrospect that this was the first experiment suggesting the existence of RNA copies of DNA.

It seemed to us that the Volkin-Astrachan observations were potentially so important that the design of an unequivocal experiment was well worth the effort. We set out, therefore, to see if bacterial cells infected with the T2 virus contained an RNA that could be specifically related to the T2 DNA. In our first experiments we sought evidence for this new type of RNA by physically isolating it from other RNA's. Two different procedures were successful. One (electrophoresis) measures the rate at which molecules migrate in an electric field; the other (sucrose-gradient centrifugation) measures their rate of migration when they are spun in a solution of smoothly varying density. Both of these methods showed that the RNA synthesized after virus infection was indeed a physically separable entity, differing in mobility and size from the bulk cellular RNA.

We found further that the ratio of the quantities of the bases (A, U, G and C) in the T2-specific RNA mimicked the ratio of the quantities of their counterparts (A, T, G and C) in the DNA of the virus. This suggested the possibility that the similarity might extend to a detailed correspondence of base sequence. A direct attack on this question by the complete determination of the sequences of bases was, and still is, too difficult.

Just at the right time, however, two groups of workers independently published experiments showing that if double-strand DNA was separated into single strands by heat (a process called denaturing), the two strands would re-form into a double-strand structure if the mixture was reheated and slowly cooled. This work was done by Julius Marmur and Dorothy Lane of Brandeis University and by Paul Doty and his colleagues at Harvard University. These investigators showed further that reconstitution of the double-strand molecule

occurs only between strands that originate from the same or closely related organisms. This suggested that double-strand hybrid structures could be formed from mixtures of single-strand DNA and RNA, and that the appearance of such hybrids could be accepted as evidence for a perfect, or near perfect, complementarity of their base sequences. It had already been shown by Alexander Rich of the Massachusetts Institute of Technology and by Doty that synthetic RNA molecules containing adenine as the only base would form hybrid structures with synthetic DNA molecules containing thymine as the only base.

With this work as background, we undertook to determine if T2 RNA would hybridize with T2 DNA. It was first necessary to solve certain technical problems. We had already devised methods for obtaining T2 RNA in a reasonable state of purity. The question was how to design the experiment so that if a hybrid structure formed, we could be certain of detecting it and identifying it as such.

All previous work on the reconstitu-

tion of two-strand DNA had involved sizable amounts of material that could form optically observable layers when it was spun in an ultracentrifuge. In our experiments the amount of hybrid material formed would probably be so small that it would escape detection by this method.

The detection method finally evolved combined several techniques. One depended on the fact that RNA has a slightly higher density than DNA; consequently RNA-DNA hybrids should have an intermediate density. Molecules of different densities can be readily separated by the density-gradient method developed by M. S. Meselson, Franklin W. Stahl and Jerome R. Vinograd at the California Institute of Technology. In this method the sample to be analyzed is added to a solution of a heavy salt, cesium chloride, and the mixture is centrifuged for about three days at more than 30,000 revolutions per minute. Under centrifugation the salt solution attains a smooth gradation in density, being most dense at the bottom of the sample tube and least dense at the top. The components of the sample migrate to layers at which their density

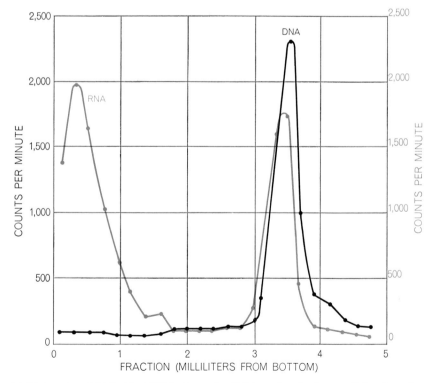

HYBRIDIZATION EXPERIMENT shows that RNA produced after a cell has been infected with the T2 virus is genetically related to the DNA of the virus. The RNA is labeled with radioactive phosphorus and the T2 DNA with radioactive hydrogen (tritium). The sample is subjected to density-gradient centrifugation (*see bottom illustration on opposite page*) and the radioactivity of the various fractions is determined. Although some of the RNA is driven to the bottom of the sample tube, much of it has hybridized with the lighter DNA fraction and thus appears between three and four milliliters above the bottom.

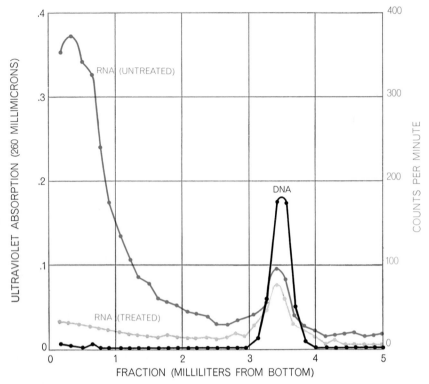

HYBRIDIZATION OF BACTERIAL RNA AND DNA is demonstrated for the bacterium *Pseudomonas aeruginosa.* Untreated RNA chiefly represents messenger RNA obtained by a special "step-down" procedure described in the text. In this experiment the presence of DNA in centrifuged fractions is determined by ultraviolet absorption. The coincident peaks in the two RNA curves represent RNA bound in RNA-DNA hybrids. "Treated RNA" refers to a portion of the sample that was treated before centrifugation with ribonuclease, an enzyme that normally destroys RNA. Although the enzyme has little or no effect on the hybridized RNA, it largely eliminates unhybridized RNA from the centrifuged sample.

exactly matches that of the salt solution. In place of the analytical ultracentrifuge we employed a centrifuge with swinging-bucket rotors, which permits actual isolation and analysis of various fractions. For this purpose the plastic sample tube is punctured at the bottom and the fractionated sample is withdrawn drop by drop for analysis [*see bottom illustration on page 62*].

To ensure a sensitive and unambiguous detection of the hybrid we labeled RNA with one radioactive isotope and DNA with another. The T2 RNA was labeled with radioactive phosphorus (P-32) and the T2 DNA with radioactive hydrogen (H-3). The beta particles emitted by P-32 have a characteristic energy different from those emitted by H-3; thus the isotopes can be assayed in each other's presence. The existence of hybrids in the centrifuged fractions would be signaled by the appearance of a layer containing the P-32 label of the RNA and the H-3 label of the DNA. Subsequently we observed that the layer of the hybrid fraction coincided closely with the layer of the unhybridized DNA. We could

therefore dispense with the radioactive label on DNA and establish its presence simply by its strong absorption of ultraviolet radiation at a wavelength of 260 millimicrons.

With these techniques we soon found that T2 RNA indeed hybridizes with T2 DNA. Furthermore, analysis of the hybrid confirmed that it was similar in overall base composition to T2 DNA. It was then necessary to show that hybrid formation occurs only between RNA and DNA that are genetically related. We exposed T2 RNA to a variety of unrelated DNA's from both bacteria and viruses. No hybrid formation could be detected, even with unrelated DNA's having an overall base composition indistinguishable from that of T2 DNA.

From these experiments one can conclude that T2 RNA has a base sequence complementary to that of at least one of the two strands in T2 DNA. Thus the similarity in base composition first noted by Volkin and Astrachan is a reflection of a more profound relatedness.

These experiments also tell us something about the events that take place when a virus invades a bacterial cell.

If precautions are taken to ensure that all the cells in a given sample are infected with the DNA virus, one finds that none of the RNA synthesized later can hybridize with the host DNA. This suggests that one of the first steps taken by a virulent virus in establishing infection is turning off production of the host's messenger RNA. Evidently RNA transcribed from the viral DNA provides the genetic messages needed for the formation of various proteins required to manufacture complete virus particles. Subsequent studies at the University of Cambridge by Sydney Brenner, François Jacob and Meselson have shown that the T2 messenger RNA is able to make use of ribosomes preexisting in the host cell for the synthesis of proteins.

We wondered next whether the transcription of the DNA code into RNA messages was a universal mechanism or whether it might be restricted to the simple mode of replication followed by viruses. The study of the flow of genetic information in normal cells is a problem of considerable difficulty. As noted above, about 95 per cent of the RNA present at any given moment is of the nontranslatable variety, consisting of ribosomal RNA and transfer, or dictionary, RNA. It is precisely because the translatable RNA molecules are so few—only about 5 per cent of the total amount of RNA—that they were overlooked for so long in normal cells. The detection of the RNA messages of T2 was made easy because the synthesis of ribosomal and transfer RNA is turned off in virus-infected cells.

We decided to look for a situation in normal cells that would imitate the advantages provided by infected ones. It had been known that the total RNA content of cells is positively correlated with rate of growth, and since most of the RNA is ribosomal RNA, a high growth rate implies a high content of ribosomes. What happens if cells are subjected to a "step-down" transfer, that is, a transfer from a rich nutrient medium to a poor one? The growth rate declines, usually by about half. More important, for a generation after they have been placed in a poorer medium the cells contain more ribosomes than they can usefully employ. We reasoned that in this period the synthesis of ribosomal RNA might stop. Since protein production continues at a low rate, however, some synthesis of RNA messages, which must be continuously replaced, should persist.

My colleague Masaki Hayashi undertook experiments to determine if this was the case. If it was, any RNA synthesized after step-down transition would be different from the ribosomal RNA. Hayashi selected three species of bacteria with DNA's of widely different base composition. In all three species the RNA synthesized after step-down transition possessed all the features that had characterized the RNA produced in virus-infected cells. These included instability, a base composition similar to that of the organisms' DNA's and a range of molecular sizes different from that of the ribosomal RNA.

Hybridization tests were carried out between the message-RNA fraction and genetically related DNA as well as with genetically unrelated DNA. The results were clear-cut. Hybrid structures were formed only when the mixture contained RNA and DNA of the same genetic origin. An experiment in hybrid formation that involved RNA and DNA from the bacterium *Pseudomonas aeruginosa* is summarized in the illustration on the opposite page.

This particular experiment illustrates an interesting and useful property of RNA-DNA hybrids. A portion of each sample of hybrid material was treated with the enzyme ribonuclease, which normally destroys RNA. One of the curves shows the amount of RNA in each fraction that was resistant to the enzyme. It can be seen that the RNA bound in the hybrid is quite resistant, whereas the free RNA is almost completely destroyed. This phenomenon turned out to be very useful for distinguishing between free and hybridized RNA. We can conclude from Hayashi's studies, and from those of others, that the flow of information from DNA to translatable RNA occurs normally in bacteria and is probably a universal mechanism in protein synthesis.

By the time these investigations were completed we were convinced that the RNA-DNA hybridization technique could be developed into an extremely powerful and versatile tool. Accordingly we decided to put it to a severe test. The problem we wanted to solve was this: Where do the nontranslatable molecules of RNA—ribosomal RNA and transfer RNA—come from?

Let us consider first the ribosomal variety. Two principal alternatives can be suggested for its mode of origin. Either it is formed on a DNA template or it is not. If it is formed on DNA, it should be complementary to some seg-

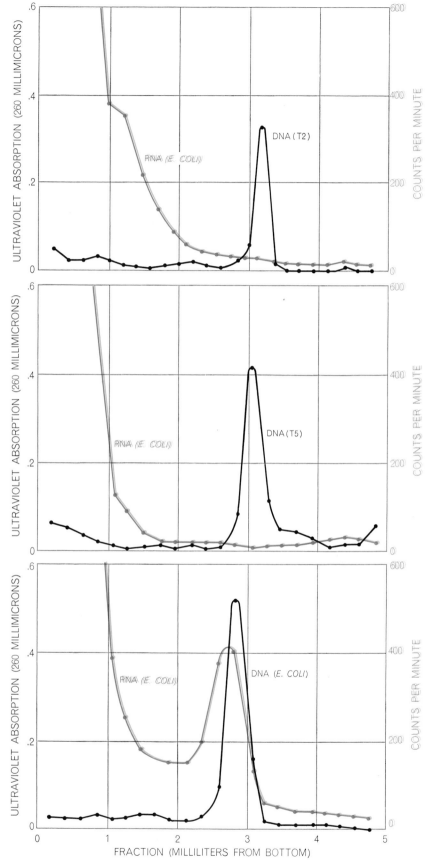

HYBRIDIZATION OF RIBOSOMAL RNA provides evidence that, like messenger RNA, it too is formed on a DNA template. In this experiment ribosomal RNA of the 23S variety was obtained from the colon bacillus (*Escherichia coli*). The top and middle curves show that no hybridization occurs when the RNA is challenged with single-strand DNA from the T2 and T5 viruses. When challenged with DNA from *E. coli*, however, hybridization is seen.

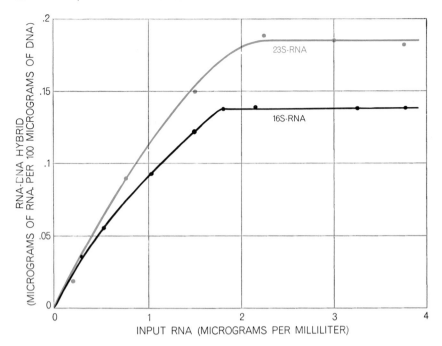

SATURATION CURVES indicate what fraction of the DNA molecule is set aside for producing the two forms of ribosomal RNA designated as 16S and 23S. The RNA and DNA samples were obtained from *Bacillus megaterium*. The results show that about .14 per cent of the DNA molecule is complementary to 16S and about .18 per cent to the 23S form.

ment of DNA and hence subject to hybridization.

It has been known for some time that the base composition of ribosomal RNA is not correlated with that of DNA found in the same cell. This, however, tells us nothing about the origin of the RNA; the DNA segment needed to serve as a template for ribosomal RNA might be so small as to constitute a nonrepresentative sample of the DNA's overall base composition.

Some three years ago one of my students, Saul A. Yankofsky, undertook the job of determining if hybridization could shed any light on this problem. The major complication was that a ribosomal RNA molecule appeared to be only about a ten-thousandth as long as the entire DNA molecule in a typical bacterial cell. We were faced, therefore, with the task of designing experiments that would detect hybridizations involving only a minute segment of DNA.

Theoretically the required sensitivity can be attained simply by labeling RNA so that it has a suitably high level of radioactivity. If no radioactivity was found in association with DNA, one could conclude that no hybrid had been formed. Experiments of this sort would require RNA labeled at a level of about one million counts per minute per microgram. The trouble with such high levels of radioactivity is that irrelevant "noise" can spoil the experiment. It is easy to detect 100 counts per minute

above the background level of radiation. Thus if as little as .0001 microgram of unhybridized RNA accidentally got into the DNA fraction, it would be detected and give a false reading. Such accidental contamination could occur in a number of ways. For example, the ribosomal RNA preparation might contain traces of radioactive translatable RNA that would hybridize with DNA. Small amounts of ribosomal RNA might be mechanically trapped by strands of DNA. Or there might be partial hybridization resulting from accidental coincidences of base complementarity over small regions.

By a variety of biological and technical stratagems it was possible to design a satisfactory experiment. Organisms were chosen with a DNA base composition far removed from that of ribosomal RNA, thereby making it possible to show that hybridized material actually contained ribosomal RNA. Contamination of the radioactive ribosomal RNA preparation by radioactive translatable RNA was eliminated by a simple trick. After the RNA in the cells was labeled with a suitable isotope the cells were transferred to a nonradioactive medium for a period long enough for the labeled RNA messages to disappear. Ribosomal RNA, being stable, retains its radioactive label. Finally, to avoid false readings from RNA that was either mechanically trapped or accidentally paired over short regions, all

suspected hybrids were treated with ribonuclease. The RNA in a genuine hybrid is resistant to this treatment.

It was noted earlier in this article that ribosomes contain two types of RNA, designated 23S RNA and 16S RNA. The outcome of a series of hybridizations between 23S RNA obtained from the colon bacillus and three different DNA preparations is presented in the illustration on the preceding page. A ribonuclease-resistant structure appears in the DNA-density region only when the DNA and the ribosomal RNA are from the same organism. These results clearly imply that ribosomal RNA is produced on a DNA template.

An extension of these studies gave us an answer to the following question: How much of the DNA molecule is set aside for turning out ribosomal RNA? To get the answer we simply add increasing amounts of ribosomal RNA to a fixed amount of DNA and determine the ratio of RNA to DNA in the hybrid at saturation. The illustration at the left shows the outcome of this experiment with the ribosomal RNA of *Bacillus megaterium*. The results indicate that approximately .18 per cent of the total DNA molecule is complementary to 23S RNA and .14 per cent to 16S RNA.

The difference in these two saturation values suggests that 23S RNA and 16S RNA are distinctively different molecules, but the evidence is not unequivocal. Although different in size, the two ribosomal RNA molecules have essentially the same base composition. There is still no direct way of telling whether they have the same or different base sequences. The similarity in base composition and the fact that the 23S RNA has about twice the weight of the 16S RNA had led, however, to the concept that the 23S-RNA molecule is a union of two 16S-RNA molecules.

To probe the matter further we designed an experiment to find out if the two kinds of ribosomal RNA compete for the same sites when they are hybridized with DNA. Hybridization mixtures were prepared that contained fixed amounts of DNA and saturating concentrations of 23S RNA labeled with P-32. To these we added increasing amounts of 16S RNA labeled with H-3, after which we determined the relative amounts of P-32 and H-3 in the hybrid structures. If the two kinds of RNA have an identical sequence, the entry of the H-3-labeled 16S RNA into the hybrid should displace an equivalent amount of P-32-labeled 23S RNA. If the sequences are different, the 16S RNA should hybridize as though the 23S material were not present. The experiment decisively supported the second alternative [*see illustration on this page*].

Following these experiments, there seemed little doubt that the third variety of RNA, transfer RNA, would also be found to originate on segments of DNA. The small size of transfer-RNA molecules made hybridization experiments even more difficult than the earlier ones. Nevertheless, the experiments were successfully carried out by Dario Giacomoni in our laboratory and by Howard M. Goodman in Rich's laboratory at M.I.T. Both workers obtained virtually identical results. They demonstrated by specific hybridization that the DNA of a cell contains sequences complementary to its molecules of transfer RNA. The amount of DNA set aside for the cell's genetic dictionary was found by both groups to be about .025 per cent, or less than a tenth of the combined space allotted to the two types of ribosomal RNA.

These experiments also ruled out an interesting possibility. The molecules of transfer RNA contain only about 80 bases (compared with about 2,000 for 16S RNA) and it was conceivable that the sequence of bases in transfer RNA's might be the same, or much the same, in the cells of different organisms. This possibility seemed more likely when Günter von Ehrenstein of Johns Hopkins University and Fritz A. Lipmann of the Rockefeller Institute showed, in a joint experiment, that transfer RNA's from the colon bacillus can serve as a dictionary in translating the RNA message for the synthesis of the protein hemoglobin from materials present in the red blood cells of the rabbit.

Giacomoni was able to show, however, that the base sequence in transfer-RNA molecules differs from organism to organism. In one such experiment a mixture of transfer-RNA molecules from two different organisms was challenged with DNA molecules obtained from one of them. For identification the genetically related transfer RNA was labeled with P-32 and the unrelated variety with H-3. Only the related RNA formed a hybrid; the genetically unrelated RNA did not [*see chart at left in illustration on next page*].

Instead of using one kind of DNA and two kinds of transfer RNA, one can reverse matters and also demonstrate specificity. For this purpose it is helpful to choose DNA preparations that migrate to different layers when they are subjected to density-gradient centrifuga-

tion. In such a mixture a hybrid will form only with radioactively labeled transfer RNA that is genetically related to one of the DNA's. In the experiment performed in our laboratory the DNA was obtained from two bacteria, *Pseudomonas aeruginosa* and *Bacillus megaterium,* and the transfer RNA was obtained only from the latter [*see chart at right in illustration on next page*].

These experiments reveal an interesting feature of the biological universe. It is assumed that only three of the 80-odd bases in a transfer-RNA molecule provide the means for "reading" the three-base code "words" in the RNA message. Although evidence is lacking on this point, it is possible that a temporary association between three bases in transfer RNA and three bases in the RNA message guarantees that the correct amino acid is deposited where it belongs in a growing protein chain [*see lower illustration on page 61*].

If this picture is accepted, what is

the role of the other 70-odd bases in transfer RNA? The function of the noncoding portion is unknown, but its presence provides an opportunity for biological individuality, from species to species, without disturbing the dictionary function of the molecule. The fact that the base sequences are different in the transfer RNA's of different organisms shows that this opportunity has not been neglected in the course of biological evolution.

We have now seen that all forms of RNA can be traced back to their point of origin on the DNA template. But the double-strand helix of DNA represents two templates, one the complement of the other. When any given segment of DNA is transcribed, two entirely different RNA molecules can be produced, depending on which strand of the DNA molecule serves as a template. Assuming that the entire length of the DNA molecule contains genetic

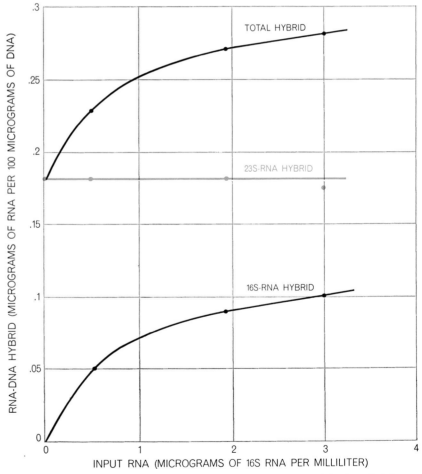

COMPETITION TEST shows that 16S and 23S ribosomal RNA form hybrids with different segments of the DNA molecule. The 16S RNA was labeled with tritium, the 23S RNA with radioactive phosphorus. Increasing amounts of 16S RNA were added to hybridization mixtures containing a saturating concentration of 23S RNA. Subsequently the relative amounts of tritium and radioactive phosphorus in the hybrids were determined. Since the two kinds of RNA hybridize without interference they must have different base sequences.

information that must be transcribed into RNA, there are three possibilities: (1) All of both strands are transcribed into complementary RNA; (2) both strands serve as templates, but in any given segment only one strand or the other is transcribed; (3) only one strand is transcribed.

Here again the hybridization test has supplied evidence to decide among the alternatives. Ideally what is required is a method of separating the two strands of the DNA molecule. If this could be done, one could test the various forms of RNA against each strand and determine if hybridization occurs.

Although the two strands of normal DNA can be separated, no way has yet been found to obtain a pure preparation containing strands of only one type. Fortunately nature provides a solution to the problem in the form of an organism that contains a single strand of DNA. The organism is the small DNA virus φX174, discovered in Parisian sewage about 30 years ago by French investigators. It is fairly easy to purify the virus particle and remove its DNA. Nature also provides a source of the complementary strand. When the virus infects a bacterial cell, the single strand of DNA serves as a template for the synthesis of a complementary strand, resulting in a normal double-strand DNA molecule. This molecule, known as the replicating form, can also be isolated for experimental purposes.

In order to run a hybridization test my co-workers Marie and Masaki Hayashi grew φX174 in infected cells in the presence of P-32 and extracted labeled molecules of translatable RNA. These molecules were then brought together with the single-strand DNA of φX174 and with a denatured sample of the double-strand form. The results obtained were satisfyingly clear. No hybrids were formed with the single-strand DNA, but excellent hybrids were produced with the DNA from the double-strand form. This implied that the RNA messages are complementary to the *other* strand in the two-strand DNA molecule, that is, the one not normally present in the φX174 particle. As a final confirmation we analyzed the base composition of the RNA that was hybridized. The results agreed with the expectation that it was complementary to only one of the two strands of the replicating form of φX174 DNA.

Using similar methods with other viruses, identical conclusions have now been drawn by two other groups: Glauco P. Tocchini-Valentini and his co-workers at the University of Chicago and Carol Greenspan and Marmur at Brandeis University. There seems little doubt that in all organisms only one strand of the DNA molecule serves as a template for RNA synthesis.

The original procedures of detecting hybrids involved lengthy high-speed centrifugations. Ekkehard K. F. Bautz of Rutgers University and Benjamin D. Hall of the University of Illinois have introduced the use of cellulose-acetate columns for hybridization experiments. Ellis T. Bolton and Brian J. McCarthy of the Carnegie Institution of Washington's Department of Terrestrial Magnetism have developed a convenient and rapid method using an agar column. Here the DNA is trapped on the agar gel and the RNA is hybridized with it. The RNA can then be removed by raising the temperature of the column and lowering the ionic strength of an eluting, or rinsing, solution.

The exploitation of the hybridization technique is still at an early stage, but it has already proved of great value in the analysis of gene function. It seems likely to play an increasingly important role in helping to illuminate many problems of molecular biology, including those pertinent to an understanding of the specialization of cells and biological evolution in general.

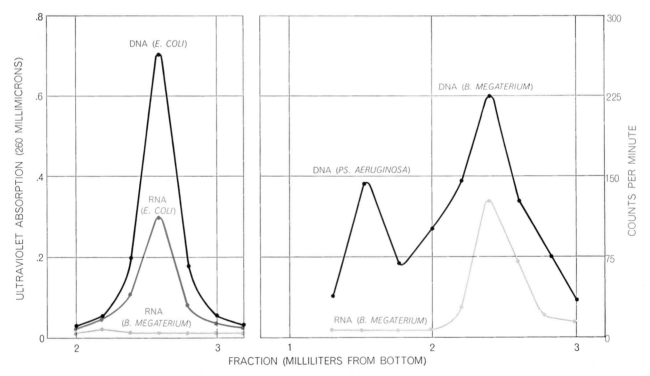

TESTS FOR GENETIC RELATIONSHIP can be carried out by challenging the RNA from two different organisms with the DNA from one of them. In one experiment (*left*) transfer RNA from *E. coli* was labeled with radioactive phosphorus; transfer RNA from *B. megaterium* was labeled with tritium. Only the former hybridizes with *E. coli* DNA. Conversely, in a second experiment (*right*), transfer RNA from *B. megaterium* hybridizes with genetically related DNA but not with DNA from *Ps. aeruginosa*.

II

MACROMOLECULAR AGGREGATES AND ORGANIZED STRUCTURES

II

MACROMOLECULAR AGGREGATES AND ORGANIZED STRUCTURES

INTRODUCTION

Not all functional units in cells are individual biopolymers; some are aggregates, or polymers of polymers. Among these are blood clots; structural, elastic, and contractile elements; chromosomes; and nerve fibers.

The association of biopolymers into functional aggregates and the delineation of hierarchies of organization are described in "Giant Molecules in Cells and Tissues," by Francis O. Schmitt. In addition to addressing the general questions involved in biopolymer aggregation, Schmitt notes the power and utility of electron microscopy as a tool for the elucidation of biological structure at the cellular and subcellular levels.

Viruses are one class of biomolecular aggregates whose structure has been particularly well studied. They are essentially complexes of protein and nucleic acid, the former acting as a protective coat for the genetic information contained in the latter. In his article "The Structure of Viruses," R. W. Horne shows how electron microscopy has been especially useful in elucidating virus structure. The negative-staining technique devised by Horne has proved particularly important. Since the viral nucleic acid has sufficient genetic information to code only a small number of proteins, and since most virus particles contain several hundreds or thousands of protein molecules, most of the proteins in a virus coat must be present in many copies. This suggests that they may be located in identical, or nearly identical, environments. Thus, virus protein coats, or capsids, have high symmetry. Horne describes the three major types of symmetry found in capsid structures—icosahedral symmetry in closed structures, helical symmetry in rodlike linear viruses, and compound symmetry in some complicated viruses.

Cell membranes serve to separate a cell from its environment, to facilitate or deny passage of chemicals between exterior and interior, and are the site of many metabolic activities. Membranes are composed of lipids, especially phospholipids, and proteins. In "The Structure of Cell Membranes," C. Fred Fox surveys current ideas on the arrangement of these components, which indicate that proteins are embedded in a two-dimensional, fluid, phospholipid bilayer. He then discusses the way in which this arrangement allows selective transport, and the mechanism of assembly of the components into intact membranes. The dynamic nature of these processes raises intriguing new questions that have stimulated much current research.

In the article "Pumps in the Living Cell," Arthur K. Solomon concentrates on one of the most important and mysterious functions of cell membranes: active transport, or the ability to transport sodium (and other ions) against adverse concentration and electro-chemical gradients. Ingenious micro-techniques have been developed to measure small concentrations and voltages at particular points on a membrane surface. The "pumps" involved in sodium transport are shown to be closely coupled with mitochondria, which provide energy for their operation. More recent work has identified these pumps as particular membrane-bound enzymes.

Living things carry out an amazing variety of specialized activities—metabolism, reproduction, movement, sensory response—with unexcelled efficiency, speed, and control. Understanding the molecular mechanisms that underlie these activities is a central task of biophysical chemistry. One of the most obvious examples of a process important to life is the conversion of chemical energy into useful work in muscular contraction. The structural basis of this process is described in "The Mechanism of Muscular Contraction" by H. E. Huxley. Muscle is found to be a complex of macromolecules—in this case the proteins myosin and actin. The organization of proteins into filaments—and interfilament bridges of proper polarity to give directed contraction—is a consequence solely of the properties of the individual protein molecules.

Other *Scientific American* articles related to the topics in this section are "Bone" by Franklin C. McLean, February, 1955 (Offprint 1064); and "Liquid Crystals" by James L. Fergason, August, 1964.

72

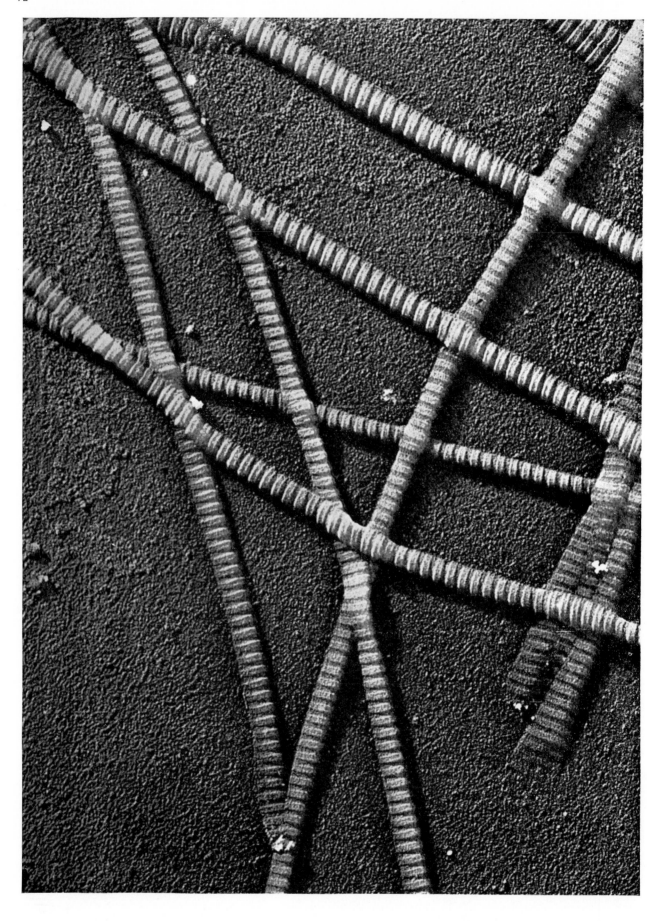

FIBRILS OF COLLAGEN, the principal protein of skin, are enlarged 60,000 diameters in this electron micrograph made by Jerome Gross of the Massachusetts General Hospital. The fibrils have been shadowed with a heavy metal to bring out their banded structure.

7

Giant Molecules in Cells and Tissues

by Francis O. Schmitt
September 1957

How are protein and nucleic acid molecules organized in protoplasm? It appears, largely on the basis of evidence supplied by the electron microscope, that these molecules are monomers of higher polymers

Two giant molecules, the proteins and the nucleic acids, provide the principal structural and dynamic constituents of the living cell. Each is made up of smaller, simpler molecular units (monomers), which are joined together, or polymerized, in the subtle chemical processes which taken together express the cell's identity as a living system. In this article it will be seen that the process of polymerization, by which proteins and nucleic acids are synthesized, provides a model for understanding how they in turn organize themselves into the higher structures of cells and tissues. This spontaneous process apparently depends upon specific properties built into these molecules. Proteins and nucleic acids, in short, may be regarded as the monomers of the living cell.

These giant monomers may be polymerized in the cell where they are made, or they may be transported in inactive form to another part of the organism. There they may be activated and polymerized as the occasion demands. For example, the soluble protein fibrinogen is always present in the blood, but it is polymerized into the insoluble fibrin of a blood clot only when bleeding must be stopped.

The function of a natural high polymer is of course reflected in its properties. The protein keratin has great tensile strength; it forms the principal structure of hair, horn and fingernail. The protein collagen serves a similar purpose in skin and tendon. Elastin is a springy protein; it occurs in ligaments and the elastic fibers of connective tissue.

These three types of polymer are more or less passive; others respond actively to changes in their chemical environment. Under the influence of such changes the protein of muscle contracts. Contraction is a property not only of muscle but also of many other biological structures, from the rapidly oscillating tail of the sperm cell to the flowing pseudopods of the amoeba. Indeed, contractility is an essential feature of all living cells, and probably employs a common molecular mechanism.

The threadlike chromosomes of the cell nucleus, made up of protein and nucleic acid, are polymers with another function: the maintenance of the specific linear sequence of the genes. We should also reserve a category for natural high polymers whose purpose is not known. A good example is the fibrous protein of nerve cells. These so-called neurofibrils are very long and less than a millionth of an inch thick. They run through the core of all nerve fibers; in the light microscope they are seen as bundles of coagulated filaments. Their presence in all nerve tissue indicates that they must serve some function, but to date no one has been able to show what it might be. Considerable effort is currently being made to isolate this protein (from the giant nerve fibers of the squid) and to determine its composition and function.

The idea that biological fibers are composed of fibrous molecules is not new. In the 19th century microscopists observed that such fibers tended to fray into finer fibers, and assumed that the hierarchy continued downward in scale. At the same time natural fibers were analyzed with polarized light, and the analysis indicated that their molecules were organized in regular structures. In

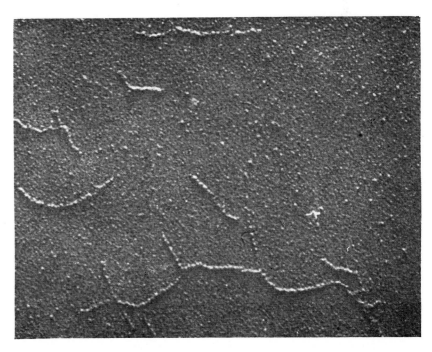

MOLECULES OF COLLAGEN, enlarged 100,000 diameters, appear as long, thin threads in this electron micrograph made by Cecil E. Hall of the Massachusetts Institute of Technology.

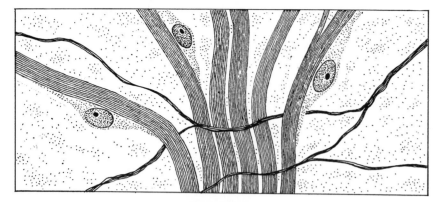

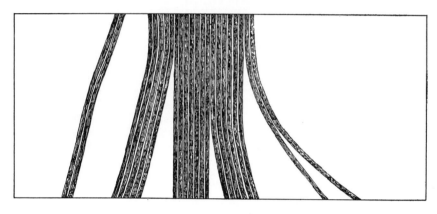

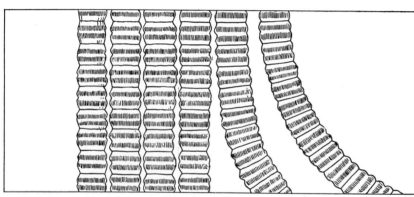

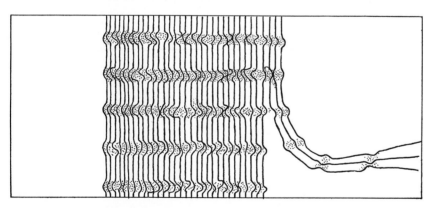

HIERARCHY OF FIBERS is outlined in these drawings of collagen as seen at four different magnifications. At top are collagen fibers in a bit of connective tissue as they appear under the light microscope. In a light microscope of higher power a single frayed collagen fiber is seen to consist of many fibrils (*second drawing from top*). Enlarged still more with an electron microscope, the same fibrils show cross-bands and other details (*third drawing from top*). X-ray diffraction methods permit even finer analyses revealing that a single fibril is made up of parallel chains of collagen molecules (*fourth drawing*).

the 1920s and 1930s polarization analysis showed that the cellular fibers were very thin. Today we know that they are the molecules themselves; they can actually be visualized in the electron microscope. Used in conjunction with X-ray diffraction and other older methods, this instrument has provided significant information on the molecular organization of structures in living cells.

When a thin section of a muscle cell is observed in the electron microscope, it is resolved into finer fibers called myofibrils, each of which consists of many thin filaments. A striking feature of the fibrils is the striations that occur regularly along their length. These cross-bands are brought out by staining the muscle tissue with compounds of heavy metals, which scatter electrons more than do the lighter elements of which the tissue is composed. Thus the bands are regions that combine well with the stains. When the fibrils are shadowed with metal so that they can be seen in relief, adjacent bands have slightly different thicknesses and hence are more visible.

The band pattern provides a kind of molecular fingerprint by which the protein can be identified in the electron microscope and which provides important clues to how giant molecules are organized into fibers. Let us consider in detail collagen, the protein of skin and tendon. Collagen, which is also present in bone, teeth and loose connective tissue, represents as much as a third of all the protein in an animal. In compact form such as tendon it will resist a pull of as much as 100,000 pounds per square inch, roughly the strength of steel wire. Intensively studied by electron microscopists, X-ray crystallographers and chemists, it is now one of the best-known fibrous proteins.

If we tease a bit of tendon or skin with fine needles, or break it up in a blendor, its fibers are split into fibrils. We can see under the light microscope that the fibrils have a fairly uniform width: usually from 200 to 1,000 Angstrom units (an Angstrom unit is a hundred millionth of a centimeter). The fibrils are composed of still finer strands: protofibrils. These are thought to consist of fibrous molecules strung end to end.

When undried native collagen fibrils are examined in the electron microscope, they show cross-bands which repeat at intervals of about 700 Angstroms. Richard S. Bear of the Massachusetts Institute of Technology has offered the following explanation for this banding. The molecular chain of collagen consists of amino acid units in specific sequence. Thus the characteristic side chains of the amino acids also occur in regular sequence. When protofibrils lie side by side, certain side chains, especially the short ones, fit together in regions of relative order. The longer side chains, on the other hand, do not fit well, and give rise to less orderly regions. Bear concluded that the regions of relative disorder correspond to bands (the regions which are slightly thicker and take on more stain); the regions of relative order, to the space between bands (interbands). The banding of collagen would thus reflect the sequence of amino acid units along the molecular chains.

Now when a collagenous tissue, such as the tendon of a rat's tail, is placed in dilute acid, it swells up and eventually dissolves; the resulting solution is as clear as water but relatively viscous. The solution can then be spun in an ultracentrifuge to remove the larger aggregates of collagen. The molecules in the remaining solution were measured by Paul Doty of Harvard University, using the methods of the physical chemist. He determined that they are 14 Angstroms wide and 2,900 Angstroms long. The molecules were visualized directly by Cecil E. Hall of M.I.T., who found that their dimensions were roughly the same [see illustration on page 73].

The dispersed molecules can be reassembled into fibrils by changing the character of the solution. The crossbands of the artificial fibrils can then be studied in the electron microscope. Depending on the nature of the treatment, the band patterns show considerable variation. First, the bands may be entirely absent. Second, the principal bands may repeat about every 700 Angstroms, as in the native fibril. Third, the bands may repeat in about a third of this distance. Fourth, they may repeat about every 2,800 Angstroms, a distance four times that of the normal spacing. This "long spacing" occurs in two forms: "fibrous" and "segment" [see illustrations on pages 76 and 77].

Reflecting on these band patterns, Jerome Gross of the Massachusetts General Hospital and John H. Highberger of the United Shoe Machinery Corporation, working in collaboration with the author, deduced that they represent different ways in which the collagen molecules can come together under different conditions. When the molecules are linked up side by side with their ends "in register," their principal bands will re-peat about every 2,800 Angstroms. In other words, the distance between the principal bands is about the same as the length of the molecule. If adjacent molecules are in register and pointing in the same direction, the pattern will be of the segment long-spacing type. If adjacent molecules are in register but pointing in opposite directions, the pattern will be fibrous long-spaced. If the adjacent molecules are pointing in the same direction, but are regularly staggered by about a fourth of their length, they will form bands with the native spacing of 700 Angstroms. If the ends of neighboring molecules have no orderly arrangement, no bands at all are formed.

These observations will explain why it seems likely that the organization of protoplasm depends on highly specific properties built into its giant monomers. The collagen molecule consists of three chains of amino acid units wound in a helix around a common axis. Side chains extending from this molecule interact with similarly placed side chains on adjacent collagen molecules. The molecules will "recognize" each other even if much foreign material is present.

This interaction is significantly influenced by the chemical environment of the molecules. Thus an artificial fibril of collagen with its band pattern repeating every 2,800 Angstroms cannot be made in a solution of pure collagen. A second constituent, such as adenosine triphosphate (ATP), must be added. Presumably the ATP, by combining with certain side chains of the collagen molecule, changes the pattern of the side chains still available for interaction with neighboring molecules.

The cross-bands of various kinds of muscle-protein also have been analyzed, though not in such detail as those of collagen. When the muscle that holds shut the shell of a clam is minced in a dilute salt solution, it breaks up into fibrils of the protein paramyosin. In the electron microscope the stained fibrils have a band pattern which repeats about every 145 Angstroms. By means of X-ray diffraction Bear has demonstrated that a unit five times this length also repeats along the fibrils.

Paramyosin fibrils can be dissolved and reconstituted so that the pattern of their bands repeats at the length of the paramyosin molecule. The Australian electron microscopist Alan J. Hodge, working at M.I.T., has shown that this distance is about twice the X-ray period, or about 1,500 Angstroms. It would be interesting to know the connection be-tween the structure of the paramyosin fibril and the long-lasting contraction of the clam's muscle.

A fast muscle of the sort that causes the blink of an eyelid has a rather different construction. The fibrils consist of parallel filaments made up of at least two kinds of protein: actin and myosin. They are segmented into alternating regions, in one of which the filaments are more highly organized than in the other. Throughout both regions a finer band pattern repeats about every 400 Angstroms. The bands of isolated actin fibrils have a similar length, as do those of light meromyosin (a derivative of myosin). How this length is related to the repeating pattern of the fibril and its constituent protein molecules is not yet clear.

In contractile structures other than muscle the filaments may be specialized. The hairlike cilia of microorganisms and higher animals, for example, consist of two fine filaments surrounded by nine thicker filaments or pairs of filaments,

SIDE CHAINS of collagen molecules lying side by side may be long or short. The long side chains sometimes end in an electrically charged group (+ or −). The short side chains tend to fit together in an orderly manner; the long, to interact in a disorderly manner. These disordered regions correspond to the bands in electron micrographs.

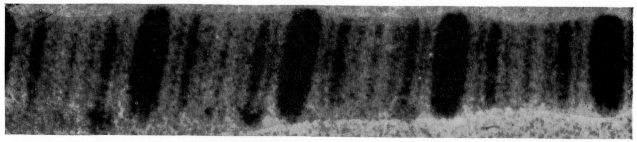

COLLAGEN RECONSTITUTED in a solution containing certain organic substances has a band pattern which repeats at intervals of about 2,800 Angstroms. The electron micrograph at left, made by Gross, shows reconstituted fibrils with "fibrous long-spacing";

RECONSTITUTED PARAMYOSIN, *i.e.*, paramyosin dissolved in salt solution and then made to reassemble, is enlarged 67,000 diameters in an electron micrograph by J. W. Jacques of M.I.T. Here the band pattern repeats every 1,600 to 1,800 Angstrom units.

LIGHT MEROMYOSIN, derived from the muscle protein myosin, is enlarged 160,000 diameters in an electron micrograph furnished by Andrew G. Szent-Gyorgyi of the Marine Biological Laboratory in Woods Hole, Mass. The pattern repeats every 420 Angstroms.

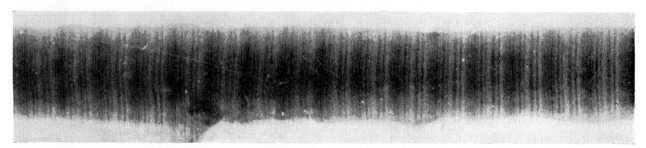

NATIVE COLLAGEN FIBRIL has a band pattern which repeats at intervals of about 640 Angstroms. Collagen fibrils dissolved in acid and reconstituted in salt solution may also have this pattern. The electron micrograph enlarges the fibril 170,000 diameters.

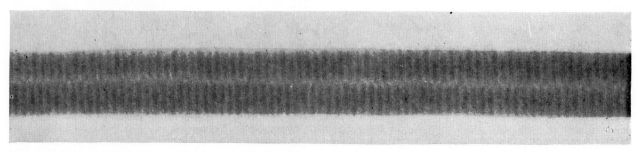

FIBRIL OF PARAMYOSIN, a protein from the muscle of a clam, is enlarged 160,000 diameters in this electron micrograph by Hall. The pattern of bands visible in the fibril repeats every 145 Angstrom units, roughly a 10th the length of the paramyosin molecule.

the micrograph at right, made by Alan J. Hodge at M.I.T., "segment long-spacing."

the whole encased in a cylindrical sheath. It is not yet known whether such filaments are composed, like those of muscle, of several types of protein.

What makes contractile polymers contract? We can only sketch the process in rough outline. In the case of muscle there must be a protein which forms filaments extending the length of the muscle. This protein transmits tension to the connective-tissue sheath of the muscle and thus to the tendon; it may be actin or a complex of actin and other proteins. In close proximity to these filaments is a second set of filaments, presumably consisting primarily of myosin. The Nobel laureate Albert Szent-Gyorgyi and his school have shown that muscular contraction and relaxation involve a quick change in the relationship between these two proteins. The relationship is influenced by ions of potassium, sodium, calcium and magnesium, and by the energy-rich substance ATP. We must discover what this relationship is if we are to understand what makes fibrous systems contract. According to some workers, contraction occurs when one set of filaments slides past the other. It is believed by others that during contraction a filament of one kind coils in a helix around a filament of the other. Whatever the mechanism, it seems likely that it is governed by complementary patterns or chemical groups built into both kinds of protein molecule. This complementarity is expressed or suppressed by the presence or absence of substances such as ATP. The function of ATP in turn is regulated by the enzyme adenosine triphosphatase (ATPase), which is an integral part of the myosin filament. This enzyme is localized in such a way that it can act on ATP only at one specific moment during the cycle of contraction and relaxation.

Another example of this subtle interaction between specific natural polymers and their chemical environment is the fibrinogen-fibrin system of blood clot-

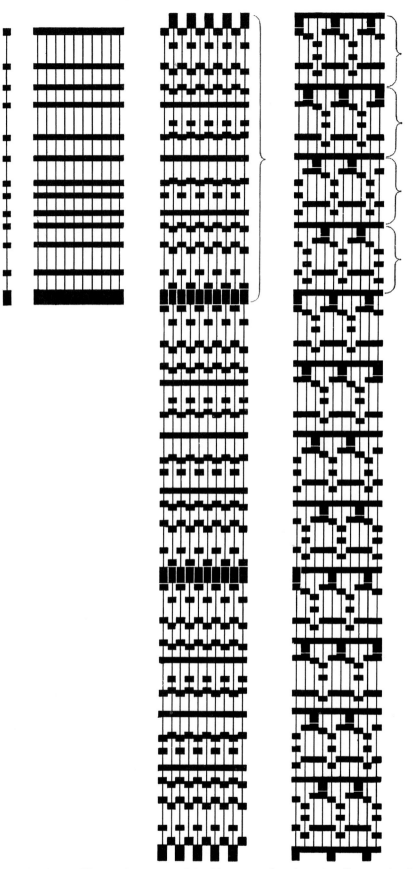

BAND PATTERNS of collagen are explained by means of a schematic collagen molecule (*upper left*) marked at regions where it interacts with other collagen molecules. If adjacent molecules point in the same direction and are in register (*second from left*), their pattern is "segment long-spaced." If adjacent molecules point in opposite directions and are in register (*third from left*), the pattern is "fibrous long-spaced." If they point in the same direction but are regularly staggered (*fourth*), the pattern repeats at a shorter interval.

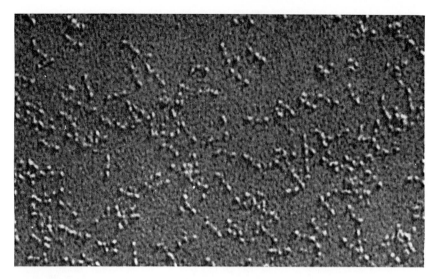

MOLECULES OF FIBRINOGEN are enlarged 130,000 diameters in this electron micrograph by Hall. Each of the molecules consists of three tiny spheres joined by a fine thread.

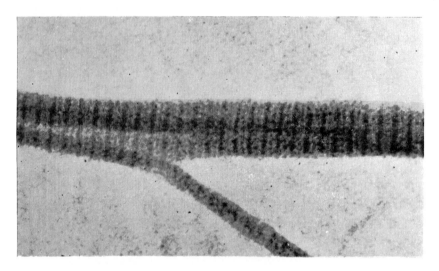

FIBRIL OF FIBRIN, the insoluble protein made up of fibrinogen, is enlarged 180,000 diameters in this electron micrograph by Hall. Its band pattern repeats every 235 Angstroms.

NERVE FILAMENTS are enlarged 26,000 diameters in this electron micrograph by Myles Maxfield of M.I.T. Filaments of this kind are found in the central core of all nerve fibers.

ting. When measured in solution, fibrinogen molecules are about 40 Angstroms in diameter and 550 Angstroms long. In the electron microscope they are about 400 Angstroms long and consist of three spheres, each 30 to 40 Angstroms in diameter, joined by a delicate thread [see illustration at left].

When blood clots, a peptide segment is split out of the fibrinogen molecules. The molecules are thus activated and combine to form a so-called intermediate polymer, the length of which is 4,000 to 5,000 Angstroms. The intermediate polymers, which have been observed in the electron microscope by B. M. Siegel of Cornell University, now come together to form fibrils of the insoluble protein fibrin. These fibrils prevent bleeding by clogging the broken blood vessels. In the electron microscope the fibrils of fibrin have a band pattern which repeats every 235 Angstroms. This distance is not clearly related to the length of the fibrinogen molecule. John D. Ferry of the University of Wisconsin has suggested that the bands may be related to activated fibrinogen molecules which are lined up side by side but staggered.

The mechanism by which soluble fibrinogen is activated and converted into the intermediate polymer is extremely complicated. Like the contraction of muscle, it involves not only the interacting protein molecules but also ions, enzymes and other substances. Some of these facilitate the reaction; others inhibit it. Only by such delicate feed-back mechanisms can the organism maintain high-polymer systems which go into action when they are needed but remain inactive when they are not.

We now come to the most remarkable high-polymer system of all. This is the system that enables the organism to reproduce itself: the nucleic acid and protein of the chromosomes.

The structure of chromosomes has been studied intensively and fruitfully under the light microscope, but reliable information about their molecular organization is scanty. In the electron microscope chromosomes of many kinds appear to have the same basic structure: a relatively dense, smooth-edged filament from 100 to 200 Angstroms wide. On this scale no bands or discontinuities are apparent.

This lack of bands or discontinuities does not mean that the chromosome threads are not made up of long molecules which specifically interact with one another. It is due to the fact that nucleic acid molecules, unlike protein molecules, have no long side chains which, by interacting with the side chains of neighboring molecules, can form bands. It is well known, however, that under the light microscope the giant chromosomes in the salivary gland of the fruit fly have pronounced bands. The darker bands represent regions in which the ratio of DNA to protein is high; the lighter bands, regions in which the ratio is low. From this it is obvious that the position of the DNA and protein molecules is precisely determined. What is perhaps more to the point, geneticists have related the hereditary traits of the fruit fly and other insects to specific segments of their chromosomes.

I should like to suggest that the lessons learned from the organization of a protein such as collagen can be applied profitably to the study of the chromosome and the gene. Let us for the moment neglect the internal structure of DNA and its protein partners in the chromosome, and merely make some reasonable deductions from other facts.

First, DNA can be dissolved out of the nucleus of the cell by salt solution. This is a very weak chemical treatment, yet it effectively breaks the bonds that link DNA molecules to protein molecules and to one another. The fact that these bonds can be broken so easily makes it seem rather unlikely that, in its replication and in exerting its biochemical effects, the chromosome is an indivisible unit with all its macromolecules in an unchanging array.

An alternative is that the genetic specificity resides ultimately in the individual giant molecules of the chromosomes and determines the manner in which they interact. Since DNA molecules preserve their chemical pattern from one generation to the next, they must be capable of precise replication. If they can perform such a difficult feat, they may also be capable of highly specific interactions with other kinds of DNA molecules and with protein molecules. Thus they might spontaneously aggregate into the specific patterns characteristic of native chromosomes.

In this picture the chromosome is an aggregation of DNA and protein molecules which is stable in a particular chemical environment in the nucleus of the cell. A change in this immediate environment may alter not only the structural relationship of the molecules but also their biochemical and genetic activity. These possibilities are now being investigated in our laboratory at M.I.T. by applying the techniques used in the study of collagen to DNA, to the protein of chromosomes and to the giant banded chromosomes of fruit flies.

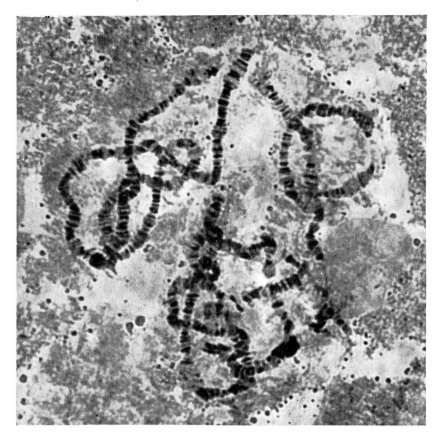

GIANT CHROMOSOMES of the fruit fly *Drosophila* are enlarged 500 diameters in this light-microscope photograph made by Herman W. Lewis of M.I.T. Cross-bands are visible.

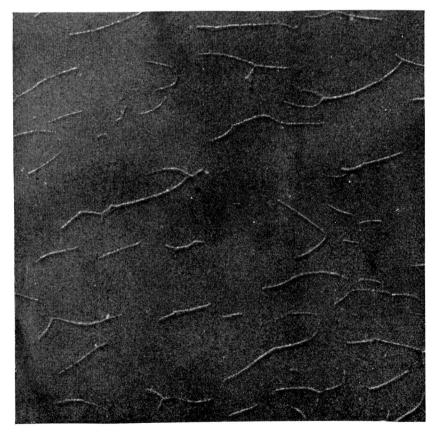

MOLECULES OF DNA are enlarged 100,000 diameters in this electron micrograph made by Hall. The material for the micrograph was extracted from the sperm cells of the salmon.

The Structure of Viruses

by R. W. Horne
January 1963

The electron microscope reveals that these infectious particles possess three principal types of symmetry. Each species of virus is ingeniously assembled from just a few kinds of building block

When the smaller members of the virus family are enlarged several hundred thousand times in the electron microscope, they are found to possess an extremely high degree of structural symmetry. In such viruses it is probable that the subunits visible in electron micrographs are individual protein molecules, often identical in kind, packed together to form a simple geometric structure. In the larger viruses the geometry is usually more complex, and a certain degree of structural flexibility begins to appear. Viewing the micrographs one has the impression of being shown how the inanimate world of atoms and molecules shades imperceptibly into the world of forms possessing some of the attributes of life.

Viruses are the smallest biological structures that embody all the information needed for their own reproduction. Essentially they consist of a shell of protein enclosing a core of nucleic acid—either ribonucleic acid (RNA) or deoxyribonucleic acid (DNA). The shell serves as a protective jacket and in some instances as a means for breaching the walls of those living cells that the virus is capable of attacking. The nucleic acid core enters the cell and redirects the cell machinery toward the production of scores of complete virus particles. When the job is done, the cell ruptures and the viruses spill out.

Most viruses fall in a size range between 10 and 200 millimicrons; in other terms, between a fortieth of a wavelength and half a wavelength of violet light. Since objects smaller than the wavelength of light cannot be seen in an ordinary microscope, viruses can be observed directly only with the aid of the electron microscope. These instruments employ a beam of electrons whose wavelength is much smaller than the dimensions of a virus. Viruses can also be studied indirectly by placing crystals of a pure virus preparation in an X-ray beam and recording the diffraction patterns produced when the X rays are reflected from the planes of atoms in the crystal. Analysis of such X-ray diffraction patterns suggested that the protein subunits forming the virus shell were arranged symmetrically. The tobacco mosaic virus, for example, showed up in early electron micrographs as a slender rod without visible subunits. When the virus was examined by X-ray diffraction, however, one could see patterns suggesting that the subunits were arranged in a helix. On the other hand, most small viruses, which looked spherical in electron micrographs, gave rise to X-ray patterns indicating that they had a cubic symmetry. This suggested that they were regular polyhedrons and also members of the group of Platonic solids: solids with four, six, eight, 12 and 20 sides.

In the light of the X-ray results, and arguing from general principles, F. H. C. Crick and James D. Watson proposed in 1956 and 1957 that the amount of nucleic acid present in the small viruses was limited, and that the information it carried would be sufficient to code for only a few kinds of protein. They suggested, therefore, that the shells of small "spherical" viruses were probably built from a number of identical protein subunits packed symmetrically. The most likely way for identical units to be packed on the surface of a sphere, Crick and Watson pointed out, would be in some pattern having cubic symmetry.

Some of the predictions of Crick and Watson were subsequently confirmed by electron micrography. There was a period, however, when the design and development of the electron microscope outpaced methods of preparing virus specimens for observation. Dehydrated virus particles are essentially transparent to an electron beam. Various techniques have had to be devised to make the particles visible. One of the earliest and simplest methods was to create "shadows" by allowing a stream of heavy-metal atoms to fall on the virus particles at an angle. This was done by placing the specimen of virus particles in a vacuum chamber and evaporating the metal atoms from a source toward the side of the chamber. The metal atoms that accumulated on the virus particle itself would block the passage of electrons, whereas electrons could pass freely through the shadows where metal atoms had not been deposited. In this way it was possible to discern the overall shape of the virus particle but not all the fine details of its surface structure.

Within the past few years a new and simple method of "staining" isolated particles such as viruses and large protein molecules has been even more successful than shadowing for revealing fine detail at the high magnifications now available in electron microscopes. It consists of surrounding the particles to be examined by an electron-dense material: potassium phosphotungstate. This is achieved by mixing the virus suspension with a solution of the phosphotungstate and spraying the mixture or depositing droplets on the specimen mounts. Since the phosphotungstate method produces images that are reversed compared with those obtained with the normal preparation procedures, it is called "negative staining" or "negative contrast." Application of this method to a large number of viruses has shown that they fall into three main symmetry groups: those with cubic symmetry, those with helical symmetry and those with complex symmetry or combined symmetries.

The class of polyhedrons that have

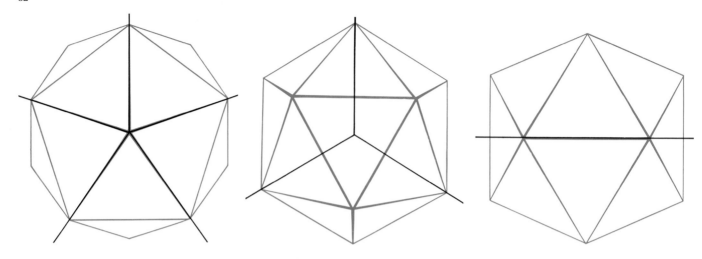

AXES OF SYMMETRY are shown for a regular icosahedron, a figure with 12 corners, 20 faces and 30 edges. Viewed along an axis at any corner, the figure can be rotated in five positions without changing its appearance (*left*). Rotated around any face axis, a regular icosahedron exhibits threefold symmetry (*middle*). Rotated around any edge axis, the figure shows twofold symmetry (*right*).

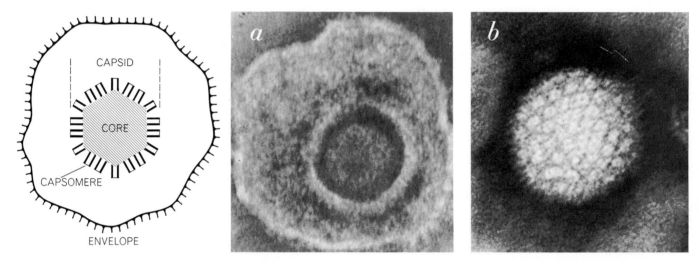

VIRUS NOMENCLATURE covers principal features observed in electron micrographs.

HERPES VIRUS sometimes has an envelope (*a*). The magnification is 310,000 diameters. The capsid (*b*) is composed of 162 capsomeres. Negative staining (*c*) indicates

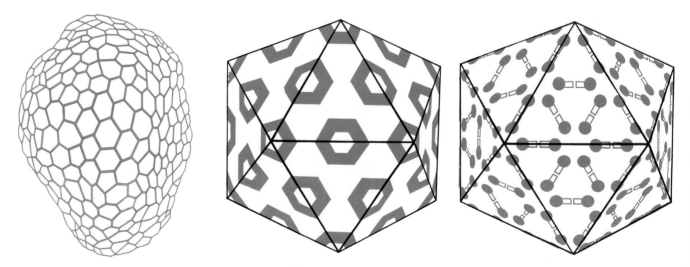

RADIOLARIANS, small marine organisms, have skeletons built of pentagons and hexagons.

ALTERNATIVE SCHEMES show how a regular icosahedron containing 42 pentagonal and hexagonal capsomeres (*left*) could be built up from 120 (or 240) small subunits.

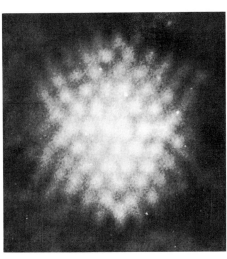

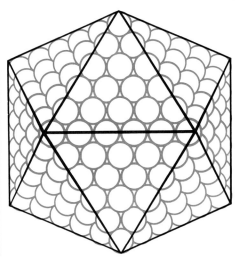

ADENOVIRUS is shown embedded in phosphotungstate, magnified about one million diameters (*left*). The drawing shows how the particle's 252 surface subunits, or capsomeres, are arranged with icosahedral symmetry. There are 12 on corners, 240 on faces or edges.

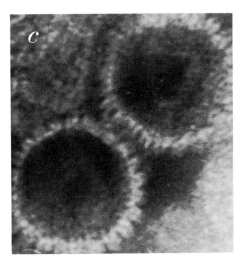

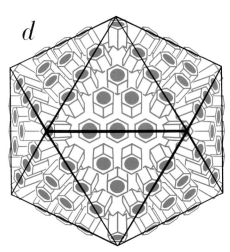

that they are hollow. The drawing (*d*) shows icosahedral arrangement. Micrographs are by P. Wildy and W. C. Russell of the Institute of Virology in Glasgow and the author.

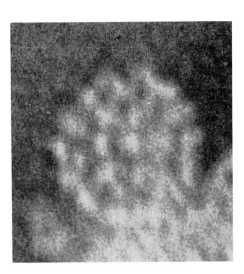

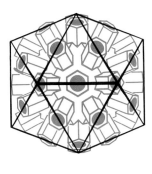

POLYOMA VIRUS is magnified one million diameters in micrograph by Wildy, M. G. P. Stoker and I. A. Macpherson of the Institute of Virology. It has 42 capsomeres (*right*).

cubic symmetry includes the regular tetrahedron (four faces), dodecahedron (12 faces) and icosahedron (20 faces). Shadowed preparations of the tipula iridescent virus, which causes a disease in the larvae of several insects, showed it to have the shape of a regular icosahedron, and the symmetry was self-evident [*see bottom illustration on page 85*]. Smaller viruses, on the other hand, do not reveal symmetry unless they are examined at very high magnification, and this requires the use of negative phosphotungstate staining.

Consider the symmetry properties of a regular icosahedron, in which each face is an equilateral triangle. If spokes are projected from the center of the icosahedron through the corners of the triangles, the spokes will represent one axis of rotational symmetry. Spokes projected from the center of the solid through the center of each face will represent a second axis. And spokes projected from the center through the midpoint of each edge will represent a third axis. (There will be 12 corner spokes, 20 face spokes and 30 edge spokes.) If the icosahedron is viewed along the spoke at any corner, one finds that the body can be rotated in five positions without changing its appearance [*see top illustration on opposite page*]. If the icosahedron is viewed along the spoke at any face, the body can be rotated in three positions without changing its appearance. And if the icosahedron is viewed along an edge spoke, it can be rotated in two positions without change of appearance. The regular icosahedron is thus said to have 5.3.2. symmetry.

Let us see now what implication this symmetry pattern has for a particle of adenovirus, which is associated with respiratory disease in man. The electron microscope shows that the surface of the particle is composed of regularly arranged structural units resembling tiny balls. Moreover, these balls are seen on the vertexes, faces and edges of an icosahedron [*see top illustration on this page*]. One can identify certain balls surrounded by five neighbors, which indicates that they are located on vertexes and therefore on axes of fivefold symmetry. Balls surrounded by six neighbors must lie on faces or edges and thus must occupy axes of either threefold or twofold symmetry. Along each edge there are six balls, including two balls occupying vertexes. To calculate the total number of balls covering the entire icosahedron one applies the simple formula $10(n-1)^2 + 2$, where n is the number of balls along one edge. Substituting 6

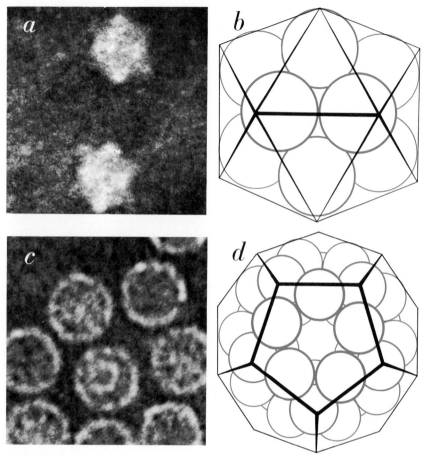

BACTERIOPHAGE ΦX174, magnified 750,000 diameters in *a*, appears to consist of 12 capsomeres arranged in icosahedral symmetry as shown in *b*. In other micrographs (*c*) smaller subunits seem to be arranged in ringlike structures. Each capsomere might actually be formed from five subunits as shown in *d*. Thirty such subunits would form a dodecahedron.

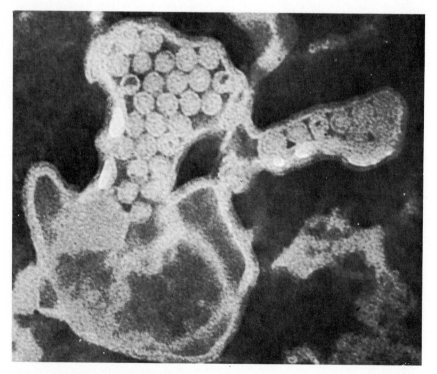

POLIOMYELITIS VIRUS PARTICLES are shown inside a fragment of an infected cell. The particles, magnified 250,000 diameters, appear to be composed of subunits smaller than the typical capsomere. The structural arrangement is not established. Electron micrograph is by Jack Nagington of the Public Health Laboratory in Cambridge and the author.

for *n* yields 252 as the number of morphological units composing the shell of the adenovirus particle.

For purposes of description (and to avoid the term "subunit," which can be applied to morphological, structural or chemical features) I shall adopt the recent terminology suggested for the various viral components [*see illustration at middle left, page 82*]. The morphological units composing the shell have been given the name "capsomeres." The shell itself is the "capsid." The region inside the capsid is the "core." The outer membrane, seen surrounding the capsid of some viruses, is the "envelope."

One merit of the negative-staining technique is that the electron-dense material is capable of penetrating into extremely small regions between, and even within, the capsomeres. A striking instance of such penetration can be seen in the electron micrograph of the herpes virus shown at the middle left on the preceding page. (In man the herpes virus causes, among other things, "cold sores.") Electron micrographs of the shadowed particle had indicated that it had the same external shape and symmetry as the adenovirus. When the two viruses were negatively stained and still further magnified, however, it could be seen on close examination that the capsomeres of the herpes virus, unlike those of the adenovirus, were elongated hollow prisms, some hexagonal in cross section and others pentagonal. In a number of particles the phosphotungstate penetrated into the central region, or core, normally containing the nucleic acid. In these "empty" particles the elongated capsomeres stand out clearly in profile at the periphery of the virus, and one can see their hollow form and the precision of their radial arrangement.

From the micrographs the number of capsomeres located on each edge was estimated to be five, giving a total of 162 capsomeres for the herpes virus. Of the 162 capsomeres, 12 are pentagonal prisms and 150 are hexagonal prisms. To satisfy the packing arrangement in accordance with icosahedral symmetry, the 12 pentagonal prisms would have to be placed at the corners and the 150 hexagonal prisms located on the edges or faces of the particle [*see drawing at middle right on preceding page*].

The need for pentagonal units goes deeper than the simple need to satisfy icosahedral symmetry. As early geometers observed, there is no way to arrange a system of hexagons so that they will enclose space. But if pentagonal units are included with hexagons, it is possible to enclose space in an almost

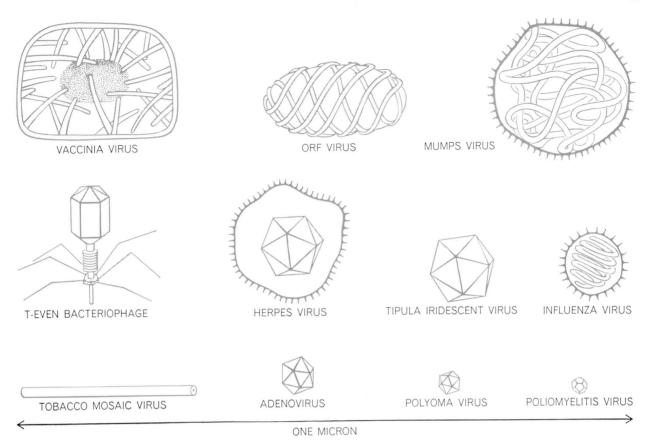

VACCINIA VIRUS

ORF VIRUS

MUMPS VIRUS

T-EVEN BACTERIOPHAGE

HERPES VIRUS

TIPULA IRIDESCENT VIRUS

INFLUENZA VIRUS

TOBACCO MOSAIC VIRUS

ADENOVIRUS

POLYOMA VIRUS

POLIOMYELITIS VIRUS

← ONE MICRON →

RELATIVE SIZES OF VIRUSES are shown in this chart. A micron, used as a measuring stick, is a thousandth of a millimeter; it is enlarged 175,000 times. The five viruses with polyhedral structures possess cubic symmetry. The tobacco mosaic virus and the internal components of influenza and mumps virus have helical symmetry. The remaining viruses exhibit complex symmetry.

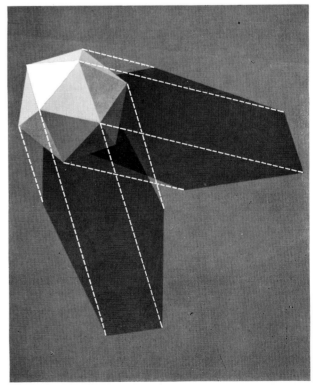

TIPULA IRIDESCENT VIRUS, an insect virus, is so large that its geometrically regular structure shows up clearly when specimens are shadowed with atoms of a heavy metal and enlarged in the electron microscope. In the doubly shadowed micrograph (*left*) the virus particles are enlarged about 58,000 diameters. The shadows indicate that each particle is a regular icosahedron (*right*). The micrograph was made by Kenneth Smith of the University of Cambridge and Robley C. Williams of the University of California.

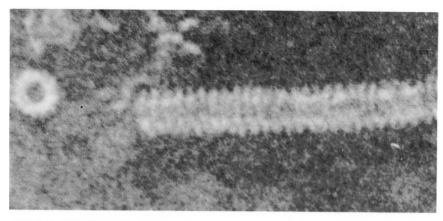

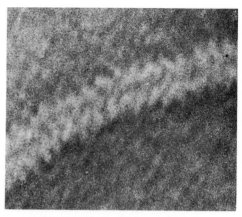

HELICAL SYMMETRY is shown in the electron micrographs of the rodlike tobacco mosaic virus, magnified 800,000 diameters, at left. The second electron micrograph, of the same magnification, shows an internal thread from a disrupted member of the myxovirus group. It too seems to possess helical symmetry. (Intact myxovirus particles are shown directly below.) The tobacco mosaic virus has

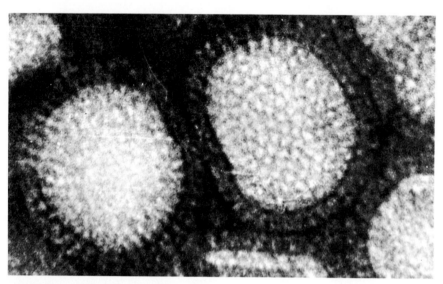

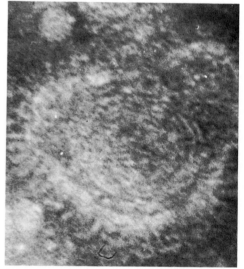

INFLUENZA VIRUS PARTICLES, members of the myxovirus family, are magnified 700,000 diameters in the electron micrograph at left. Although the particles are irregular in both size and shape, they appear to bristle with regularly spaced surface projections. In the second micrograph, which has a magnification of 600,000 diameters, phosphotungstate has penetrated the core of a particle, reveal-

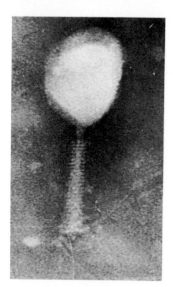

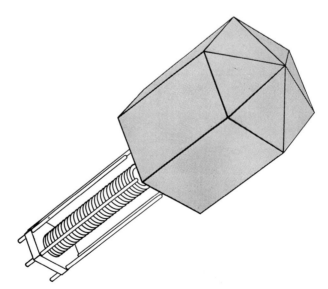

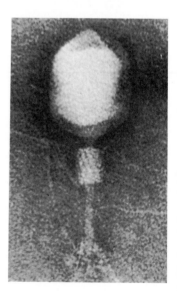

COMPLEX SYMMETRY is displayed by the T2 bacteriophage and other members of the "T even" family. Electron micrographs, in which the particle is magnified 300,000 diameters, clearly show that T2 exists in "untriggered" and "triggered" forms. The untriggered form is shown in the first pair of illustrations. The head of the phage is a bipyramidal hexagonal prism. The tail is a tube-like structure surrounded by a helical sheath. An end plate carries six tail fibers. When triggered, as shown in the second pair of illus-

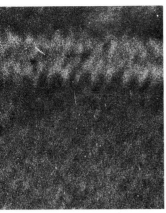

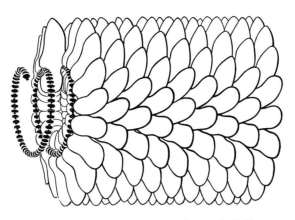

2,130 elongated capsomeres, consisting of protein molecules, arranged around a hollow core, as shown in the diagram at far right. The helical coil embedded in the capsomeres represents viral nucleic acid. The micrographs are by Nagington, A. P. Waterson and the author.

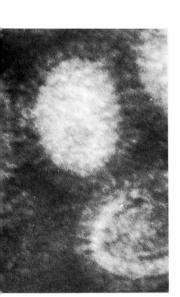

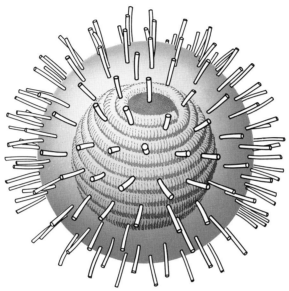

ing a coiled structure inside. The diagram at right shows a possible arrangement of the components in a typical myxovirus. The diagram follows a model built by L. Hoyle, Waterson and the author. The micrographs are by Waterson, Wildy, A. E. Farnham and the author.

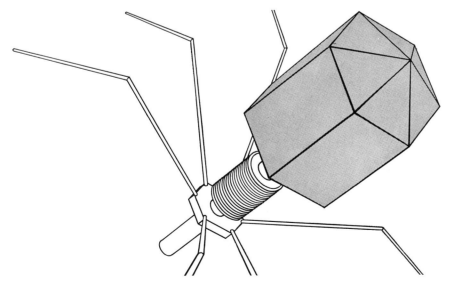

trations, the sheath contracts away from the end of the tail and the tail fibers are released. Presumably this coincides with the ejection of the DNA core (not shown), which previously had been coiled up in the head of the particle. The micrographs are by Sydney Brenner, George Streisinger, S. Champe, Leslie Barnett, Seymour Benzer, M. Rees and the author.

endless variety of ways, with forms both regular and irregular. The radiolarians, a group of marine protozoa, provide a fascinating example of varied structures assembed from pentagonal and hexagonal units [see illustration at bottom left on page 82].

Viruses smaller than the herpes virus usually have fewer capsomeres, but the relation between size and capsomere number is somewhat variable. The polyoma virus, which produces tumors in rodents and has stimulated a search for viruses in human cancer, appears to be almost spherical when examined by the shadowing technique. Nevertheless, negative staining shows that the outer shell is probably composed of 42 elongated angular capsomeres arranged in icosahedral symmetry [see bottom illustration, page 83]. Such a shell can be constructed by placing 12 pentagonal prisms at the corners of an icosahedron and 30 hexagonal prisms on the 30 edges. In this case the 20 faces have no capsomeres of their own, which helps to explain the nearly spherical appearance of the virus.

In the electron microscope the turnip yellow mosaic virus, which causes a disease of the leaves in the turnip and related plants, appears to have 32 capsomeres arranged in accordance with cubic symmetry. Crystals of the same virus studied by X-ray diffraction also show cubic symmetry, but this method indicates that there are 60 subunits instead of 32. Strictly speaking, neither number can be used to construct an icosahedron. But both numbers of subunits can be disposed symmetrically on the surface of an icosahedron. The smaller number can be distributed by placing 12 subunits on corners, 20 on faces and none on edges. (The 32 capsomeres could also be placed on the 32 vertexes of a pentakis dodecahedron or a rhombic triacontahedron.) The larger number can be distributed according to strict icosahedral symmetry by placing two subunits on each of the 30 edges and none on corners or faces. It is evident that if the two figures were transparent, one could be fitted over the other and the subunits of one would fall precisely in between the subunits of the other without overlapping. This suggests that the 60 subunits inferred from X-ray diffraction patterns may combine in some fashion to give the appearance of 32 subunits when the virus particle is observed in the electron microscope.

It has therefore been suggested that in the small spherical viruses the morphological features resolved as pentagons

VIRUS	SYMMETRY	NUMBER OF CAPSOMERES	SIZE OF CAPSID (ANGSTROM UNITS)	NUCLEIC ACID
TIPULA IRIDESCENT	CUBIC	812	1,300	DNA
ADENOVIRUS	CUBIC	252	700–750	?
GAL (GALLUS ADENO-LIKE)	CUBIC	252	950–1,000	?
INFECTIOUS CANINE HEPATITIS	CUBIC	252	820	?
HERPES SIMPLEX	CUBIC	162	1,000	DNA
WOUND TUMOR	CUBIC	92	?	RNA
POLYOMA	CUBIC	42	450	DNA
WARTS	CUBIC	42	500	?
TURNIP YELLOW MOSAIC	CUBIC	32	280–300	RNA
ΦX174	CUBIC	12	230–250	DNA
TOBACCO MOSAIC	HELICAL	2,130	3,000 × 170	RNA
MUMPS	HELICAL	—	170 (DIAMETER)	RNA
NEWCASTLE DISEASE	HELICAL	—	170 (DIAMETER)	RNA
SENDAI	HELICAL	—	170 (DIAMETER)	?
INFLUENZA	HELICAL	—	90–100 (DIAMETER)	RNA
T-EVEN BACTERIOPHAGE	COMPLEX	—	1,000 × 800 (HEAD)	DNA
CONTAGIOUS PUSTULAR DERMATITIS (ORF)	COMPLEX	—	2,600 × 1,600	?
VACCINIA	COMPLEX	—	3,030 × 2,400	DNA

TABLE OF VIRUSES shows the symmetry classification, number of capsomeres and capsid size of some of the principal families. (An angstrom unit is a ten-millionth of a millimeter; the wavelength of violet light is 4,000 angstrom units.) Nucleic acid (*column at far right*) is the genetic material of the virus. DNA is deoxyribonucleic acid; RNA, ribonucleic acid.

and hexagons may actually be built up from smaller structural subunits. These subunits may not all be identical, but they may be of two or three different molecular species. The diagram at the bottom right on page 82 indicates how such subunits might be assembled to produce pentagonal and hexagonal units, in strict accordance with icosahedral symmetry. The arrangement illustrated, one of several possible combinations, was proposed by A. Klug, D. L. D. Caspar and J. Finch of the University of London. It shows how 42 capsomeres could be formed from 120 (or 240) smaller subunits. Recent evidence suggests that the capsomeres in some of the larger viruses are linked together by small structures that may well correspond to the subunits.

High-resolution electron micrographs have revealed that structures originally identified as capsomeres in one very small virus are indeed composed of still smaller subunits. The virus, known as φX174, has been intensively studied because it contains an unusual single-stranded form of DNA [see "Single-stranded DNA," by Robert L. Sinsheimer; SCIENTIFIC AMERICAN Offprint 128]. When first examined in the elec-tron microscope, the virus appeared to have a shell composed of 12 spherical capsomeres, the minimum number needed for icosahedral symmetry. More recent electron micrographs indicate that each capsomere is formed from five subunits, but since each capsomere may be shared with a neighbor, the number of subunits is 30 [see *top illustration on page 84*]. If they are not shared and each capsomere is composed of five subunits, the total would be 60 and the shape would be that of a dodecahedron. Similar subunits smaller than capsomeres have been observed in electron micrographs of the virus of poliomyelitis, but it has not yet been possible to count them accurately [see *bottom illustration on page 84*].

The second broad group of viruses I shall discuss are those that have helical symmetry. Far and away the best known of this group is the virus that causes the mosaic disease of tobacco. Its helical structure was originally inferred from X-ray diffraction data. These data, combined with evidence from other physical and chemical observations, have led to a detailed knowledge of the tobacco mosaic virus' architecture. The subunits appear to be elongated structures so arranged that about 16 subunits form one turn of a helix. The subunits project from a central axial hole that runs the entire length of the virus. The nucleic acid of the virus does not occupy the hole, as might be expected, but is deeply embedded in the protein subunits and describes a helix of its own. The virus is composed of 2,130 identical protein subunits. Each subunit is a large molecule formed by the joining together of 168 amino acid molecules. The diagram of the virus' structure at the top of this page is based on a model by R. E. Franklin, Klug, Caspar and K. Holmes of the University of London.

Until recently helical symmetry was observed only in plant viruses. Now it has also been found in the complex animal viruses that are members of the influenza, or myxovirus, group. The group includes the viruses of mumps, Newcastle disease (a respiratory ailment of fowl), fowl plague and Sendai disease (a form of influenza). Electron micrographs produced by the shadow-casting technique showed these viruses to be of various shapes and sizes. Some were roughly spherical, some were filaments and others were complex and irregular. Thin sections of purified virus and particles seen at the surface of infected cells suggested the existence of an internal component in the form of ringlike structures surrounded by an outer membrane.

Recent studies using the negative staining method have shown that the internal component, or capsid, has the same dimensions and appearance as the rods of tobacco mosaic virus but is more flexible. This is particularly evident in electron micrographs of mumps virus, which show that the helical capsid forms coils or loops after being released. The particles of influenza and fowl plague are more structurally compact than the mumps virus and, unless subjected to special chemical treatment, are rarely observed releasing their internal components.

The envelopes of influenza virus and fowl plague virus carry surface projections that evidently contain the protein known as hemagglutinin, so named because it causes red blood cells to agglutinate. If these two viruses are treated with ether, the internal helix is released and can be separated from the hemagglutinin in a centrifuge. When this inner component is studied by electron microscopy, it is found to be of smaller diameter than that in the viruses of mumps, Newcastle disease virus and Sendai disease virus. The precise length of the helical components in the various myxoviruses is not yet known, nor the way they are packed within their envelopes. A possible arrangement for a typical myxovirus is shown in the diagram at the middle on the preceding page.

The last of the three broad groups of viruses are those whose symmetry is complex. This category includes the large bacterial viruses, such as the T2 virus that infects the bacterium *Escherichia coli,* and the large pox viruses. The T2 virus and several of its "T even" relatives are particularly remarkable because they contain some sort of contractile mechanism, a feature that has not been discerned in any other family of viruses. The electron micrographs at the bottom of page 86 show that the T2 virus has a head shaped in the form of a bipyramidal hexagonal prism. Attached to one end of the prism is a tail sructure consisting of a helical contractile sheath surrounding a central hollow core. At the extreme end of the core there is a curious hexagonal plate carrying six slender tail fibers. The plate structure and tail fibers probably make initial contact with the wall of the bacterium that is being attacked. After contact has been made the helical sheath contracts, allowing the nucleic acid

core of the virus to enter the bacterium.

The contraction of the T2 sheath raises many fascinating questions. The entire T2 virus appears to contain only a few different kinds of protein molecule. If these are allocated to the construction of the different structures—head, sheath, tail plate and tail fibers—one must conclude that the contractile sheath is composed of only two or at most three different kinds of protein. How can so few kinds of building block produce a sheath with contractile ability? What substances trigger the contraction? And how is the contraction related to the ejection of the long DNA molecule that is tightly packed in the T2 core?

Still larger viruses having complex symmetry are several important members of the pox virus family: the viruses of variola, vaccinia, cowpox and ectromelia. They are among the few viruses large enough to be seen in the light microscope. In early shadowed electron micrographs the vaccinia virus appeared to have a three-dimensional bricklike shape with a spherical dense central region. More detailed studies of the virus seen in infected cells after staining and thin sectioning revealed morphological features not observed in

other viruses. The central dense region appeared to be surrounded by a number of layers, or membranes, of varying opacity to the electron beam. In some micrographs tubelike structures could be seen between the outer membranes and the central region. The electron micrographs below illustrate the structural variations that exist between two members of the pox group. In the particles of the virus that causes orf, or contagious pustular dermatitis, the tubular components form a definite crisscross pattern. It is difficult to say whether the tubular structures should be described as capsids or as capsomeres, nor can one say just where the nucleic acid is located in relation to them.

The electron microscope, together with other methods, has greatly contributed to the study of viruses, and it has shown that they come in a surprising variety of mathematically ordered families. It has been understood for many years, of course, that proteins are versatile building blocks and that they account for the tremendous diversity of living forms. But it required the electron microscope to reveal directly what intricate and exquisite structures can be created by putting together only a few kinds of protein molecule.

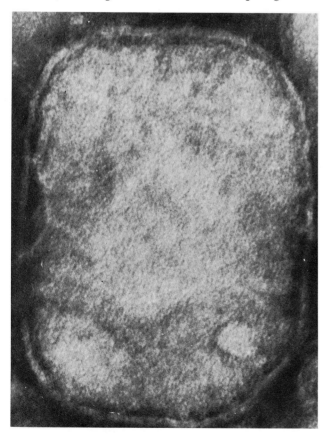

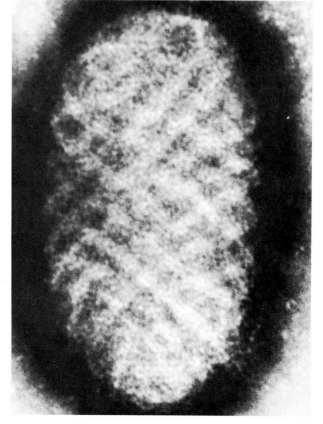

VACCINIA VIRUS, one of the giant pox viruses, is about twice the diameter of the smallest living cells, which are known as pleuropneumonia-like organisms. The magnification is 400,000 diameters.

ORF VIRUS, another pox virus, has components wound in a crisscross pattern. The magnification is 450,000 diameters. Micrographs of the orf and vaccinia viruses are by Nagington and the author.

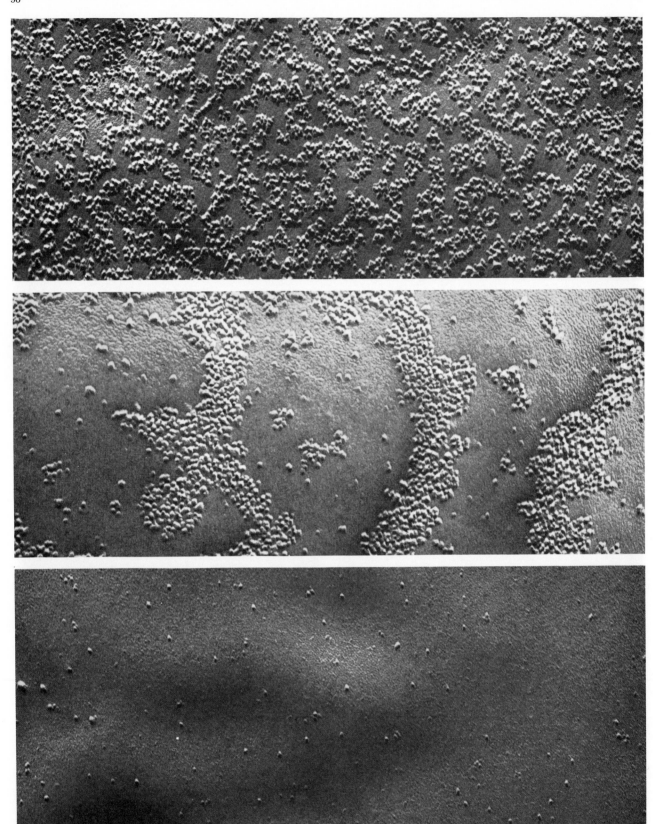

EVIDENCE FOR PROTEINS within the bilayer structure of cell membranes is provided by freeze-etch electron microscopy. A suspension of membranes in water is frozen and then fractured with a sharp blade. The fracture will often split a membrane in the middle along a plane parallel to the surface. After platinum and carbon vapors are deposited along the fracture surface the specimen can be studied in the electron microscope. The micrograph at the top shows many particles 50 to 85 angstroms in diameter embedded in a fractured membrane from rabbit red blood cells. The other two views show how the number of particles is greatly reduced if the membrane is first treated with a proteolytic enzyme that digests 45 percent (*middle*) or 70 percent (*bottom*) of the original membrane protein. The missing particles have presumably been digested by the enzyme. The membrane preparations are enlarged some 95,000 diameters in these micrographs made by L. H. Engstrom in Daniel Branton's laboratory at the University of California at Berkeley.

9

The Structure of Cell Membranes

by C. Fred Fox
February 1972

The thin, sturdy envelope of the living cell consists of lipid, phosphate and protein. The proteins act as both gatekeepers and active carriers, determining what passes through the membrane

Every living cell is enclosed by a membrane that serves not only as a sturdy envelope inside which the cell can function but also as a discriminating portal, enabling nutrients and other essential agents to enter and waste products to leave. Called the cytoplasmic membrane, it can also "pump" substances from one side to the other against a "head," that is, it can extract a substance that is in dilute solution on one side and transport it to the opposite side, where the concentration of the substance is many times higher. Thus the cytoplasmic membrane selectively regulates the flux of nutrients and ions between the cell and its external milieu.

The cells of higher organisms have in addition to a cytoplasmic membrane a number of internal membranes that isolate the structures termed organelles, which play various specialized roles. For example, the mitochondria oxidize foodstuffs and provide fuel for the other activities of the cell, and the chloroplasts conduct photosynthesis. Single-cell organisms such as bacteria have only a cytoplasmic membrane, but its structural diversity is sufficient for it to serve some or all of the functions carried out by the membranes of organelles in higher cells. It is clear that any model formulated to describe the structure of membranes must be able to account for an extraordinary range of functions.

Membranes are composed almost entirely of two classes of molecules: proteins and lipids. The proteins serve as enzymes, or biological catalysts, and provide the membrane with its distinctive functional properties. The lipids provide the gross structural properties of the membrane. The simplest lipids found in nature, such as fats and waxes, are insoluble in water. The lipids found in membranes are amphipathic, meaning that one end of the molecule is hydrophobic, or insoluble in water, and the other end hydrophilic, or water-soluble. The hydrophilic region is described as being polar because it is capable of carrying an ionic (electric) charge; the hydrophobic region is nonpolar.

In most membrane lipids the nonpolar region consists of the hydrocarbon chains of fatty acids: hydrocarbon molecules with a carboxyl group (COOH) at one end. In a typical membrane lipid two fatty-acid molecules are chemically bonded through their carboxyl ends to a backbone of glycerol. The glycerol backbone, in turn, is attached to a polar-head group consisting of phosphate and other groups, which often carry an ionic charge [*see illustration on next page*]. Phosphate-containing lipids of this type are called phospholipids.

When a suspension of phospholipids in water is subjected to high-energy sound under suitable conditions, the phospholipid molecules cluster together to form closed vesicles: small saclike structures called liposomes. The arrangement of phospholipids in the walls of both liposomes and biological membranes has recently been deduced with the help of X-ray diffraction, which can reveal the distance between repeating groups of atoms. An X-ray diffraction analysis by M. F. Wilkins and his associates at King's College in London indicates that two parallel arrays of the polar-head groups of lipids are separated by a distance of approximately 40 angstroms and that the fatty-acid tails are stacked parallel to one another in arrays of 50 or more phospholipid molecules.

The X-ray data suggest a structure for liposomes and membranes in which the phospholipids are arranged in two parallel layers [*see illustrations on page 93*]. The polar heads are arrayed externally on the bilayer surfaces, and the fatty-acid tails are pointed inward, perpendicular to the plane of the membrane surface. This model of phospholipid structure in membranes is identical with one proposed by James F. Danielli and Hugh Davson in the mid-1930's, when no precise structural data were available. It is also the minimum-energy configuration for a thin film composed of amphipathic molecules, because it maximizes the interaction of the polar groups with water.

Unlike lipids, proteins do not form orderly arrays in membranes, and thus their arrangement cannot be assessed by X-ray diffraction. The absence of order is not surprising. Each particular kind of membrane incorporates a variety of protein molecules that differ widely in molecular weight and in relative numbers; a membrane can incorporate from 10 to 100 times more molecules of one type of protein than of another.

Since little can be learned about the disposition of membrane proteins from a general structural analysis, investigators have chosen instead to study the orientation of one or a few species of the proteins in membranes. In the Danielli-Davson model the proteins are assumed to be entirely external to the lipid bilayer, being attached either to one side of the membrane or to the other. Although information obtained from X-ray diffraction and high-resolution electron microscopy indicates that this is probably true for the bulk of the membrane protein, biochemical studies show that the Danielli-Davson concept is an oversimplification. The evidence for alternative locations has been provided chiefly by Marc Bretscher of the Medical Research Council laboratories in Cambridge and by Theodore L. Steck, G. Franklin and Donald F. H. Wallach of the Harvard Medical School. Their results suggest that certain proteins penetrate the lipid bilayer and that others extend all the way through it.

Bretscher has labeled a major protein of the cytoplasmic membrane of human red blood cells with a radioactive substance that forms chemical bonds with the protein but is unable to penetrate the membrane surface. The protein was labeled in two ways [*see illustration on pages 94 and 95*]. First, intact red blood blood cells were exposed to the label so that it became attached only to the portion of the protein that is exposed on the outer surface of the membrane. Second, red blood cells were broken up before the radioactive label was added. Under these conditions the label could attach itself to parts of the protein exposed on the internal surface of the membrane as well as to parts on the external surface.

The two batches of membrane, labeled under the two different conditions, were treated separately to isolate the protein. The purified protein from the two separate samples was degraded into definable fragments by treatment with a proteolytic enzyme: an enzyme that cleaves links in the chain of amino acid units that constitutes a protein. A sample from each batch of fragments was now placed on the corner of a square of filter paper for "fingerprinting" analysis. In this technique the fragments are separated by chromatography in one direction on the paper and by electrophoresis in a direction at right angles to the first. In the chromatographic step each type of fragment is separated from the others because it has a characteristic rate of travel across the paper with respect to the rate at which a solvent travels. In the electrophoretic step the fragments are further separated because they have char-acteristic rates of travel in an imposed electric field.

Once a separation had been achieved the filter paper was laid on a piece of X-ray film so that the radioactively labeled fragments could reveal themselves by exposing the film. When films from the two batches of fragments were developed, they clearly showed that more labeled fragments were present when both the internal and the external surface of the cell membrane had been exposed to the radioactive label than when the outer surface alone had been exposed. This provides strong evidence that the portion of the protein that gives rise to the additional labeled fragments is on the inner surface of the membrane.

Steck and his colleagues obtained similar results with a procedure in which they prepared two types of closed-membrane vesicle, using as a starting material the membranes from red blood cells. In one type of vesicle preparation (right-side-out vesicles) the outer membrane surface is exposed to the external aqueous environment. In the other type of preparation (inside-out vesicles) the inner surface of the membrane is exposed to the external aqueous environment. When the two types of vesicle are treated with a proteolytic enzyme, only those proteins exposed to the external aqueous environment should be degraded. Steck found that some proteins are susceptible to digestion in both the right-side-out and inside-out vesicles, indicating that these proteins are exposed on both membrane surfaces. Other proteins are susceptible to proteolytic digestion in right-side-out vesicles but not in inside-out vesicles. Such proteins are evidently located exclusively on only one side of the membrane. This information lends credence to the concept of sidedness in membranes. Such sidedness had been suspected for many years because the inner and outer surfaces of cellular membranes are thought to have different biological functions. The development of a technique for preparing vesicles with right-side-out and inside-out configurations should be extremely useful in determining on which side of the membrane a given species of protein resides and thus functions.

Daniel Branton and his associates at the University of California at Berkeley have developed and exploited the technique of freeze-etch electron microscopy to study the internal anatomy of membranes. In freeze-etch microscopy a suspension of membranes in water is frozen rapidly and fractured with a sharp blade. Wherever the membrane surface runs parallel to the plane of fracture much of the membrane will be split along the middle of the lipid bilayer. A thin film of platinum and carbon is then evaporated onto the surface of the fracture. This makes it possible to examine the anatomy of structures in the fracture plane by electron microscopy.

The electron micrographs of the fractured membrane reveal many particles, approximately 50 to 85 angstroms in diameter, on the inner surface of the lipid bilayer. These particles are not observed if the membrane samples are first treated with proteolytic enzymes, indicating that the particles probably consist of protein [*see illustration on page 90*]. From quantitative estimates of the number of particles revealed by freeze-etching, Branton and his colleagues have suggested that between 10 and 20 percent of the internal volume of many biological membranes is protein.

Somewhere between a fifth and a quarter of all the protein in a cell is physically associated with membranes. Most of the other proteins are dissolved in the aqueous internal environment of the cell. In order to dissolve membrane proteins in aqueous solvents detergents must be added to promote their dispersion. One might therefore expect membrane proteins to differ considerably from soluble proteins in chemical composition. This, however, is not the case.

The amino acids of which proteins are composed can be classified into two groups: polar and nonpolar. S. A. Rosenberg and Guido Guidotti of Harvard University analyzed the amino acid composition of proteins from a number of membranes and found that they contain about the same percentage of polar and nonpolar amino acids as one finds in the soluble proteins of the common colon

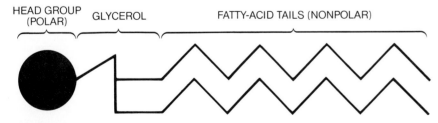

HEAD GROUP (POLAR)   GLYCEROL   FATTY-ACID TAILS (NONPOLAR)

TYPICAL MEMBRANE LIPID is a complex molecular structure, one end of which is hydrophilic, or water-soluble, and the other end hydrophobic. Such a substance is termed amphipathic. The hydrophilic, or polar, region consists of phosphate and other constituents attached to a unit of glycerol. The polar-head group, when in contact with water, often carries an electric charge. The glycerol component forms a bridge to the hydrocarbon tails of two fatty acids that constitute the nonpolar region of the lipid. In this highly schematic diagram the zigzag lines represent hydrocarbon chains; each angle is occupied by a carbon atom and two associated hydrogen atoms. The terminal carbon of each chain is bound to three hydrogen atoms. Phosphate-containing amphipathic lipids are called phospholipids.

bacterium *Escherichia coli.* Thus differences in amino acid composition cannot account for the water-insolubility of membrane proteins.

Studies conducted by L. Spatz and Philipp Strittmatter of the University of Connecticut indicate that the most likely explanation for the water-insolubility of membrane proteins is the arrangement of their amino acids. Spatz and Strittmatter subjected membranes of rabbit liver cells to a mild treatment with a proteolytic enzyme. The treatment released the biologically active portion of the membrane protein: cytochrome b_5. In a separate procedure they solubilized and purified the intact cytochrome b_5 and treated it with the proteolytic enzyme. This treatment also released the water-soluble, biologically active portion of the molecule, together with a number of small degradation products that were insoluble in aqueous solution. The biologically active portion of the molecule, whether obtained from the membrane or from the purified protein, was found to be rich in polar amino acids. The protein fragments that were insoluble in water, on the other hand, were rich in nonpolar amino acids. These observations suggest that many membrane proteins may be amphipathic, having a nonpolar region that is embedded in the part of the membrane containing the nonpolar fatty-acid tails of the phospholipids and a polar region that is exposed on the membrane surface.

We are now ready to ask: How do substances pass through membranes? The nonpolar fatty-acid-tail region of a phospholipid bilayer is physically incompatible with small water-soluble substances, such as metal ions, sugars and amino acids, and thus acts as a barrier through which they cannot flow freely. If one measures the rate at which blood sugar (glucose) passes through the phospholipid-bilayer walls of liposomes, one finds that it is far too low to account for the rate at which glucose penetrates biological membranes. Information of this kind has given rise to the concept that entities termed carriers must be present in biological membranes to facilitate the passage of metal ions and small polar molecules through the barrier presented by the phospholipid bilayer.

Experiments with biological membranes indicate that the hypothetical carriers are highly selective. For example, a carrier that facilitates the transport of glucose through a membrane plays no role in the transport of amino acids or other sugars. An interesting experimental

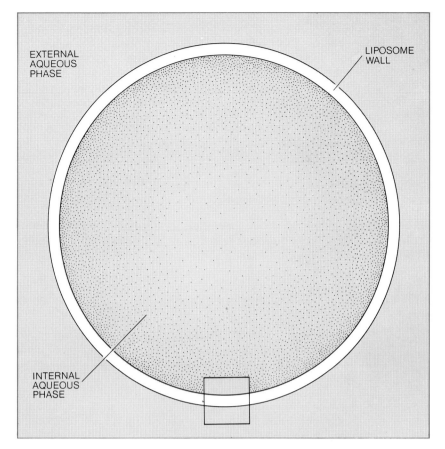

ARTIFICIAL MEMBRANE-ENCLOSED SAC, known as a liposome, is created by subjecting an aqueous suspension of phospholipids to high-energy sound waves. X-ray diffraction shows that the phospholipids in the liposome assume an orderly arrangement resembling what is found in the membranes of actual cells. Area inside the square is enlarged below.

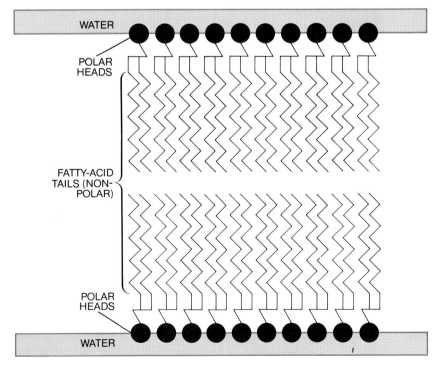

CROSS SECTION OF LIPOSOME WALL shows how the membrane is formed from two layers of lipid molecules. The polar heads of amphipathic lipids face toward the aqueous solution on each side while the nonpolar fatty-acid tails face inward toward one another.

system for measuring selective ion transport was developed by A. D. Bangham, M. M. Standish and J. C. Watkins of the Agricultural Research Council in Cambridge, England, and by J. B. Chappell and A. R. Crofts of the University of Cambridge. As a model carrier they used valinomycin, a nonpolar, fat-soluble antibiotic consisting of a short chain of amino acids (actually 12); such short chains are termed polypeptides to distinguish them from true proteins, which are much larger. Valinomycin combines with phospholipid-bilayer membranes and makes them permeable to potassium ions but not to sodium ions.

The change in permeability is conveniently studied by measuring the change in electrical resistance across a phospholipid bilayer between two chambers containing a potassium salt in aqueous solution. The experiment is performed by introducing a sample of phospholipid into a small hole between the two chambers. The lipid spontaneously thins out until the chambers are separated by only a thin membrane consisting of a phospholipid bilayer. Electrodes are then placed in the two chambers to measure the resistance across the membrane.

The resistance across a phospholipid bilayer in the absence of valinomycin is several orders of magnitude higher than the resistance across a typical biological membrane: 10 million ohms centimeter squared compared with between 10 and 10,000. This indicates that phospholipid-bilayer membranes are essentially impermeable to small hydrophilic ions. If a small amount of valinomycin (10^{-7} gram per milliliter of salt solution) is intro-

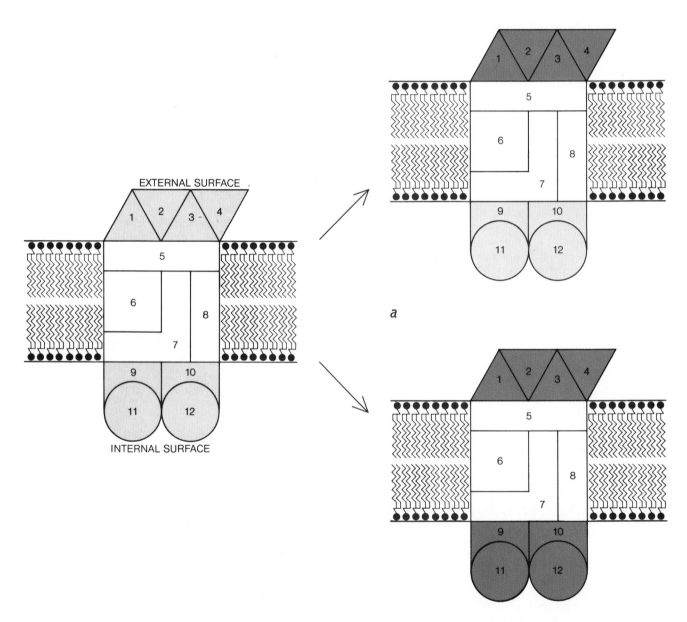

LOCATION OF PROTEINS IN MEMBRANES can be inferred by attaching radioactive labels to the proteins. These diagrams depict an experiment in which a major protein in the membrane of red blood cells was labeled (*a*). When intact cells (*top sequence*) are exposed to the radioactive substance, only the portion of the protein on the outside wall picks up the label (*color*). When the cells are broken before labeling (*bottom sequence*), the radioactive label is able to reach portions of the protein that are exposed to the internal as well as to the external surfaces of the membrane. This can be demonstrated by isolating and purifying the protein labeled under the two conditions. The protein is then broken up into defined fragments (*numbered shapes*) by treating it with a proteolytic enzyme (*b*). Portions of the two batches of fragments are spotted on the corners of filter paper for "fingerprinting" (*c*). This is a

duced into the chambers containing the potassium solution, the resistance falls by five orders of magnitude and the permeability of the phospholipid bilayer to potassium ions rises by a like amount. The permeability of the experimental membrane now essentially duplicates the permeability of biological membranes.

If the experiment is repeated with a sodium chloride solution in the chambers, one finds that the addition of valinomycin causes only a slight change in resistance. Hence valinomycin meets two of the most important criteria for a biological carrier: it enhances permeability and it is highly selective for the transported substance. The question that now arises is: How does valinomycin work?

First of all, valinomycin is nonpolar. Thus it is physically compatible with and can dissolve in the portion of the bilayer that contains the nonpolar fatty-acid tails. Second, valinomycin can evidently diffuse between the two surfaces of the bilayer. S. Krasne, George Eisenman and G. Szabo of the University of California at Los Angeles have shown that the enhancement of potassium-ion transport by valinomycin is interrupted when the bilayer is "frozen" by lowering the temperature. Third, valinomycin must bind potassium ions in such a way that the ionic charge is shielded from the nonpolar region of the membrane. Finally, valinomycin itself must have a selective binding capacity for potassium ions in preference to sodium or other ions.

With valinomycin as a model for carrier-mediated transport, one can postulate three essential steps: recognition of the ion, diffusion of the ion through the membrane, and its release on the other

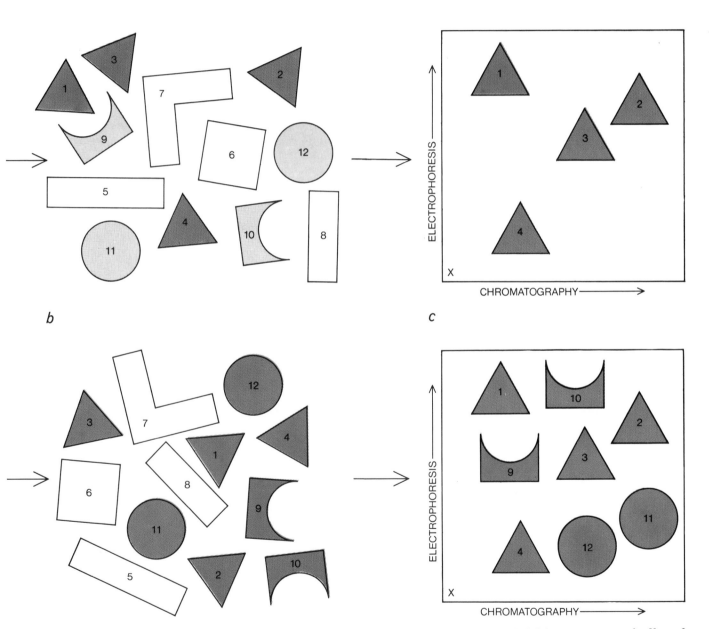

b

c

technique that combines chromatography with electrophoresis. By chromatography alone protein fragments would migrate at different rates depending primarily on their solubility in the solvent system. Electrophoresis involves establishing an electric-potential gradient along one axis of the filter paper. Since various fragments have different densities of electric charge they are further separated. A piece of X-ray film is then placed over each sheet of filter paper. Radiation from the labeled fragments exposes the film and reveals where the various fragments have come to rest. A comparison of the X-ray films produced in the parallel experiments shows that more protein fragments are labeled when the red blood cells are broken before labeling and that the additional fragments (9, 10, 11, 12) must represent portions of the original protein that extend through the membrane and penetrate the inner surface.

side. In the first step some part of the valinomycin molecule, embedded in the membrane, "recognizes" the potassium ion as it approaches the surface of the membrane and captures it. In the second step the complex consisting of valinomycin and the potassium ion diffuses through the membrane. Finally, on reaching the opposite surface of the membrane the potassium ion is dissociated from the complex and is released.

The argument to this point can be summarized in a few words. The fundamental structure of biological membranes is a phospholipid bilayer, the phospholipid bilayer is a permeability barrier and carriers are needed to breach it. In addition, the membrane barrier must often be breached in a directional way. In a normally functioning cell hundreds of kinds of small molecule must be present at a higher concentration inside the cell than outside, and many other small molecules must be present at a lower concentration inside the cell than outside. For example, the concentration of potassium ions in human cells is more than 100 times greater than it is in the blood that bathes them. For sodium ions the concentrations are almost exactly reversed. The maintenance of these differences in concentration is absolutely essential; even slight changes can result in death.

Although the model system based on valinomycin provides considerable insight into the function and selectivity of carriers, it sheds no light on the transport mechanism that can pump a substance from a low concentration on one side of the membrane to a higher concentration on the other. Our understanding of concentrative transport (or, as it is usually termed, active transport) owes much to the pioneering effort of Georges Cohen, Howard Rickenberg, Jacques Monod and their associates at the Pasteur Institute in Paris. The Pasteur group studied the transport of milk sugar (lactose) through the cell membrane of the bacterium *Escherichia coli*. Genetic experiments suggested that the carrier for lactose transport was a protein. Studies of the rate of transport revealed that the transport process behaves like a reaction catalyzed by an enzyme, giving further support to the idea that the carrier is a protein. The Pasteur group also found that the lactose-transport system is capable of active transport, producing a lactose concentration 500 times greater inside the cell than outside. The active-transport process depends on the expenditure of metabolic energy; poisons that block energy metabolism destroy the ability of the cell to concentrate lactose.

A model that accounts for many (but not all) of the properties of the active-transport system that are typified by the lactose system postulates the existence of a carrier protein that can change its shape. The protein is visualized as resembling a revolving door in the membrane wall [*see illustration on opposite page*]. The "door" contains a slot that fits the target substance to be transported. The slot normally faces the cell's external environment. When the target substance enters the slot, the protein changes shape and is thereby enabled to rotate so that the slot faces into the cell. When the target substance has been discharged into the cell, the protein remains with its slot facing inward until the cell expends energy to rotate the protein so that the slot again faces outward.

Working with Eugene P. Kennedy at the Harvard Medical School in 1965, I succeeded in identifying the lactose-transport carrier. We found, as we had expected, that it is a protein with an enzyme-like ability to bind lactose. Since then a number of other transport carriers have been identified, and all turn out to be proteins. The lactose carrier resides in the membrane and is hydrophobic; thus it is physically compatible

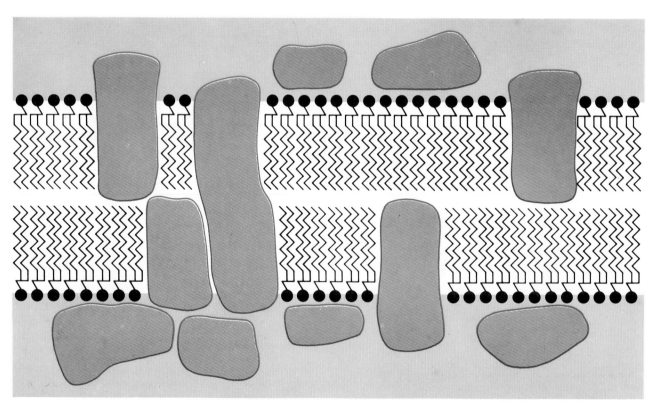

ANATOMY OF BIOLOGICAL MEMBRANE is suggested in this schematic diagram. Phospholipid molecules stacked side by side and back to back provide the basic structure. The gray shapes represent protein molecules. In some cases several proteins (for example the five at the left) are bound into a single functional complex. Proteins can occupy all possible positions with respect to the phospholipid bilayer: they can be entirely outside or inside, they can penetrate either surface or they can extend through the membrane.

with the nonpolar-lipid phase of the membrane.

In 1970 Ron Kaback and his associates at the Roche Institute of Molecular Biology observed that the energy that drives the active transport of lactose and dozens of other low-molecular-weight substances in *E. coli* is directly coupled to the biological oxidation of metabolic intermediates such as D-lactic acid and succinic acid. How energy derived from the oxidation of D-lactic acid can be used to drive active transport is one of the more interesting unsolved problems in membrane biology.

Since transport carriers must be mobile within the membrane in order to move substances from one surface to the other, one might guess that the region of the membrane containing the fatty-acid tails should not have a rigid crystalline structure. X-ray diffraction studies indicate that the fatty acids of membranes in fact do have a "liquid crystalline" structure at physiological temperature, that is, around 37 degrees Celsius. In other words, the fatty acids are not aligned in a rigid crystalline lattice. The techniques of electron paramagnetic resonance and nuclear magnetic resonance can be used to study the flexibility of the fatty-acid side chains in membranes. Several investigators, notably Harden M. McConnell and his associates at Stanford University, have concluded that the fatty acids of membranes are quasi-fluid in character.

Membranes incorporate two classes of fatty acids: saturated molecules, in which all the available carbon bonds carry hydrogen atoms, and unsaturated molecules, in which two or more pairs of hydrogen atoms are absent (with the result that two or more pairs of carbon atoms have double bonds). The fluid character of membranes is largely determined by the structure and relative proportion of the unsaturated fatty acids. In phospholipids consisting only of saturated fatty acids the fatty-acid tails are aligned in a rigidly stacked crystalline array at physiological temperatures. In phospholipids consisting of both saturated and unsaturated fatty acids the fatty acids are packed in a less orderly fashion and thus are more fluid. The double bonds of unsaturated fatty acids give rise to a structural deformation that interrupts the ordered stacking necessary for the formation of a rigid crystalline structure [*see illustration on next page*].

My colleagues and I at the University of Chicago (and later at the University of California at Los Angeles) and Peter Overath and his associates at the Uni-

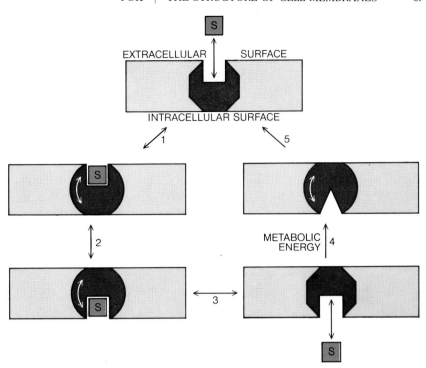

MECHANISM OF "ACTIVE" TRANSPORT may involve a carrier protein (*dark gray*) with the properties of a revolving door. A carrier protein can capture a substance, S, that exists outside the membrane in dilute solution and transport it to the inside of the cell, where the concentration of S is greater than it is outside. When S is bound to the protein, the protein changes shape (*1*), thus enabling it to rotate (*2*). When S becomes detached and enters the cell (*3*), the protein returns to its immobile form. Metabolic energy must be expended (*4*) to alter the protein's shape so that it can rotate and again present its binding site to the cell exterior (*5*). Other protein carriers have the capacity to transport substances from low concentration inside the cell to solutions of higher concentration outside the cell.

versity of Cologne have varied the fatty-acid composition of biological membranes to study the effects of fatty-acid structure on transport. When the membrane lipids are rich in unsaturated fatty acids, transport can proceed at rates up to 20 times faster than it does when the membrane lipids are low in unsaturated fatty acids. These experiments show that normal membrane function depends on the fluidity of the fatty acids.

The temperature at which cells live and grow can have a pronounced effect on the amount of unsaturated fatty acid in their membranes. Bacteria grown at a low temperature have membranes with a greater proportion of unsaturated fatty acid than those grown at a higher temperature. This adjustment in fatty-acid composition is necessary if the membranes are to function normally at low temperature. A similar adjustment can take place in higher organisms. For example, there is a temperature gradient in the legs of the reindeer; the highest temperature is near the body, the lowest is near the hooves. To compensate for this temperature gradient the cells near the hooves have membranes whose lipids are enriched in unsaturated fatty acids.

Although, as we have seen, phospholipids can spontaneously form bilayer films in water, this process only provides a physical rationale as to why the predominant structure in membranes is a phospholipid bilayer. The events leading to the assembly of a biological membrane are far more complex. The cells of higher organisms contain a number of unique membrane structures. They differ widely in lipid composition, and each type of membrane has its own unique complement of proteins. The diversity in protein composition and in the location of proteins within membranes explains the functional diversity of different types of membrane. Rarely does a single species of protein exist in more than one type of membrane.

Since all membrane proteins are synthesized at approximately the same cellular location, what is it that determines that one type of protein will be incorporated only into the cytoplasmic membrane and that another type will turn up only in a mitochondrial membrane? At present this question can be answered only by conjecture tinctured with a few facts. Two general hypotheses for membrane assembly can be offered. One pos-

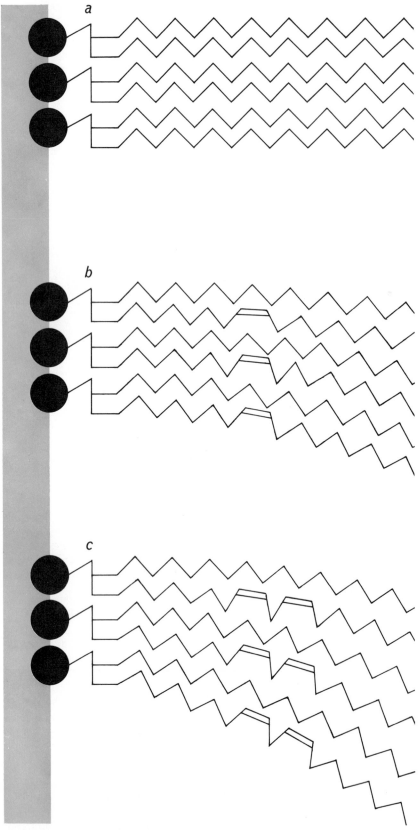

VARIATION IN FATTY-ACID COMPOSITION can disrupt the orderly stacking of phospholipids in a biological membrane. In a lipid layer composed entirely of saturated fatty acids (a) the fatty-acid chains contain only single bonds between carbon atoms and thus nest together to form rigid structures. In a lipid layer containing unsaturated fatty acids with one double bond (b) the double bonds introduce a deformation that interferes with orderly stacking and makes the fatty-acid region somewhat fluid. When fatty acids with two double bonds are present (c), the deformation and the consequent fluidity are greater still.

sibility is that new pieces of membrane are made from scratch by a self-assembly mechanism in which all the components of a new piece of membrane come together spontaneously. This new piece could then be inserted into an existing membrane. A second possibility is that newly made proteins are simply inserted at random into a preexisting membrane.

Recent studies in my laboratory at the University of California at Los Angeles and in the laboratories of Philip Siekevitz and George E. Palade at Rockefeller University support the second hypothesis. That is all well and good, but what determines why a given protein is incorporated only into a given kind of membrane? Although this must be answered by conjecture, it is known that many proteins are specifically bound to other proteins in the same membrane. Such protein-protein interactions are not uncommon; many of the functional entities in membranes are complexes of several proteins. Thus the proteins in a membrane may provide a template that is recognized by a newly synthesized protein and that helps to insert the newly synthesized protein into the membrane. In this way old membrane could act as a template for the assembly of new membrane. This might explain why different membranes incorporate different proteins.

Why, then, do different membranes have different lipid compositions? The answers to this question are even more obscure. In general lipids are synthesized within the membrane; the enzymes that catalyze the synthesis are part of the membrane. Some lipids, however, are made in one membrane and then shuttled to another membrane that has no inherent capacity to synthesize them. Since there is an interchange of lipids between various membranes, it seems unlikely that the variations in lipid composition in different membranes can be explained by dissimilarities in the synthetic capacity of a given membrane for a given type of lipid. There are at least two possible ways of accounting for differences in lipid composition. One possibility is that different membranes may destroy different lipids at different rates; another is that the proteins of one species of membrane may selectively bind one type of lipid, whereas the proteins of another species of membrane may bind a different type of lipid. It is obvious from this discussion that concrete evidence on the subject of membrane assembly is scant but that the problems are well defined.

Pumps in the Living Cell

by Arthur K. Solomon
August 1962

One of the main activities of most animal cells and many animal tissues is the excretion of sodium. How is this "active transport" accomplished? Here the answer is sought in the kidney of the amphibian Necturus

One of the main tasks of the living cell is simply to maintain its physical integrity: to keep itself from bursting open. In its normal liquid environment the cell is subject to an inexorable force acting to push water across its enclosing membrane from outside to inside and thereby increase the pressure inside. Some plant cells resist the force by structural means. The strong outer wall of such a cell contains the pressure, which, incidentally, furnishes the rigidity that gives the cell its shape. Animal cells have evolved a different answer. In effect they hold back the water with pumps.

Biological pumping, or "active transport," plays other fundamental roles in the functioning of both individual cells and multicellular organisms. The process has attracted the interest of biophysicists in many laboratories. They have undertaken to isolate its mechanism and to identify the source of the energy that drives it.

In preventing the cell from bursting, the pump is working against the force of osmosis. This force results from the uni-versal tendency of solutions on each side of a semipermeable membrane to reach equal concentration. The membrane of the cell is freely permeable to the molecules of water and to other small molecules, but not to large molecules such as those of proteins. In general the concentration of large molecules is higher inside the cell than it is in the surrounding liquid; therefore water tends to flow in to equalize the concentration. Almost all animal cells are thought to offset the effect by pumping out sodium. They expend a substantial fraction of their total energy budget in maintaining a lower sodium concentration inside the cell than outside. The result is that the total concentration of large and small molecules is the same inside the cell and out, and therefore there is no osmotic pressure.

Since the sodium is tranferred as positively charged ions, its outflow leaves the interior of the cell electrically negative with respect to the outside. The potential difference across the membrane is only about a tenth of a volt, but the membrane is only about a millionth of a centimeter thick. Hence the field amounts to 100,000 volts per centimeter, which is quite a respectable figure. Among its various physiological roles the electrical potential difference makes possible the transmission of impulses along nerve fibers and the triggering of contraction in muscle.

Under certain circumstances a system of cells pumps sodium to move water rather than to prevent it from moving. For example, approximately seven quarts of water enter the human intestinal tract every day, mostly in saliva and secretions of the stomach and pancreas. Here the problem is to return most of the water to the body fluids. It appears that the lining of the gut contains a pump that moves sodium, and water along with it, out of the intestine and back to the bloodstream.

In our laboratory at the Harvard Medical School we have been studying a somewhat similar system in the vertebrate kidney. The kidney is the organ that regulates the composition of the blood plasma and hence the extracellular environment of all the cells in the body. In an adult human being some 50 gallons of blood a day pass through the little capsules in the kidney called glomeruli. Here the proteins are filtered out, remaining behind in the blood, and the rest of the solution passes through the filter into the kidney tubules. In the first, or proximal, section of the tubule 80 per cent of the salt and water and all the glucose are pumped out and returned to the bloodstream. The rest of the material, which includes all the waste, travels on to the more distant segments of the tubule, which exercise fine control over the composition of the urine. Our interest has centered on the proximal section of the tubule, where the bulk of the work is done.

The experimental animal we have been using is the large fresh-water amphibian called *Necturus* or the mud puppy. The kidney tubule of *Necturus* is relatively large: its internal diameter of about a two-hundredth of an inch is big enough to admit a special micropipette. Moreover, the proximal section of the tubule starts and finishes near the surface of the kidney, providing access to both its ends even though its middle portion descends into the kidney and cannot be seen [*see illustrations on the next page*]. These advantages have attracted a number of workers. The pioneers were Alfred N. Richards and his group at the University of Pennsylvania School of Medicine, who carried out a

AUTHOR'S NOTE

The author wishes to acknowledge the collaboration of his associates in the work reported in this article. They are: Irwin B. Hanenson, Joseph C. Shipp, Hisato Yoshimura, Erich E. Windhager, Hans J. Schatzmann, Guillermo Whittembury, Donald E. Oken, W. J. Flanigan, Nobuhiro Sugino, Francisco C. Herrera, Raja N. Khuri, Charles J. Edmonds and David L. Maude.

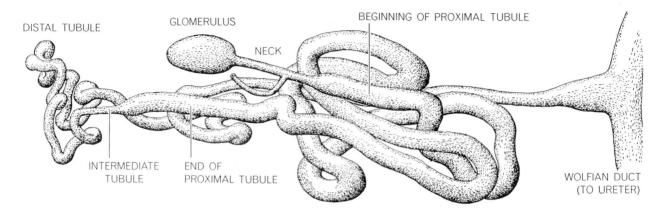

DISTAL TUBULE   GLOMERULUS   BEGINNING OF PROXIMAL TUBULE

NECK

INTERMEDIATE TUBULE   END OF PROXIMAL TUBULE

WOLFIAN DUCT (TO URETER)

PROXIMAL TUBULE of the *Necturus* kidney runs from the glomerulus at one end to the narrower intermediate tubule at the other. For most of its length it is within the kidney. Both ends, however, are near the surface and are accessible to the experimenter.

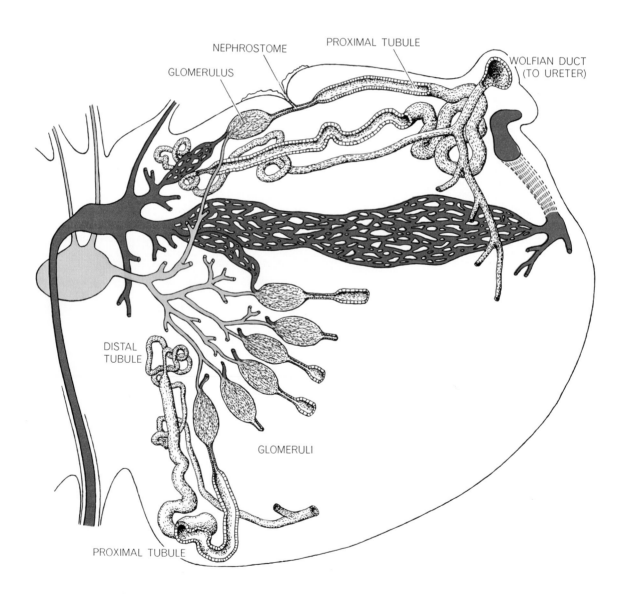

NEPHROSTOME   PROXIMAL TUBULE

GLOMERULUS

WOLFIAN DUCT (TO URETER)

DISTAL TUBULE

GLOMERULI

PROXIMAL TUBULE

NECTURUS KIDNEY is estimated to contain approximately 800 tubules. In this schematic drawing the venous blood vessels are shown in dark red and the arterial in light red. Much of the water and salt flowing through the kidney returns to blood via tubules.

classic series of studies on the *Necturus* kidney in the 1930's.

Taking up the investigation in 1953, we hoped to extend Richards' results with the help of certain new theoretical ideas on active transport and new methods, particularly the use of radioactive tracers. At first we were interested in the outward movement of solutions through the wall of the proximal tubule. It should be emphasized that we were not then studying the activity of the membrane of the individual cell but of a wall composed of a layer of cells.

By means of a technique known as stopped-flow microperfusion we can measure precisely the amount of water and dissolved molecules that are absorbed from a single kidney tubule. Working with the exposed kidney of an anesthetized living animal, we insert a micropipette into a glomerulus and inject colored oil until it enters the upper part of the proximal tubule. Then a second pipette is inserted into the drop of oil and a solution containing a radioactive tracer is injected, splitting the oil drop in two. As fluid begins to fill the tubule, it pushes the "front" part of the oil drop; this disappears from view, descending into the kidney, and finally reappears at the end of the proximal tubule. We can identify the end because at this point the tubule abruptly narrows, and the drop assumes a characteristic shape. Now the tubule is filled with fluid in the process of being perfused. We isolate the fluid completely by injecting a little more oil through the glomerulus to seal the hole by which the tubule was filled.

In 20 minutes a measurable amount of salt and water has disappeared from the tubule. Then another micropipette is inserted into the beginning of the tubule and a second drop of oil is sent through, pushing the perfusion fluid into a collecting pipette. The amount of fluid we collect is only about a ten-thousandth of a cubic centimeter, a quantity just visible to the naked eye. We measure the volume by transferring the fluid under the microscope to a small capillary tube of constant bore, a technique devised by Richards. Evaporation presents a problem in dealing with such small amounts of solution, and we have to be careful to keep the droplet away from the extreme ends of the capillary.

To determine how much water has been lost from the tubule, we put into a perfusion fluid the inert substance inulin, the molecules of which are too large to pass through the tubule wall. The increase in the concentration of inulin in the fluid collected from the tubule is then a measure of the decrease in water volume. Fortunately we were able to obtain inulin labeled with the radioactive isotope carbon 14. As a result we can measure changes in inulin concentration by radioactive techniques, which are simpler than chemical ones.

Richards had found that as fluid moves down along the tubule, some of it passing out through the wall, the concentration of dissolved salts (chiefly sodium chloride) remains almost constant. Evidently salt and water move together. But which is actively pumped and which follows along passively? It had generally been assumed that salt, or rather sodium, is the actively transported material in situations of this kind, but we wished to take nothing for granted.

It took three and a half years before we mastered the techniques sufficiently to conduct a successful experiment. We found that 27 per cent of our perfusion fluid water was absorbed in 20 minutes, a figure that agrees fairly well with the observation of Richards' group that 19 per cent is absorbed when fluid passes through the tubule under normal conditions.

We attacked the problem of identifying the actively and passively transported materials by varying the salt concentration of the perfusion fluid. To offset osmotic-pressure effects from changes in salt concentration, any sodium chloride removed was replaced with mannitol, another inert substance. When the sodium chloride concentration inside the tubule was reduced (but the osmotic pressure held constant), the outflow of water also decreased. A further reduction stopped all water movement; at a still lower concentration water moved into the tubule. These results indicated

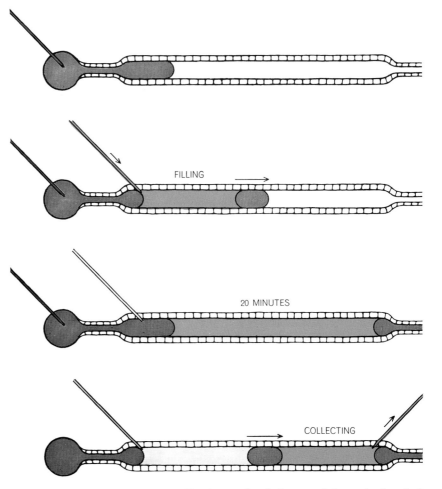

STOPPED-FLOW MICROPERFUSION technique, by which many of the results described in this article were obtained, is illustrated schematically. In drawing at top a micropipette is inserted into a glomerulus (*left*) and oil is injected until it fills the narrow neck and enters the proximal tubule. In second drawing from top perfusion fluid is injected through a second pipette into middle of oil, forcing a droplet ahead of it. Tubule is full when droplet reaches the far end of the tubule (*third from top*). After about 20 minutes the fluid is collected (*bottom*) by injecting a second liquid behind the oil remaining at near end of tubule.

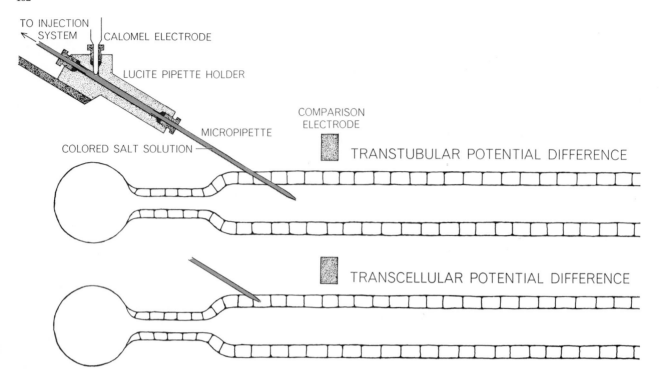

TO INJECTION SYSTEM

CALOMEL ELECTRODE

LUCITE PIPETTE HOLDER

MICROPIPETTE

COLORED SALT SOLUTION

COMPARISON ELECTRODE

TRANSTUBULAR POTENTIAL DIFFERENCE

TRANSCELLULAR POTENTIAL DIFFERENCE

ELECTRICAL POTENTIAL DIFFERENCES across the inner and outer sections of the membranes of the tubule wall are measured by inserting microelectrodes into the tubule (*top*) and into the cells (*bottom*). Colored salt solution furnishes electrical contact through calomel electrode. After measurement colored solution is injected into tissue, where it marks the point of measurement.

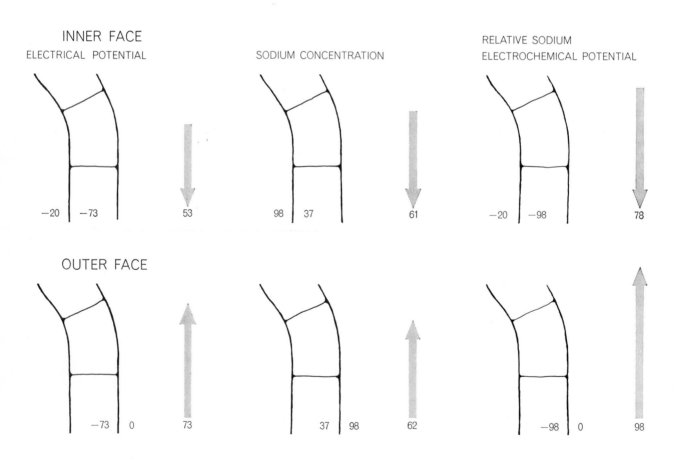

INNER FACE

ELECTRICAL POTENTIAL

SODIUM CONCENTRATION

RELATIVE SODIUM ELECTROCHEMICAL POTENTIAL

−20 | −73   53

98 | 37   61

−20 | −98   78

OUTER FACE

−73 | 0   73

37 | 98   62

−98 | 0   98

POTENTIAL DIFFERENCES, both electrical and chemical, that sodium ion encounters at inner face (*top*) and outer face (*bottom*) of cells of tubule wall are indicated by gray arrows. Numbers are units appropriate to each type of potential or concentration.

that water movement is dependent on salt movement.

Although sodium chloride was by far the most abundant substance dissolved in the perfusion fluid, the fluid also contained other molecules, in particular mannitol, that might leak back and forth across the tubule wall. This or any other molecule would transport some water. To establish that water movement is completely passive we had to take into account the leakage of all molecular species, known and unknown. The simplest way to do this is to measure the freezing point, which is lowered by an amount proportional to the concentration of independent dissolved particles. In a series of experiments we varied the salt concentration of the perfusion fluid, causing water and dissolved molecules to move into the tubule in some cases and out of it in others. In each case we measured the small change in the freezing point of the fluid before and after the experiment, and the total water transferred. We were able to show that water movement is strictly proportional to solute movement: when there is no net transfer of dissolved substances, there is no flow of water. Therefore the movement of water through the wall of the tubule is entirely passive.

We were now ready to look for the pump; that is, to find the substance that is actively transported. Before describing the experiments let us consider in more detail what is meant by the term "active transport."

If a weight is moved uphill, it is moved from a lower gravitational potential to a higher one. Such a motion, which is opposite to the natural tendency of the weight, requires active transport. Now think of a sugar solution that is more highly concentrated at one side of its container than it is at the other. It is obvious that the transfer of sugar from the more concentrated to the less concentrated side is a downhill process. If sugar is seen to move en masse the other way, it must be actively transported against a chemical potential difference. In the case of charged atoms or molecules (ions) the electrical potential must also be considered. A positive ion moves downhill from a region of higher to a region of lower electrical potential; for it to move the other way implies active transport. The laws of physical chemistry show how to combine the two effects: at 37 degrees centigrade an electrical potential difference of 61 millivolts (thousandths of a volt) is equivalent to a 10-fold concentration difference. It is just as hard to move a positive ion such

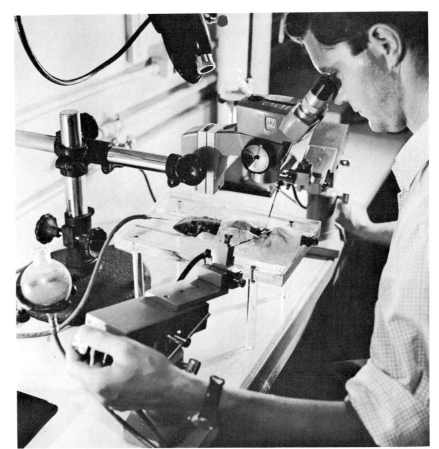

INSERTION OF MICROPIPETTE into a kidney tubule is accomplished with the aid of a pair of micromanipulators located at each side of a tray on which the anesthetized animal is placed (*center*). The experimenter monitors the operation through binocular microscope.

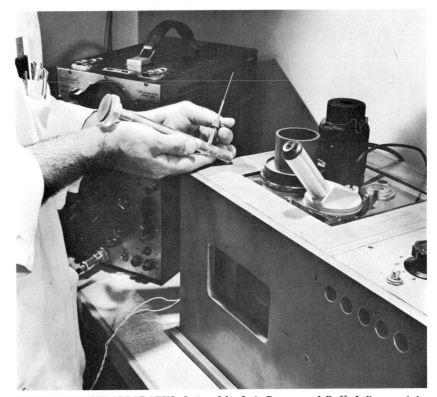

FREEZING-POINT APPARATUS, designed by J. A. Ramsay and R. H. J. Brown of the University of Cambridge, is contained in the box at right. The experimenter is about to insert the sample-holder. Freezing is observed through microscope eyepiece at top of box.

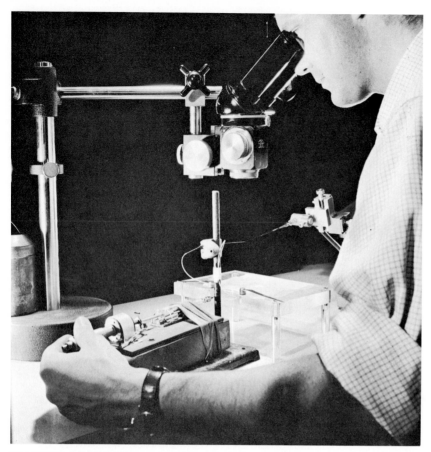

TRANSFER OF FLUID from micropipette, extending from apparatus at right, to measuring capillary tube is also carried out under microscope with aid of a micromanipulator.

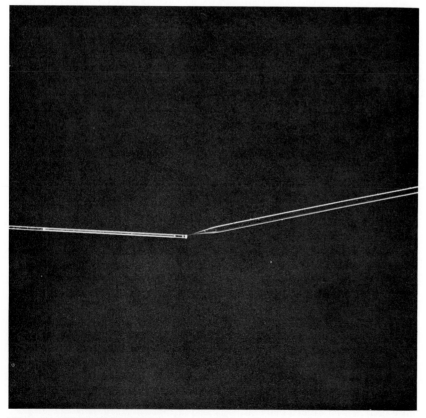

PIPETTE AND CAPILLARY are photographed in close-up, magnified three diameters. Bore of capillary (*left*) is uniform, so length of liquid column is measure of its volume.

as sodium "upward" across a 100-fold concentration difference as across a 10-fold concentration difference and an electrical potential difference of 61 millivolts. One student of active transport, Thomas Rosenberg of Denmark, defines it as net movement up an electrochemical potential gradient.

In view of what is known, or at least suspected, about sodium in many cellular systems, we first turned our attention to this element. Our first perfusion experiments, in which we progressively lowered the concentration of sodium chloride, had shown that water and presumably sodium moved out of the tubule even when the inside sodium concentration was only 67 per cent of the outside concentration. Evidently sodium is transported up a chemical concentration gradient.

Next we had to know about the electrical potential difference between the inside and the outside of the tubule. Gerhard H. Giebisch of the Cornell University Medical College had already measured the potential and found it to be 20 millivolts more positive outside than inside. Hence it appeared that the sodium ion also travels up an electrical potential gradient. Giebisch's determination, however, was made under conditions somewhat different from those of our experiments, so we thought we should make the measurement ourselves. We adopted his technique, inserting microelectrodes into the tubules [*see top illustration on page 102*]. The tips of the electrodes are much smaller than those of the micropipettes and cannot be seen under the microscope. Therefore we had no direct way of telling whether or not they were correctly placed in the tubule. To check their location we filled the hollow microelectrodes with colored salt solution (the solution served as an electrical conductor). When a measurement had been completed, we forced the colored liquid out of the electrode and into the tubule, a process that requires considerable pressure to overcome the capillary forces in the minute electrode tips. If the colored solution could be seen in the tubule, the measurement was accepted; if not, the figure was discarded. To our gratification we obtained a potential difference of 20 millivolts. We could now say unequivocally that sodium is actively transported through the tubule wall up an electrochemical potential gradient, in agreement with Rosenberg's definition.

A more thoroughgoing analysis of active transport has been made in the past few years by Hans H. Ussing of the University of Copenhagen. He recognized that processes other than net uphill transport require an immediate energy supply. One example is downhill motion at faster than equilibrium rate. An automobile coasts downhill at a speed governed by the steepness of the grade and the various frictional forces involved. If it goes down faster, it must be taking power from the engine. To take this kind of motion into account Ussing derived an equation in which the ratio of uphill to downhill movement

is compared with the known electrochemical potential difference.

In an attempt to apply Ussing's equation, we have tried to measure the backward flow of sodium in the tubule. By adding some radioactive sodium to the fluid injected into the tubule and determining the amount of radioactivity remaining in the fluid collected at the end of the experiment, we hope to get a direct measure of the outflow of sodium. Combining this figure with the net change in sodium content in the tubule would enable us to calculate a net return flow. But

this method turns out to be incapable of showing whether or not individual ions have accumulated in the cells of the wall and then leaked back into the tubule several times during the course of the experiment, and so we cannot really say what the total uphill or downhill motion is. At present, therefore, we cannot subject our results to Ussing's more fine-grained analysis.

Although the results were disappointing, the quantitative methods required to deal with such small amounts of material are worth mentioning. It is easy

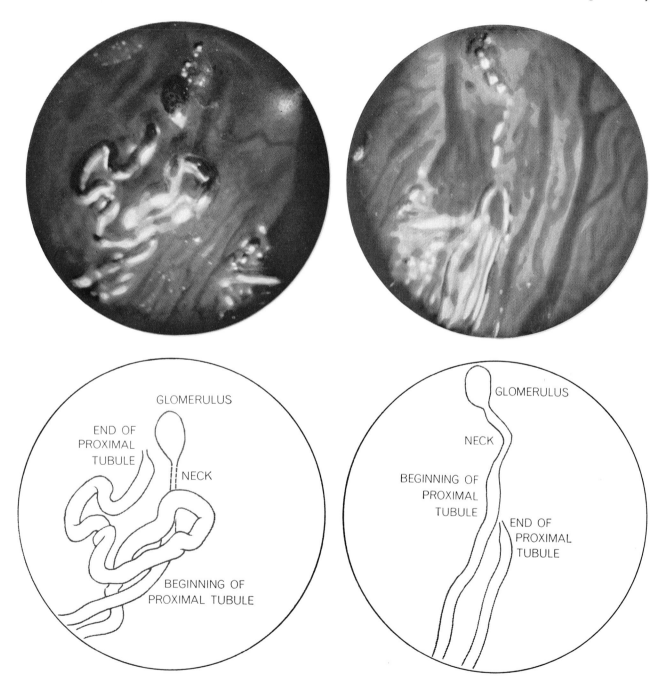

KIDNEY TUBULES in *Necturus* have been injected with mercury to make them visible in the two photomicrographs at top. The two drawings at bottom indicate what parts are seen. Photomicrographs were made by Bradford D. Pearson of Harvard Medical School.

enough to assay radioactive sodium by ordinary counting techniques; determining total sodium content less than a millionth of a gram is another matter. The measurements were made in a flame photometer specially adapted for ultra-micro quantities. The solution is placed in a flame and the yellow sodium line is isolated by a spectroscope. The intensity of this line, which gives an indication of the amount of sodium present, is then measured with a photomultiplier tube.

The experiment did confirm, in any case, that the net movement of sodium in the *Necturus* kidney tubule is upward against an electrochemical gradient. It remained to find out if any of the other important ions in the system are also actively pumped. The migration of chloride ions has been studied by Giebisch together with Erich E. Windhager, formerly of our group. They found no need for active transport. The 20-millivolt potential difference across the wall is in the right direction to pull the negatively charged chloride ion out of the tubule.

In our laboratory we tackled the rather difficult problem of potassium. The concentration of this element in the blood of *Necturus* is about a thirtieth that of sodium. Like sodium, it passes through the glomerular filter and forms a part of the normal tubular fluid. Because of its much lower concentration, and because photomultiplier tubes are much less sensitive to the purple light it emits than they are to yellow light, potassium is harder to measure than sodium. Even with the flame photometer tuned to its highest efficiency, we were unable to determine the potassium concentration in

the fluid collected from a single tubule; we had to pool several samples in order to obtain enough material for analysis. We found that as water passed out of the tubule, the potassium concentration inside rose in exactly the same ratio as the inulin concentration, indicating no net movement of potassium. We therefore concluded that potassium transport is also passive. This left sodium as the active component.

Now we could go on to the exciting task of localizing the transport process within the cells of the tubule wall. We assumed that the pumping is done somewhere on the cell membrane, either on the part that faces the inside of the tubule or on the part that faces outward. Giebisch had already measured the electrical potential difference across the separate faces of the cell in living animals. We now began to work with slices of kidney in which we could determine the concentration of sodium, potassium and chloride inside the cell. In this work the two groups collaborated very closely. When we checked the electrical potentials in the cells of our kidney slices, we obtained results in striking agreement with Giebisch's.

Combining the electrical and chemical measurements gave the following picture [*see bottom illustration on page 102*]. In the case of sodium the concentration inside the cell is much less than the normal tubular fluid concentration. In addition, the 53-millivolt potential difference across the inside face of the cell membrane is in the direction to pull sodium ions from the tubule into the cell. Therefore sodium moves into the

cell down both an electrical and a chemical concentration gradient, and the process appears passive. At the outside face of the cell the electrical potential difference is 73 millivolts, but in a direction opposed to the motion of the sodium. The movement also goes against a chemical concentration gradient. Therefore sodium must be actively transported. The sodium pump, the only active process so far identified, appears to be located at the outside face of the cell. The beauty of this system is that it accounts for a host of apparently unrelated phenomena. The sodium pump at the outside face provides the force to pump the water out of the tubule and incidentally to preserve the cell's osmotic integrity. At the same time it controls the sodium concentration within the cell and the observed electrical potential differences.

Electron microscope studies have recently produced a picture consistent with our hypothesis. It has long been known that a major energy source for cellular reactions of all kinds are the small bodies called mitochondria. Since the sodium pump requires energy, our model suggests that there should be a concentration of mitochondria at the outer face in kidney proximal tubule cells. Recent electron micrographs show, most dramatically in frogs and to a lesser extent in *Necturus,* that this is indeed the case. Modern anatomy places the energy source just where it is needed according to our biophysical studies.

By way of conclusion I might point out that the term "active transport" in a sense betrays ignorance. A naïve member of a primitive society, seeing an automobile climb a hill, might conclude that the body of the car is somehow endowed with the energy to move it. We can trace the energy to the engine, to the cylinder and to the chemical bonds in the fuel. The body of the automobile moves passively under the power of the drive shaft. The drive shaft is in turn passively actuated by the cylinders. They are passively pushed by the expansion of burning gases, and so on.

Our results, then, represent only a first step. The localization of the sodium pump within the cell is exciting, but the major problem still faces us. What is the detailed molecular mechanism by which sodium is transported, and how is this mechanism coupled to its energy supply? When these questions are finally answered, we shall be able to drop the phrase "active transport" from our vocabulary.

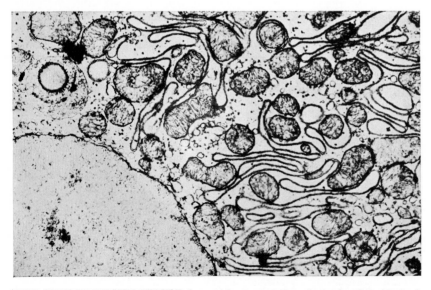

CELL OF FROG KIDNEY TUBULE shows rich concentration of mitochondria (*dark objects*) near the nuclear membrane (*curved line at bottom left*). This electron micrograph, magnification 18,000 diameters, was made by Morris J. Karnovsky of Harvard Medical School.

The Mechanism of Muscular Contraction

11

by H. E. Huxley
December 1965

*When a muscle contracts, one kind of filament within
it slides past another kind. Electron microscopy and
other techniques have begun to disclose how the
filaments exert a force on each other*

An outstanding characteristic of all animals is their ability to move voluntarily by contracting their muscles. When I summarized our understanding of muscle contraction seven years ago [see "The Contraction of Muscle," by H. E. Huxley; SCIENTIFIC AMERICAN Offprint 19]. it had already been determined that during contraction two kinds of filament in voluntary muscle—thick filaments and thin ones—slide past each other so as to produce changes in the length of the muscle. At that time one could offer only a hypothetical description of contraction at a more detailed level; it was assumed that a relative force is somehow exerted between the thick and thin filaments at sites where they are connected by tiny cross-bridges. Now, thanks to advances in electron microscopy and allied techniques, we have been able to substantiate that hypothesis and to learn considerably more about the nature of the interaction of the thick filaments (composed mainly of the protein myosin) and the thin ones (composed of another protein, actin). It appears that at each site where the proteins of the two kinds of filament are in contact one of them (probably myosin) acts as an enzyme to split a phosphate group from adenosine triphosphate (ATP) and thus provide the energy for contraction. The basic problem is to understand how the conversion of chemical into mechanical energy takes place.

Let us briefly review what is known about the structure and function of muscle. Under the microscope voluntary muscles—for example those that can move the leg of a frog—appear regularly striated at right angles to their length. The muscles responsible for the slow and regular movements of organs that work involuntarily, such as the gut,

appear smooth. For reasons of technical convenience most investigations of muscle have dealt with striated muscle, and so our discussion will refer specifically to muscle of that type. A good deal of what has been learned about striated muscle, however, may apply to smooth muscle as well.

Striated muscle can shorten at speeds equal to several times its length per sec-

ond; it can generate a tension of some 40 pounds per square inch of its cross section; it can contract or relax in a very small fraction of a second. A muscle consists of individual fibers with a diameter of between 10 and 100 microns (a micron is a thousandth of a millimeter); the fibers run the length of the muscle, or a good part of it. Each fiber is surrounded by an electrically po-

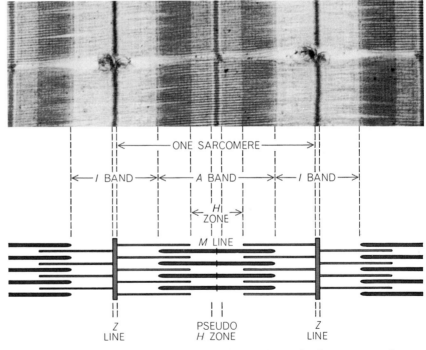

STRIATED MUSCLE from the leg of a frog is shown in longitudinal section in an electron micrograph (*top*) and the overlap of filaments that gives rise to its band pattern is illustrated schematically (*bottom*). Parts of two myofibrils (long parallel strands organized into muscle fiber) are enlarged some 23,000 diameters in the micrograph. The myofibrils are separated by a gap running horizontally across the micrograph. The major features of the sarcomere (a functional unit enclosed by two membranes, the *Z* lines) are labeled. The *I* band is light because it consists only of thin filaments. The *A* band is dense (and thus dark) where it consists of overlapping thick and thin filaments; it is lighter in the *H* zone, where it consists solely of thick filaments. The *M* line is caused by a bulge in the center of each thick filament, and the pseudo *H* zone by a bare region immediately surrounding the bulge. The electron micrograph and others illustrating this article were made by the author.

larized membrane, the inside of which is generally a tenth of a volt negative with respect to the outside. Contraction is signaled by an impulse that travels down a nerve to a motor "end plate" in contact with the fiber. The arrival of the impulse depolarizes the membrane and causes the release throughout the fiber of an activating substance, probably calcium. It is this activation that enables one of the muscle proteins to act as an enzyme and split a phosphate group from ATP. The muscle stays contracted until nerve impulses cease (or until it becomes exhausted), at which point the activating substance withdraws, probably by being bound to the sarcoplasmic reticulum, a network of tiny channels within the fiber.

An individual fiber is made up of a number of parallel elements called myofibrils, each about a micron in diameter. Each myofibril itself consists of parallel actin and myosin filaments that, when they are viewed from the end, are seen to lie a few hundred angstrom units apart in a remarkably regular array. (An angstrom unit is a ten-thousandth of a micron.) The myosin filaments are some 160 angstroms in diameter but often appear somewhat thinner when fixed for electron microscopy. They are about a micron and a half in length. The actin filaments are only about a micron in length and 50 to 70 angstroms in diameter. The overlap between the arrays of thick and thin filaments gives rise to the pattern of striations visible in the microscope. The pattern is characterized by a succession of dense bands (called A bands) and light bands (I bands). In the A bands the myosin filaments lie in register in hexagonal array and are responsible for the bands' high density. The actin filaments are attached in register on each side of a narrow, dense structure that traverses the I band: the Z line. In a relaxed muscle the distance between Z lines—one sarcomere—is such that about half of the length of a thin filament and two-thirds of the length of an adjacent thick filament overlap. In the region of overlap in a relaxed fiber the array contains twice as many thin filaments as thick ones. The thin filaments terminate at the edge of the H zone, a region of low density in the center of the A band. In the center of the H zone lies the "pseudo H zone," a region of even lower density that maintains its width no matter how the length of the muscle changes. This light zone surrounds a thin, dark strip known as the M line, which is now thought to be caused by a slight bulge in the center of each thick filament.

When a longitudinal section of muscle is viewed in the electron microscope, it can be seen that the cross-bridges between a given pair of thick and thin filaments come at fairly regular intervals. The cross-bridges are the only mechanical linkage between the filaments, and they are responsible for the structural and mechanical continuity along the whole length of a muscle. It is the cross-bridges that must generate or sustain the tension developed by a muscle. As the sarcomere changes its length, either actively during contraction or passively (stretching or shortening while at rest), the filaments themselves do not perceptibly change in length but slide past one another; the thin filaments move farther into the A bands during shortening and farther out of them during stretching.

Since normal contractions involve changes in the length of the sarcomere of 20 percent or more, the thin filaments in each half of the A band must move distances of at least a quarter of a micron while maintaining tension. It seems physically impossible that the cross-bridges could remain attached to the same point on the actin filament throughout this process. We supposed, therefore, that they are attached to one site on the filament for part of the contraction, then detach and reattach themselves at a new site farther along. Moreover, we assumed that at each site

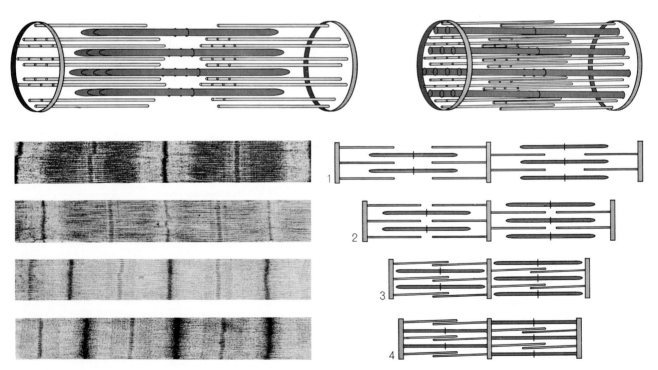

CONTRACTION OF MUSCLE entails change in relative position of the thick and thin filaments that comprise the myofibril (top left and right). The effect of contraction on the band pattern of muscle is indicated by four electron micrographs and accompany- ing schematic illustrations of muscle in longitudinal section, fixed at consecutive stages of contraction. First the H zone closes (1), then a new dense zone develops in the center of the A band (2, 3 and 4) as thin filaments from each end of the sarcomere overlap.

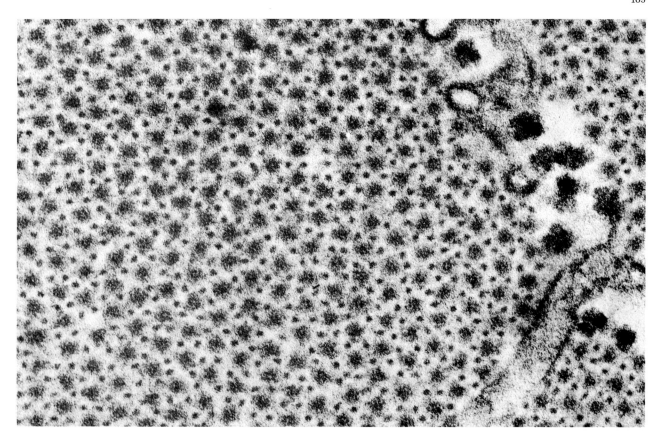

TRANSVERSE SECTION through a frog's leg muscle in its uncontracted state shows how thick and thin filaments are arrayed in a regular hexagonal pattern. Breaks in the pattern at the right side of the micrograph are channels of the sarcoplasmic reticulum. From the end thick and thin filaments look like large and small dots. This electron micrograph enlarges them some 200,000 diameters.

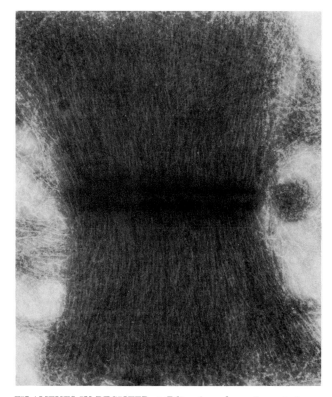

FILAMENTS IN REGISTER at Z line (*membrane in center*) are the thin filaments of actin, which alone comprise the I segment. This sample was obtained by homogenizing muscle from the back of a rabbit in the Waring blendor; it was prepared for the micrograph by negative staining. Magnification is some 47,000 diameters.

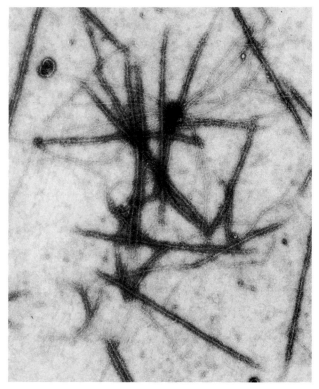

SEPARATED FILAMENTS are from rabbit muscle that has been homogenized in a Waring blendor. The dark, thick strands are filaments of myosin. The very faint thin strands are filaments of actin. Thin filaments are still attached to remnant of Z line (*dark patch at top center*). Filaments are enlarged some 35,000 diameters.

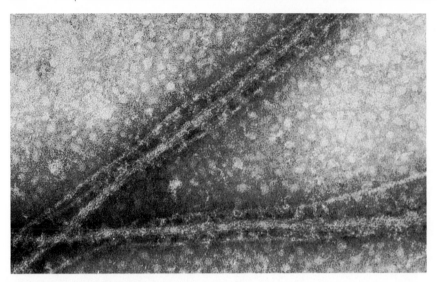

CROSS-BRIDGES between thick and thin filaments are enlarged 180,000 diameters in electron micrograph made by "negative staining." Technique involves surrounding very small objects with a dense salt (*white substance in background*) so that they stand out by contrast.

where cross-bridge and filament interact, one molecule of ATP is split to generate a sliding force between the two kinds of filament (and hence between arrays of filaments).

This general description of the structural changes associated with contraction is the sliding-filament hypothesis put forward a decade ago by Jean Hanson of the Medical Research Council unit at King's College in London and me, and independently by A. F. Huxley and R. Niedergerke of the University of Cambridge. Our hypothesis was partly based on observations of muscle prepared by what is called the thin-sectioning technique. That method involved steeping a chemically fixed (for preservation) and stained (for contrast) piece of muscle in liquid plastic and then cutting the solidified plastic into slices as thin as 100 angstroms. It turned out, however, that the thin-sectioning technique was not adequate to the task of illuminating many details of the hypothesis. In order to ascertain how a force might be developed between thick and thin filaments we needed information about the detailed structure of actin and myosin, and such information was not forthcoming until the arrival of the technique of electron microscopy known as negative staining.

This new method, in which the specimen under examination is embedded in a thin film of some very dense material such as uranyl acetate, has in recent years revealed much about the structure of small spherical viruses and particles of similar size. As adapted in our

laboratory at the Medical Research Council unit in Cambridge, the technique involves applying a drop in which particles of muscle are suspended to an electron microscope grid covered by a thin film of carbon. Many particles adhere to this film; the excess is washed away with a few drops of solvent. Before the preparation dries a shallow drop of the negative-staining material—a heavy metal salt in dilute solution—is applied. It is allowed to dry around the particles. The regions of the particles that are not penetrated by the stain show up clearly by negative contrast because they consist of protein and are much less dense than the salt that surrounds them.

The negative-staining method brings to light far more detail than the conventional positive-staining technique (which artificially increases the density of objects with respect to their background). Its disadvantage is that it can only be applied to very thin specimens; thick ones and the associated thick deposit of negative stain would impair the resolution of the electron microscope image. Thus the method is not directly applicable to whole pieces of tissue such as muscle. The muscle must first be broken down into fragments of suitable thickness (such as individual filaments), which is not easily accomplished.

The usual method of breaking down muscle tissue for purposes of investigation is to homogenize it in the Waring blendor; under this treatment it disperses readily into its constituent myofibrils but no further. In fact, the myofibrils strongly resist further breakdown, probably because the cross-bridges between thick and thin filaments bind the whole structure together in a very robust fashion. Making the assumption that the cross-bridges are the sites where actin and myosin combine, we wondered if we might weaken the structure by suspending the fibrils in certain salt solutions that tend to dissociate the two proteins. We were delighted to find that if muscles, either freshly isolated or preserved in a deep freezer in a solution of water and glycerol, were placed in the appropriate salt solution and then homogenized in the blendor, they indeed broke down into their constituent filaments. Thus they could provide excellent material for examination by the negative-staining technique.

The first specimens we prepared by this method consisted of filaments from the psoas muscle in the back of a rabbit. In the electron micrographs the layer of negative stain was thickest in the region immediately surrounding the filaments; accordingly the filaments have a dense outline [see illustrations at bottom of preceding page]. We could at once recognize the thick filaments by their resemblance to the thick filaments in earlier preparations of striated muscle. The diameter of the filaments was the expected 160 angstroms; their length was apparently about 1.5 microns. (Longer structures were never observed but shorter ones, presumably fragments, were.) Small projections, extending sideways from the filaments along most of their length, seemed to correspond to the cross-bridges. Thinner filaments 50 to 70 angstroms in diameter could also be seen, and in places we noticed a large group of such filaments extending for about a micron on each side of a Z line, to which they were still attached [see illustration at bottom left on preceding page]. These observations of thick and thin filaments of characteristic size, lying side by side and sometimes still connected by cross-bridges, confirmed the conclusions about the structure of the myofibrils reached earlier by X-ray diffraction techniques and by conventional light microscopy and electron microscopy. We subsequently considered the appearance of the individual filaments more closely.

A regular feature of the thick filaments is a short region, midway along their length, from which the projections we believe to be cross-bridges are absent. This differentiated, projection-free area, some .15 to .2 micron long, can be seen not only in negatively stained material but also in sectioned specimens. It is now apparent that the absence of cross-bridges from this region is responsible for the mysterious pseudo *H* zone. This zone maintains its uniform size at various muscle lengths because it is a structural feature of the filaments themselves and is not created by their pattern of overlap.

At first sight the projection-free middle region of the thick filaments did not seem particularly significant. It was conceivable that the region was composed of some other protein constituent of muscle. The situation was transformed, however, when we found that filaments of virtually the same appearance could be synthesized from purified solutions

of myosin, the protein that is the main component of the thick filament.

The myosin molecule is known to be an elongated structure with a length of about 1,500 angstroms and a diameter of 20 to 40 angstroms. It can be split (by the enzyme trypsin) into two well-defined fragments; the fragments were named light meromyosin and heavy meromyosin by Andrew G. Szent-Györgyi of the Institute for Muscle Research in Woods Hole, Mass. The heavy-meromyosin fragment has the ability to split a phosphate group from ATP and the ability to combine with actin. The light-meromyosin fragment possesses neither of these attributes but retains

the solubility properties that enable it to form the same kind of structure that intact myosin does. The molecule of heavy meromyosin appears to be more globular than the molecule of light meromyosin. The dimensions of the fragments suggest that before cleavage of the myosin molecule they are arranged in simple end-to-end fashion.

Isolated myosin molecules have been examined under the electron microscope, first by Robert V. Rice of the Mellon Institute in Pittsburgh and subsequently by other workers, by means of the technique known as shadow-casting. This entails treating particles on a film in a vacuum by spraying them at an angle with a vaporized heavy metal.

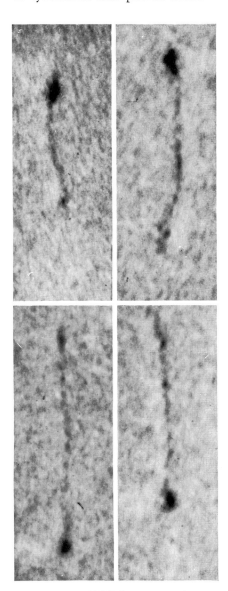 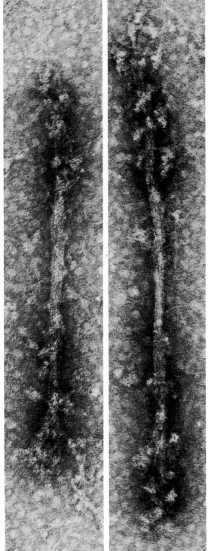 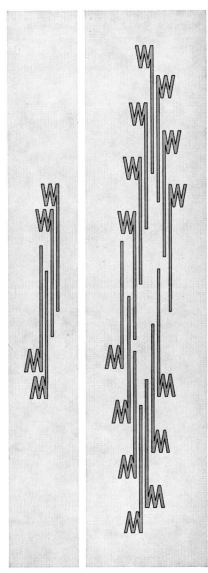

MYOSIN MOLECULES appear in electron micrographs prepared by shadow-casting method. The wide head has enzymatic properties and combines with actin. The straight tail can aggregate with other myosin molecules. Magnification is 300,000 diameters.

AGGREGATIONS of several molecules from a precipitate of pure myosin were negatively stained and magnified 175,000 diameters to reveal their characteristic appearance: a thick strand with projections near the ends and a bare region in the middle.

MOLECULAR STRUCTURE of myosin makes it aggregate in the manner shown here. Head of molecule is schematically represented by zigzag line, tail by straight line. Tails join in center; heads extend as projections at ends, oppositely pointed at each end.

In forming a layer over the sample the metal builds up the particles on the near side and leaves a shadow on the far side, where it is blocked from landing on the underlying film. When myosin molecules are prepared by this method for viewing in the electron microscope, they appear as linear structures with a globular region at one end [*see illustration at left on preceding page*]. Heavy-meromyosin fragments are seen to consist of a large globular head with a short tail. Light-meromyosin fragments appear as simple linear strands. It therefore seems that the intact myosin molecule is asymmetric—a molecule with a head and a tail. The sites (perhaps a single site) responsible for its enzymatic activity and its affinity for actin are located in its globular head, and the sites responsible for its affinity for other myosin molecules are in its tail. The head, which is 40 angstroms in diameter, accounts for about a sixth of the length of the molecule; the tail, 20 angstroms in-diameter, accounts for the rest.

It is known that under certain conditions purified myosin in potassium chloride solution will precipitate. When we examined such a precipitate by the negative-staining technique, we were delighted to find that it consisted entirely of filaments. They varied somewhat in length and diameter but generally bore a most remarkable resemblance to the thick filaments prepared directly from muscle. Systematic examination of these synthetic filaments, first of short filaments only two or three times the length of a single myosin molecule and then of longer ones, turned up an even more remarkable feature. The shortest filaments were straight rods some 1,500 to 2,500 angstroms long, with clusters of globular projections at both ends. It occurred to us that we were looking at a small number of myosin molecules arranged in two opposite directions, with their globular heads forming the projections and their linear tails overlapping [*see middle illustration on preceding page*]. Longer filaments had longer clusters of projections, but the projection-free region in the middle of each filament was the same length as the corresponding region in the shorter filaments. The longest synthetic filaments we observed closely imitated the appearance of thick filaments extracted from muscle. It seems clear that myosin filaments grow by the addition of molecules parallel to the molecules that have already aggregated. The molecules are oriented in one of two opposite directions, depending on which

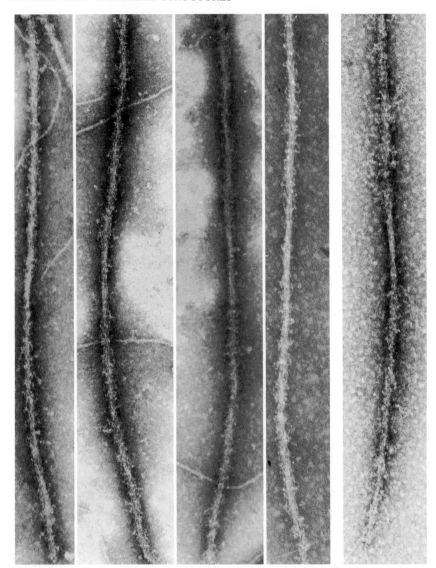

MYOSIN FILAMENTS were obtained directly from muscle homogenized in a blendor (*four electron micrographs at left*) for comparison with a synthetic filament from precipitate of pure myosin (*micrograph at right*). The thick filaments from muscle and the synthetic filament have the same form, characterized by the bridge-free zone in the center and the projections clustered at each end. The filaments are enlarged some 105,000 diameters.

end of the filament a given molecule is joining. It is this method of construction that gives rise to the projection-free region in the synthetic filaments and of course to the same feature in the natural filaments.

This study of myosin molecules and the way they aggregate impressed on us two features that explain the role of these molecules in muscle. First, the head of the molecule has the enzymatic and actin-binding properties we have long assumed the cross-bridges must have. Second, because the molecules aggregate with their heads pointed in one direction along half of the filament and in the opposite direction along the other half, they have an inherent direction-ality. The first observation leads us to conclude that the heads of myosin molecules serve as the cross-bridges connecting the thick and thin filaments in muscle. The second is important because it explains a crucial feature of the sliding-filament hypothesis at the molecular level.

In a sliding-filament system in which a relative force is developed between actin and myosin molecules located in the two types of filament, it is essential that the appropriate directionality of sliding be built into the filaments in some way. In a striated muscle the thin filaments move toward each other in the center of the A bands, so that it is required that all the elements of force generated by the cross-bridges in one

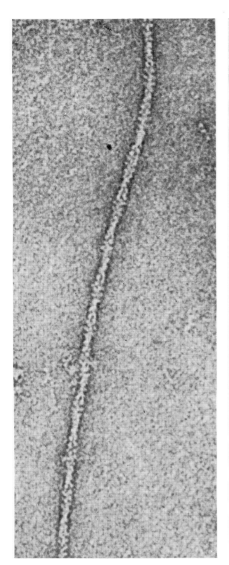

ACTIN FILAMENT has a characteristic structure, visible in micrograph in which filament is enlarged some 420,000 diameters. The filament has the appearance of two coils of globular units wound in a double helix.

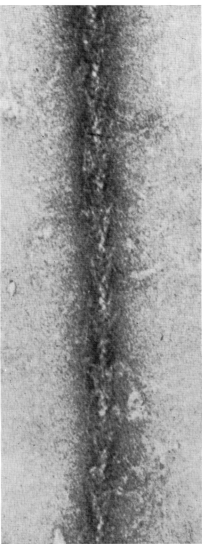

"ARROWHEADS" point in one direction along each filament of actin labeled with heavy meromyosin (extract of the globular halves of myosin molecules), implying that actin has an inherent polarity of its own.

half of the A band be oriented in the same direction, and that the direction of the force be reversed in the other half. The direction of the force developed as a result of the interaction of actin and myosin would depend either on the orientation of the myosin molecules, the orientation of the actin molecules or both. Our electron microscope observations suggest strongly that all or part of this directionality is achieved by the fact that the myosin molecules are arranged so that they point in the same direction in half of each thick filament (and hence in each A band) and in the opposite direction in the other half [see illustration at right on page 111]. Moreover, we have shown that filaments with this essential reversal of polarity at their midpoint will assemble themselves auto-

matically in vitro from purified preparations of myosin; this finding has obvious relevance to problems of how muscle develops its structure.

Let us now turn to the thin filaments. It was first noticed by Jean Hanson and J. Lowy of the Medical Research Council unit in London that the thin filaments from the smooth muscle of clams had a characteristic beaded appearance. They were able to show that the filament had the form of a double helix consisting of two chains of roughly globular subunits, the chains twisted around each other so that viewed from a given direction the crossover points were about 360 angstroms apart [see illustration at left on this page].

The thin filaments from striated

muscle show an identical structure; it can often be seen even when they are still attached to a Z line. Filaments made from actin prepared by standard biochemical techniques again show the same pattern. Thus we can confirm that the thin filaments of striated muscle do contain actin, as we had supposed. We can also deduce that the globular subunits are molecules of actin that aggregate to build up the filament. The structure itself might resemble two strings of beads twisted around each other; its alternating high points and low points suggest a general arrangement for the successive active sites on the filament to which the cross-bridges may attach themselves (assuming that each globular unit has one site). We cannot directly view enough of the internal structure or shape of the subunits to make any deductions about their directionality. To reveal such polarity we have used a natural marker, namely heavy meromyosin, the fragment of myosin that combines with actin.

When actin filaments are treated with a solution of heavy meromyosin and examined in the electron microscope by negative staining, they assume a complex appearance that we do not yet understand in full detail. Nevertheless, one salient feature stands out immediately: the filaments of the resulting compound have a well-defined structural polarity that manifests itself in an obvious arrowhead pattern [see illustration at right on this page]. The arrows always point in the same direction over the length of a given filament, even when only dilute solutions of heavy meromyosin have been applied and the arrow pattern is interrupted by long stretches of normal uncombined actin. If the polarity were imposed by some local condition such as the direction along the actin filament at which a series of heavy meromyosin molecules were attached during the formation of the compound filament, one would expect the pattern of arrowheads to lack such consistency. Therefore it would seem that it is the underlying structure of the actin that imposes the pattern. Precisely which feature of the myosin-actin combination gives rise to the arrowhead effect is unclear; it may well be that the pattern reveals the actual orientation of some part of the heavy-meromyosin fragments. A general feature can be deduced, however: all the actin molecules in a filament will combine with heavy meromyosin in precisely the same way [see top illustration on page 114]. We can conclude that all the actin molecules in a given thin filament are oriented in the same sense and that they can all interact in identical fashion with a given myosin crossbridge.

We have used the same technique to investigate the way in which the actin filaments are attached at the Z lines. As I have mentioned, preparations of thin filaments from homogenized muscle frequently contain groups of filaments still connected to both sides of a Z line. We find, in examining such assemblies after treatment with heavy meromyosin, that the arrows on all filaments always point away from the Z lines. The filaments forming the *I* substance on one side of a Z line are all similarly oriented; on the opposite side of the Z line the orientation is reversed.

This is exactly the arrangement we require in order for the same relative orientation of the actin and myosin molecules to obtain in the two halves of the A bands but for the absolute orientation to be reversed. The direction of the forces developed will consequently be reversed and the actin filaments can move in opposite directions, that is, toward each other in the middle of the sarcomere.

We conclude that both the thick and the thin filaments in a striated muscle are assembled and oriented in such a way that if a relative force were developed between a given actin and myosin molecule in either filament, all the elements of force in the whole system would be added together in the appropriate manner to give rise to the organized behavior we have observed. Several years ago we tentatively proposed the analogy of a ratchet to describe the interaction of sliding filaments. Now our understanding of the way in which cross-bridges of myosin seem to hook onto consecutive active sites on the actin filament makes the analogy seem even more appropriate.

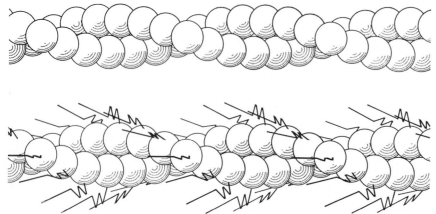

STRUCTURE OF ACTIN is represented by two chains of beads twisted into a double helix (*top*). The way in which actin might combine with heavy-meromyosin fragments to give rise to arrowheads apparent in micrograph at right on page 158 is suggested at bottom.

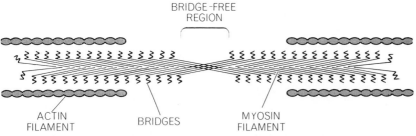

CONTACT OF ACTIN AND MYOSIN in muscle might be made in the manner schematically illustrated here. The thin actin filaments at top and bottom are so shaped that certain sites are closest to thick myosin filament in the middle. The heads of individual myosin molecules (*zigzag lines*) extend as cross-bridges to the actin filament at these close sites.

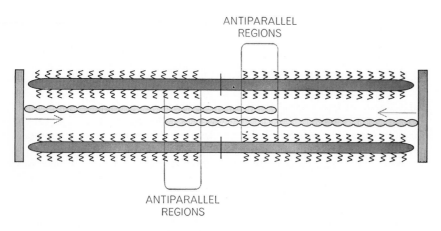

DOUBLE OVERLAP of thin filaments from each side of the sarcomere would result if the sliding-filament hypothesis is essentially correct. It is now assumed that muscle generates maximum tension when thin filaments reach center of A band, and that tension falls when thin filaments cross the center and interact with improperly oriented cross-bridges.

Recently we have examined with our improved electron microscope techniques sections of muscle fixed at various stages of contraction. The filament lengths, measured by Sally G. Page of University College London, appear to remain constant and equal to the corresponding lengths in resting muscle (discounting small changes in length from tension during fixation and other preparative steps). The most interesting feature of the contraction sequences we have studied is that at the shorter sarcomere lengths a dense zone appears in the center of the A band; the zone progressively increases in width as the muscle shortens. This zone first appears after the H zone has closed up completely. We had shown previously that the closing of the H zone during shortening is caused by the ends of the thin filaments sliding toward each other in the center of the A band. Now when we measured the distance from the Z line to the opposite end of the new dense zone, we found that it was still equal to the length of the thin filaments. We therefore suspected that the new zone might correspond to a region where the thin filaments from each end of the sarcomere overlap [*see illustration on page 108*].

This view was confirmed when we examined cross sections of muscle cut through the region of supposed double

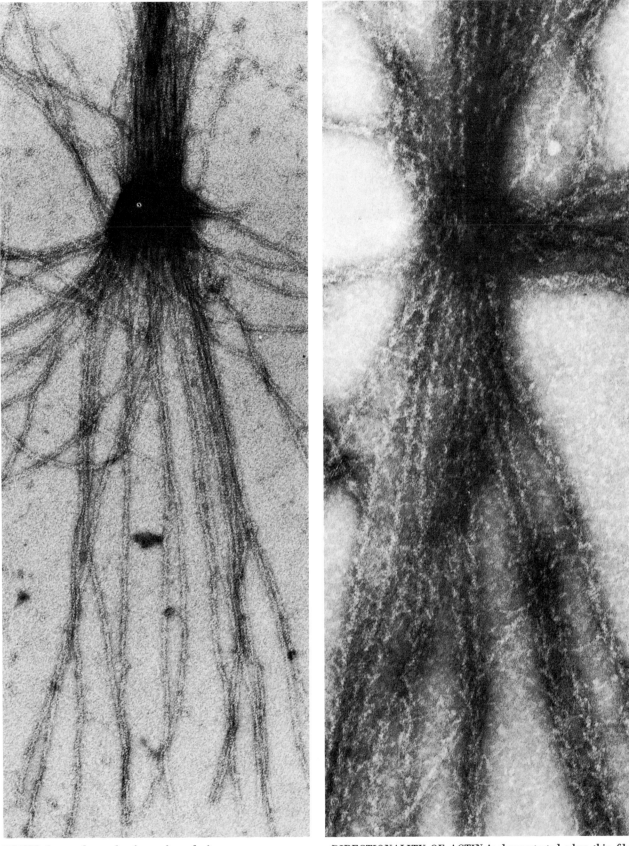

Z LINE, the membrane that forms the end of a sarcomere, appears as dark region from which strands radiate. The strands are thin actin filaments that comprise the *I* band. At times, as in this instance, they remain attached even after muscle has been homogenized. This micrograph and one at right were made by negative staining. They both have magnification of some 165,000 diameters.

DIRECTIONALITY OF ACTIN is demonstrated when thin filaments attached at the *Z* line are labeled with heavy meromyosin. Arrowheads form, pointing away from the *Z* line on each side. In this micrograph they point up at top of *Z* line, down at bottom. Opposite orientation of the two *I* segments of a sarcomere enables filaments from left and right *I* segments to converge on center.

overlap. Instead of the normal pattern of thick and thin filaments in regular hexagonal array with twice as many thin filaments as thick ones, the pattern was less regular and there were four times as many thin filaments as thick ones [see illustration below]. Apparently we were seeing the thin filaments from both ends of the sarcomere at the same time, and these must have slid past one another during the shortening. This finding confirmed that the simple sliding process describes the behavior of the thin filaments under all conditions. (Objectors had proposed, for instance, that the thin filaments might coil up within the A bands.) The finding also suggested to our group and to A. F. Huxley, who is now at University College London, a possible explanation for the observable decrease in tension generated by striated muscles at sarcomere lengths shorter than resting length. Tension would fall off in the double-overlap

region because there is a progressively increasing penetration of the thin filaments from one Z line into the "wrong" end of the A bands. We know that actin molecules in part of a thin filament penetrating the center of the A band would have an abnormal orientation with respect to the adjacent cross-bridges [see bottom illustration on page 114]. Such a region would not be expected to contribute to the development of tension by the muscle, and by interfering mechanically and chemically with the interaction of the correctly oriented actin and myosin molecules it might reduce the tension.

Can we extend our investigations to consider changes in the arrangement or configuration of the actin and myosin molecules in muscles that are actually contracting? This goal has indeed been attained, thanks to the sophistication of a technique long used

in the study of muscle: X-ray diffraction. Untreated muscle reflects X rays in a regular pattern. We can compare the way in which contracted and relaxed muscles reflect X rays and thus determine if activation and the development of tension are associated with appreciable changes in the length of the repeating units of pattern formed by the arrangement of actin and myosin molecules within the filaments (and hence with possible changes in filament length). Two groups of workers—W. Brown, K. C. Holmes and I in Cambridge, and G. F. Elliott, Lowy and B. M. Millman at the Medical Research Council Biophysics Research Unit at King's College in London—have independently conducted such studies, and both groups report that no such changes in length occur during contraction.

Another exciting finding, reported by our group, is that the relative intensity of some of the X-ray reflections associated with myosin filaments changes greatly during contraction. (Subsequently the London group reported observations consistent with our findings.) These effects have still to be analyzed in detail, but they indicate a substantial movement of the cross-bridges during contractile activity. Very recently members of our group and a group of investigators under J. W. S. Pringle at the University of Oxford have demonstrated a movement of the cross-bridges associated with the contraction of insect flight muscle. These latest findings open up new possibilities. Now that we know that measurable changes in the X-ray reflections do in fact occur during contraction, we have a method of distinguishing steps in the process by which energy for contraction is obtained.

A contracting muscle offers a uniquely favorable system for studying the outstanding problems of protein structure and function. In muscle we have now clearly identified the interacting protein molecules, the high concentration in which they are present and their regularity of arrangement. The major unsolved question about contractility is a general question of biochemistry: how do proteins act as catalysts for biochemical reactions, and what happens to them in the process? It is interesting in this regard to recall that ATP itself was first identified as the source of energy in the contraction of muscle and subsequently as the universal carrier of chemical energy in the living cell. We expect that the study of the precise basis of contractility will also lead to broadly applicable results.

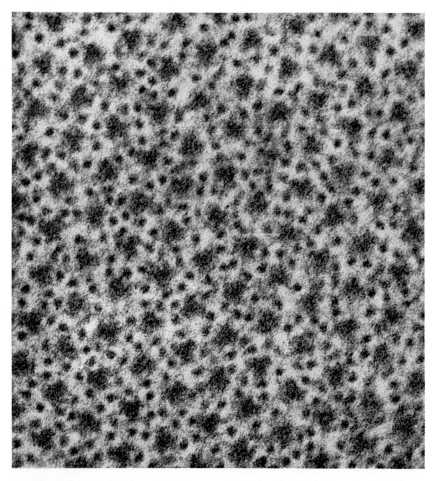

CONTRACTED MUSCLE viewed end on in this electron micrograph has four times as many thin filaments (*small dots*) as thick (*large dots*). The regular array of thick filaments is well preserved; the array of thin filaments is not. Since the ratio of thin to thick filaments in relaxed muscle (*evident in micrograph at top of page 109*) is two to one, it appears that actin filaments from each end of the sarcomere overlap during contraction. This transverse section, made by cutting through the center of an A band of a muscle from the leg of a frog (the method used in making section at top of page 109), is enlarged 250,000 diameters.

ENZYMES: MACROMOLECULAR CATALYSTS

III

ENZYMES: MACROMOLECULAR CATALYSTS

INTRODUCTION

Of all the intricate processes that have evolved in living cells, none is more striking or more essential than enzymatic catalysis. Due to the action of enzymes under the mild conditions of temperature and acidity prevailing in living systems, the myriad chemical reactions that contribute to cellular metabolism occur with enormous speed and specificity. Chemical reaction rates are commonly accelerated a billionfold, and the absence of unwanted byproducts is the envy of the synthetic chemist.

As Earl Frieden explains in "The Enzyme-Substrate Complex," enzymes exert their catalytic influence by forming geometrically specific complexes with their substrates, thereby lowering the free energy required to overcome the barrier to reaction. Substantial direct evidence for such complexes had been obtained by 1959, when Frieden wrote his article. Although recent work has provided even greater understanding of their properties, Frieden's article provides an excellent overview of the major features of enzyme-substrate complexes.

One very important recent contribution to understanding the mechanism of action of enzymes and other proteins is discussed by Daniel E. Koshland, Jr., in "Protein Shape and Biological Control." It has become apparent that the enzymesubstrate complex is not a geometrically static entity. Rather, the substrate and regulatory molecules can induce in the enzyme subtle conformational changes that help to regulate activity and coordinate complex biochemical pathways. Koshland surveys the occurrence, nature, and magnitude of these changes in a variety of enzyme systems, showing how they lead to cooperative binding and catalytic properties. He also indicates that the principle of conformational control of activity applies to many non-enzymatic systems, such as antibodies, receptors for light, taste and smell, hormonal action, and control of cell differentiation.

Among the most convincing and detailed demonstrations of enzyme interactions with substrates or substrate analogs are those provided by X-ray crystallography. By the early 1970s about a dozen enzyme structures had been determined by this technique—a body of work that is one of the great triumphs of biophysical chemistry. Lysozyme, the first enzyme deciphered by X-ray crystallography, is described by David C. Phillips in "The Three-Dimensional Structure of an Enzyme Molecule." Phillips's description is beautifully aided by the illustrations of Irving Geis. In addition, Phillips reviews the principles of X-ray diffraction and isomorphous replacement techniques in a way that nicely complements the articles by J. C. Kendrew ("The Three-Dimensional Structure of a Protein Molecule," p. 25) and M. F. Perutz ("The Hemoglobin Molecule," p. 40). The structure of lysozyme has some features in common with myoglobin and hemoglobin: charged and polar groups on the exterior in contact with the polar solvent, water, and hydro-carbon side chains clustered away from water in the interior. However, in lysozyme the alpha helix is not as prominent a structural feature as in myoglobin or hemoglobin; many other sorts of helixes and nonhelical regions are found. On the basis of the structure of lysozyme, Phillips proposes a mechanism for lysozyme folding—the way in which the primary structure determines the tertiary structure of the protein. Finally, using an inhibitor, a structural analog of the substrate of lysozyme, Phillips is able to construct a plausible model for the enzyme-substrate complex and the mechanism of catalysis. The kind of intimate, atomic-level understanding of enzymes reported here is the ideal for all biophysical chemical investigations.

Striving for simplification, biochemists have tried when possible to study enzymes in dilute solutions, separated from all other cell constituents. However, as Klaus Mosbach points out in his article "Enzymes Bound to Artificial Matrixes," perhaps the majority of enzymes function while attached to membranes or in other complex environments. The microenvironments of such enzymes may be quite different from those in vitro. Enzymes in a metabolic pathway may be fixed in position to shuttle substrates and products from one catalytic site to the next in an efficient, nonrandom manner. Many of these effects can be mimicked and studied by attaching enzymes to insoluble supports. As Mosbach reports, such matrix-bound enzymes are becoming increasingly valuable in analytical, clinical, and industrial research.

Three useful articles related to the topics in this section are "Catalysis" by Vladimir Haensel and Robert L. Burwell, Jr., December, 1971; "Enzymes" by John E. Pfeiffer, December, 1948; and "The Control of Biochemical Reactions" by Jean-Pierre Changeux, April, 1965 (Offprint 1008). Articles describing additional biochemical control processes are "Hormones and Genes" by Eric H. Davidson, June, 1965 (Offprint 1013); and "The Specificity of Antibodies" by S. J. Singer, October, 1957 (also recommended in connection with Section I).

The Enzyme-Substrate Complex

12

by Earl Frieden
August 1959

The fleeting union of the enzyme and the substance on which it acts holds a key to our understanding of life processes. Many ingenious techniques are in use today to isolate it for study

We seek him here, we seek him there,
Those Frenchies seek him everywhere.
Is he in heaven?—Is he in hell?
That demmed, elusive Pimpernel?

Biochemists have an elusive Pimpernel in the enzyme-substrate complex: the transitory combination of the enzyme and the substance upon which it acts. Like the protean hero of Baroness Orczy's romantic novel, it turns up here, there and everywhere in the chemical reactions of life, acts its decisive part and then disappears. For 30 years the complex was more than elusive; it was a figment of the biochemist's imagination. Nevertheless many biochemists believed an enzyme could work only by joining briefly with its substrate. Two decades ago they demonstrated that the complex exists, but they could not isolate it; its existence is too fleeting. Today they are closing in on the quarry; capture appears imminent.

Without enzymes biochemical processes would be far too slow to carry on what we know as life. Enzymes are catalysts: substances that accelerate reactions without themselves being chemically changed. All the enzymes isolated thus far are proteins: huge molecules made up of chains of amino acid units. Most enzymes are highly specific, acting either on a single substrate, on a group of closely related substrates, or on a characteristic region in various substrate molecules. But just how does the enzyme act upon the substrate? The momentary union of enzyme and substrate conceals the answer. If one could trap the complex long enough to elucidate its structure, one might hope to understand how an enzyme promotes the transformation of the substrate into its end product.

The systematic study of enzymes began in 1835, when the Swedish chemist Jöns Jakob Berzelius included biological reactions among the chemical changes he termed "catalytic." The fact that a mixture of enzymes from potatoes breaks down starch faster than sulfuric acid does made a great impression on Berzelius. With remarkable insight he predicted that it would eventually be found that all substances in living organisms are made under the influence of catalysts. A century later chemists were still making discoveries that confirmed his pre-diction, and were finding enzymes where one might think them unnecessary.

For example, few chemical reactions are as rapid as the decomposition of carbonic acid into carbon dioxide and water. This reaction is what we see whenever we open a bottle of carbonated beverage. Indeed, it occurs in our lungs every time we exhale. In 1928 O. M. Henriques of Denmark observed that the unaided reaction is not fast enough to account for the evolution of carbon dioxide in the lungs. He deduced that some enzyme in the blood must catalyze this reaction. In 1932 two British biochemists, N. U. Meldrum and F. J. W. Roughton, found the enzyme (named carbonic anhydrase) in red blood cells, again verifying Berzelius's prediction.

Despite the contention of Berzelius, many chemists doubted that enzymes were catalysts. Some doubted their very existence, even after James B. Sumner of Cornell University succeeded in crystallizing the enzyme urease in 1926. For one thing, enzymes seemed to disobey the chemical law of mass action, which states that the velocity of a chem-

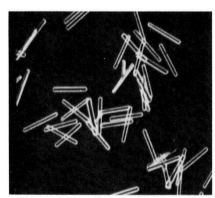
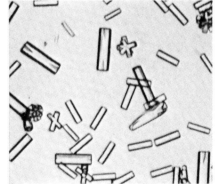

ENZYME, INHIBITOR AND COMPOUND of the two are seen in these photomicrographs. At left are crystals of the enzyme trypsin. In center are those of a substance that inhibits or stops the action of trypsin by combining with it. At right are crystals of the compound. Moses Kunitz and John H. Northrop at the Rockefeller Institute isolated enzyme and inhibitor from beef pancreas extract.

ical change is proportional to the active masses, or molecular concentrations, of the reacting substances. The speed of an enzyme reaction is proportional to the amount of substrate only at very low substrate concentrations; higher concentrations do not affect its rate. Furthermore, it was observed that a tiny amount of enzyme transforms an unbelievably large quantity of substrate.

On the other hand, the enzyme concept found support in the mainstream of chemical theory and in a growing body of experimental evidence. Before the end of the 19th century chemists understood that a molecule must obtain extra energy from another molecule before it can react. This may happen when one molecule collides with another (a frequent occurrence in gases and liquids, made more frequent by heat). The molecules with extra energy are said to be "activated." Thus in a reaction that converts substrate molecules (S) to product (P) the molecules must be in the activated state (designated S°) in order to yield P. The more S° molecules there are, the faster the reaction will go. In 1888 the Swedish chemist Svante Arrhenius published a simple equation for calculating the activation energy for a given reaction.

How do catalysts fit into this scheme? Arrhenius himself suggested that the most sensible way to explain the action of any catalyst is to suppose that it forms an intermediate compound with its substrate. From this intermediate the reaction proceeds at a lower energy of activation. The catalyst contributes no energy, but alters the reaction path so that S goes to P by a different route. With an enzyme (E) the reaction is $E + S \rightleftharpoons ES^\circ \rightarrow E + P$. As this equation shows, the catalyst is regenerated unchanged and can participate in the reaction again and again. Thus a small amount of enzyme can cause reaction in a prodigious volume of substrate.

Of course not all catalysts are enzymes. But enzymes generally reduce activation energies far more effectively than inorganic catalysts do, and they permit reactions to take place at lower temperatures. For example, to decompose hydrogen peroxide into water and oxygen requires an activation energy of 18,000 calories per gram molecule of hydrogen peroxide. Catalytic iron brings this figure down to 13,000 calories; platinum, to 12,000. But the liver enzyme catalase reduces the activation energy to less than 5,000 calories.

As early as two years after Arrhenius had proposed the existence of a catalyst-substrate complex, Cornelius O'Sullivan and F. W. Tompson of England found evidence for it in a reaction involving an enzyme. They employed a stratagem still used today to protect an enzyme when separating it from other protein. To a solution containing the yeast enzyme invertase they added its substrate, sugar. When the solution was heated, the unwanted proteins coagulated and precipitated, leaving the invertase to be recovered intact. The addition of the substrate had apparently protected the enzyme from the effects of heat. O'Sullivan and Tompson believed this was evidence for the existence of an enzyme-substrate complex. We have since learned that enzymes can be stabilized against heat by other combining substances, such as coenzymes and inhibitors. The degree of stabilization correlates well with the protective agent's affinity for the enzyme.

By the early 1900's Victor Henri of France and Adrian J. Brown of England, presenting the results of independent studies of invertase action, proposed the formation of an enzyme-substrate complex as the best explanation of that enzyme's action. Historically, however, the concept of the enzyme-substrate complex is usually credited to Leonor Michaelis and Maude Menton of Berlin. They published a remarkably clear and complete paper on the subject in 1913, and we often refer to the complex as the Michaelis complex. Since then the idea has played a central role in the interpretation of numerous studies of the kinetics, or rate, of enzyme reactions. However usefully the enzyme-substrate complex has served the purpose of these investigations, the investigations themselves have yielded no direct evidence to establish the existence of the complex. The evanescent compound must somehow be captured so that we can examine it, even if only briefly.

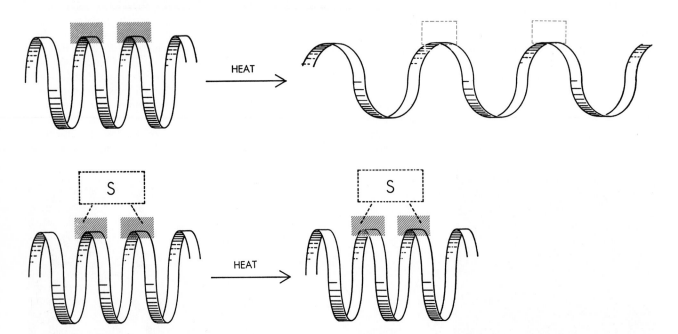

PROTECTION OF AN ENZYME against heat is effected by a substrate. At top left a bit of the helical enzyme molecule is shown, its active site hatched in color. Heat damages the helix and inacti-vates the site, as at top right. When substrate (S) is present, presumably attached in a complex at the active site, the same amount of heat leaves the enzyme and active site intact, as shown at bottom.

The first concrete evidence for the complex came in 1936, when Kurt Stern, then at Yale University, and David Keilin and Thaddeus Mann of the University of Cambridge simultaneously reported that the mixing of certain enzymes with their substrates in solution brought a change of color in the solution. Stern's enzyme was catalase, and Keilin and Mann used peroxidase isolated from horse-radish. Peroxidase catalyzes the reaction between hydrogen peroxide (or simple peroxide derivatives) and many reducing agents to form water and an oxidized substance. In doing this it first joins hydrogen peroxide in a complex; this enzyme-substrate complex is then capable of oxidizing a reducing agent such as ascorbic acid or malachite green. A solution of peroxidase is brown, but when hydrogen peroxide or another substrate is added to the solution, it turns green. Then, a few seconds later, it turns red. The addition of ascorbic acid regenerates the original brown color of the free enzyme. Keilin and Mann attributed the color changes to the formation of a complex. Their work and Stern's constituted the first qualitative capture of the enzyme-substrate complex.

In a series of brilliant papers beginning in 1940 Britton Chance of the Johnson Foundation at the University of Pennsylvania gave the complex further respectability by measuring it quantitatively. He took advantage of the fact that in the peroxidase and catalase reactions the enzyme and substrate each have light-absorbing properties that differ from those of the complex. Using very rapid mixing and flow techniques and a highly sensitive spectrophotometer, he followed the growth and decay of the enzyme-substrate complex by reading the changes in the color of the solution. He found that the complex forms rapidly, reaches a maximum and a steady state for a brief period and then declines slowly to zero [see illustration at right]. Chance's measurements of the kinetics agree so closely with theoretical predictions that we presently accept his work as the most convincing proof of the existence of the union of enzyme and substrate.

In addition to studying the complex directly, we can investigate analogous combinations. For example, certain enzymes are formed by the more or less temporary combination of an apoenzyme (a protein molecule) and a coenzyme (a small nonprotein molecule); the activity characteristic of the enzyme appears only when the two are combined. Some of these combinations are structurally

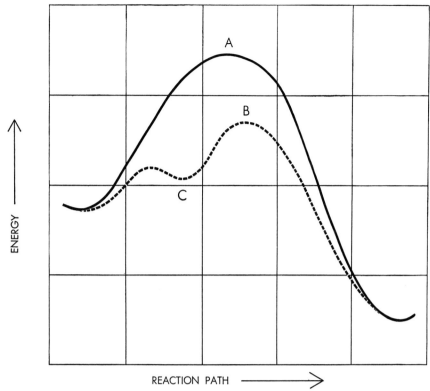

SUBSTRATE REACTION PATH is changed by an enzyme. Solid line represents reaction path without the enzyme, where high energy of activation (A) is required. The path with an enzyme present (*broken line*) shows lower activation energy (B) due to complex at C.

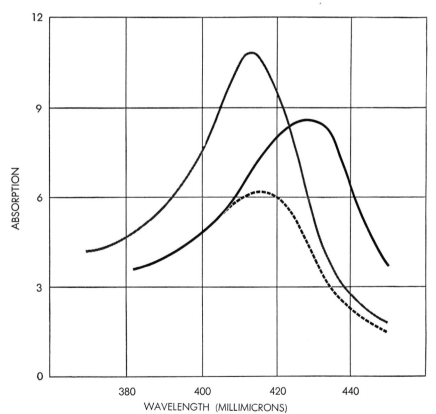

LIGHT-ABSORPTION CURVES provided first demonstration of a complex. Absorption by the enzyme peroxidase gives rise to brown color (*gray line*), but complex of the enzyme and the substrate hydrogen peroxide is green (*broken line*). Change to red color (*black line*) quickly follows. The complete visible spectrum of these compounds is not shown.

similar to the enzyme-substrate complex. In 1951 Hugo Theorell and Roger K. Bonnichsen of Sweden investigated the complex formed by the protein and the coenzyme that together comprise alcohol dehydrogenase, a liver enzyme that catalyzes the oxidation of alcohol. They found that a significant change in the absorption of light—a shift of the point of maximum absorption 30 millimicrons toward the ultraviolet—attends the combination of the coenzyme and the protein. As in the case of the colored enzymes, the close correspondence of the kinetic and spectral data confirms the presence of the elusive protein-coenzyme complex. Helmut Beinert of the Institute for Enzyme Research in Madison, Wis., has observed very rapid color changes in several enzymes containing the bright yellow vitamin riboflavin, a coenzyme. The transient intermediate that appears during enzymic activity probably represents a highly reactive or free-radical form of the coenzyme.

To trap the intermediate of a substrate or a coenzyme Sidney F. Velick of the Washington University Medical School devised the unique method of mixing the protein of alcohol dehydrogenase with its coenzyme or a substrate, and spinning them at 60,000 revolutions per minute in an ultracentrifuge. The heavy protein concentrates at the bottom of the centrifuge cell, along with any complex, while the lighter coenzyme or substrate molecules not combined with the protein remain above in the solution [see illustration at left]. Upon measuring the amounts of coenzyme or substrate carried to the bottom of the cell in transitory combination with the protein, Velick found them to be in full accord with the Michaelis-Menton equations.

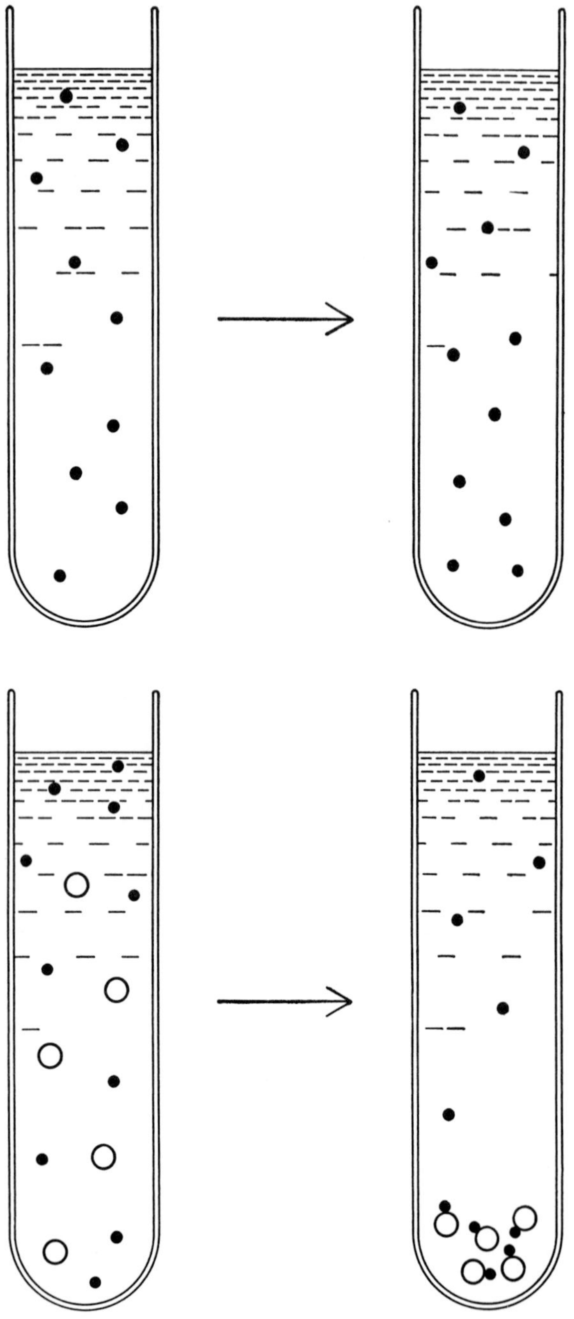

Another substitute for the substrate that may be employed to form stable complexes with enzymes are enzyme inhibitors. These are substances that block the normal activity of the enzyme without destroying its chemical integrity; the enzyme can be regenerated intact from the combination with its inhibitor. We believe that the inhibitor becomes attached to the enzyme at its "active" site, that is, at the very section of the molecular chain which forms the transitory bond with the substrate. The enzyme-inhibitor complex thus provides an excellent analogy with the enzyme-substrate complex.

David G. Doherty and Fred Vaslow have conducted a fruitful investigation on this line at the Oak Ridge National Laboratory. They have employed dialysis, one of the older techniques of biochemistry, to detect the enzyme-inhibitor complex. When a solution containing small and large molecules is placed in a bag permeable only to the small molecules, the small molecules tend to migrate into the fluid outside the bag until an osmotic equilibrium is reached [see illustration on opposite page]. Ideally, in the detection of the enzyme-substrate complex, the concentration of small substrate molecules showing up outside the bag would provide an indication of the concentration of free substrate remaining inside. Unfortunately a true substrate is always changing to product, and the system does not reach equilibrium. Accordingly Doherty and Vaslow used an inhibitor instead of a substrate. They labeled the inhibitor with radioactive atoms, calculated the amount of enzyme that formed a complex with one molecule of inhibitor, and showed that there

ULTRACENTRIFUGE SEDIMENTATION traps the complex. This drawing shows tubes from ultracentrifuge, at left before application of forces 100,000 times gravity, at right after spinning in the machine. At top substrate molecules (dots) do not sediment. At bottom, with enzyme (circles), some substrate is held in complex in sediment after spinning. Comparison of substrate in solutions (top and bottom right) gives quantity of complex.

is only one active site on each enzyme molecule. By repeating the dialysis at different temperatures and calculating the heat given off during the formation of complex they deduced some important predictions about the behavior of this enzyme. They were even able to conceive of a brand-new substrate and to show that the enzyme would act upon it.

The use of inhibitors to identify the active site in the structure of various enzymes now constitutes a major line of attack in the field of enzyme chemistry. Biochemists frequently employ the so-called nerve gases for this purpose because they form some of the most interesting complexes. The nerve gas diisopropylfluorophosphate (DFP) combines with nerve enzyme cholinesterase to yield a "DFP-enzyme" and release hydrogen fluoride. The gas is lethal because, by this reaction, it inhibits cholinesterase and thus prevents the transmission of nerve impulses. Antidotes to the nerve gas free the enzyme because they form compounds more readily with the DFP molecule.

DFP is a good research tool because it joins the enzyme at its active site. The chemist must purify the DFP-enzyme compound and obtain it, if possible, in crystalline form. He then gently breaks down the compound with other enzymes and uses paper chromatography to separate the resulting pieces of the molecule. From the fragments he determines the order in which the amino acids are linked up in the intact molecule and establishes the site at which the DFP molecule is attached. The DFP is labeled with radioactive phosphorus for easy detection. This technique has been employed by J. A. Cohen of the Netherlands with the enzyme chymotrypsin, by Hans Neurath of the University of Washington with trypsin, by Daniel E. Koshland, Jr., of Brookhaven National Laboratory with phosphoglucomutase, and by Robert Schaffer of the Army Chemical Center with cholinesterase.

All of these workers find that DFP reacts with the amino acid serine in the structure of all four enzymes. They have made an even more remarkable discovery. All four enzymes seem to have the same amino acids in exactly the same order at the DFP-reactive site: glycine-aspartic acid-serine-glycine. The sequence beyond the serine-glycine group is glutamic acid-valine-alanine for chymotrypsin and phosphoglucomutase, but it is proline-valine for trypsin. The finding that enzymes with such different catalytic activities have such closely related active sites suggests that the spe-

cificity of their action must depend upon the molecular structure adjacent to the active site, rather than only upon the small active area. We are only at the threshold of this realm of study.

Many enzymes have built-in markers at their active site, and it is not necessary to use DFP or some other inhibitor to detect it. At Florida State University we are trying to isolate the active site in a group of copper-protein enzymes. We know that copper is involved in the catalytic activity because the enzyme becomes inactive when we remove it. We can restore the activity by adding more copper ion. The identification of the copper-containing amino acid groups separated from these enzymes should lead to the pin-pointing of the active sites.

Though the theory that enzymes

function by means of an active site is now backed by considerable experimental evidence, biochemists still do not know whether the full catalytic activity of an enzyme resides in the structure as a whole or in a small part associated with the active site. The molecules of certain enzymes have been trimmed down to somewhat smaller size without appreciable loss in catalytic activity. For example, the red respiratory enzyme cytochrome C has been considerably degraded, and the resulting smaller molecule has retained some of the catalytic properties of the original enzyme. The enzyme ribonuclease loses little of its activity when its molecule is partially degraded. The molecule has also been broken into a large part and a small part that are inactive separately but become active when they are put into solution together.

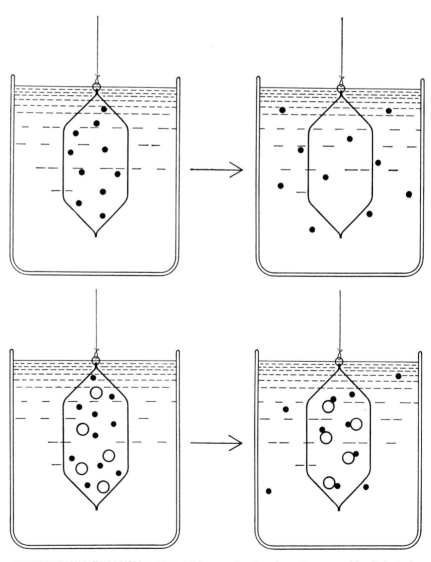

EQUILIBRIUM DIALYSIS with inhibitor molecules alone in permeable dialysis bag (*upper left*) ends with inhibitor distributed uniformly in the solution inside and outside the bag (*upper right*). Enzyme molecules (*circles in bag at lower right*) attach to inhibitor, retaining a larger number of its molecules inside the dialysis bag at the end (*lower left*).

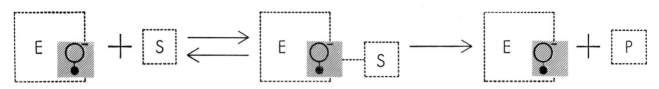

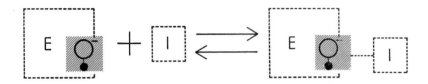

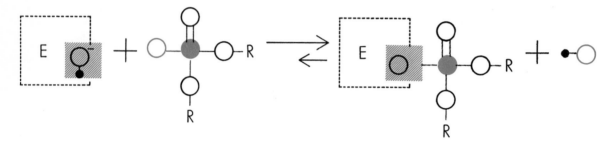

- ● HYDROGEN
- ○ OXYGEN
- ⬤ PHOSPHORUS
- ○ FLUORINE

THREE TYPES OF ENZYME reaction are depicted in these diagrams. In the top diagram an enzyme (E) and a substrate (S) form a complex at a hydroxyl (*oxygen and hydrogen atoms*) active site of the enzyme, then yield the enzyme and the product (P). In the middle diagram a competitive inhibitor (I) forms a complex with an enzyme. If a substrate were pres-

ent, the inhibitor would slow but not stop the enzyme-substrate reaction. In the bottom diagram the nerve gas diisopropylfluorophosphate (DFP) forms a complex with an enzyme, liberating hydrogen fluoride (*hydrogen and fluorine atoms*) and ending the catalytic activity of the enzyme. The letter R denotes radicals, or groups of atoms, on the DFP molecule.

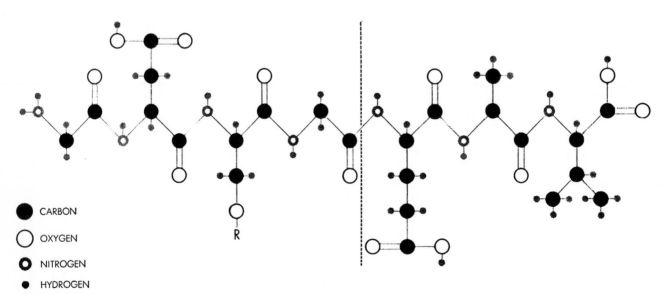

- ● CARBON
- ○ OXYGEN
- ◎ NITROGEN
- • HYDROGEN

SEQUENCE OF AMINO ACID UNITS in the vicinity of the active site is identical in four enzymes. An inhibitor, the nerve gas DFP (*labeled* R), was used to discover this. Here it is attached to active site. To left of the broken line the enzymes have the same sequence:

glycine-aspartic acid-serine-glycine. The active site of the DFP reaction is at the serine. The enzymes chymotrypsin and phosphoglucomutase both have the glutamic acid-alanine-valine sequence shown to the right of the line; trypsin and cholinesterase differ on the right.

Yet the effort to locate the catalytic activity of enzymes at a small spot on a large protein molecule may be doomed to failure. The extraordinary success of proteins as selective catalysts may depend on much more than the simple order of the amino acid units of which they are made. The chains of these units are arranged in helixes folded in various and intricate ways. The geometrical relationship of one helix to others may be crucial to catalytic activity. Heat or other relatively mild treatment easily inactivates many enzymes, though the effect of heat must be limited to changing the three-dimensional orientation of the chains and could not affect the basic order of amino acid units. Some chemists have suggested that, when the enzyme and substrate form their complex, the enzyme structure changes extensively before it exerts any catalytic activity. It is difficult to imagine such gross spatial changes in simple chains of a limited number of amino acid units.

We may find, nonetheless, that the catalytic activity of some enzymes does depend upon a relatively simple section of the long chain of the molecule. We are now discovering numerous examples of short-chain molecules that resemble a section of an enzyme molecule and have specific biological activity. Vincent du Vigneaud of the Cornell University Medical College has determined the structure of two peptide hormones of the pituitary gland: oxytocin and vasopressin. Each hormone has nine amino acid units in a chain, and differs from the other in only two of its component units. Other active short-chain molecules are the antibiotics gramicidin and tyrocidin and the growth factor streptogenin. As our knowledge of these smaller active mole-

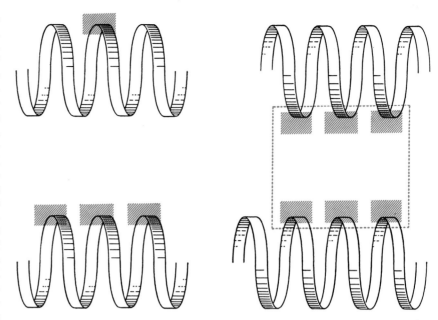

ACTIVE SITE OF ENZYME MOLECULES may be confined to limited area of the helix, as in color hatching at upper left. It may overlap several coils of the helix, as at lower left. Or perhaps two or more helixes of the enzyme, as at right, would have to be intact in order to preserve the active site of the molecule in its proper three-dimensional relationship.

cules increases, we may learn a great deal from them about the formation of complexes.

What will the capture of the enzyme-substrate complex mean? Consider the part that enzymic catalysis plays in the processes of life. The many thousands of biological reactions are mediated by virtually an equal number of highly specific enzymes. The reactions vary from relatively simple ones, such as the decomposition of hydrogen peroxide or carbonic acid, to the complicated, step-by-step synthesis of huge molecules like the nucleic acids that control heredity. Yet all this variety of activity resides in

essentially one type of molecule, assisted in some cases by other substances of low molecular weight. The versatility of enzymes is a source of wonder to the chemist.

As the biochemist Ernest Borek has written: "We live because we have enzymes. Everything we do—walking, thinking, reading these lines—is done with some enzymic process." Life is essentially a system of cooperating enzyme reactions. A unifying theme in all this diversity may be discovered with the help of a close look at the enzyme-substrate complex, now that it is no longer so elusive.

13

Protein Shape and Biological Control

by Daniel E. Koshland, Jr.
October 1973

The processes of life are turned on and off by means of a universal control mechanism that depends on the ability of protein molecules to bend flexibly from one shape to another under external influences

A living system must have both the capacity to act and the capacity to control its actions. We humans, for example, must be able to digest food, but we cannot be eating all the time. Hence we need both a positive control to turn on the process (a desire to eat when food is needed) and a negative control to turn off the process (a desire to stop eating when we have had enough). Similarly, the process of blood clotting must be turned on when we bleed from a wound but must be turned off afterward so that it does not lead to coronary thrombosis. We must also be able to turn muscles on in order to move and to turn them off in order to relax. In short, every biological system has built-in controls to initiate or accelerate a process under some conditions and to terminate or decelerate it under other conditions.

Considering the diversity of processes that must be regulated and the diversity of environmental conditions to which an organism must react, it might be expected that the controls in biological systems are enormously complex. That is true, and yet when one examines the processes more closely, it appears that the fundamental elements of control are remarkably simple and universal.

The fundamental control element in all living systems—from the smallest bacterium to man—is the protein molecule. Enzymes, the biological catalysts that control all the chemical processes of living systems, are proteins. Sensory receptors, which enable us to see, hear, taste and smell, are proteins. Antibodies, which provide immunity against infection, are proteins. Recent experiments have established that it is the ability of these proteins to change shape under external influences that provides the "on-off" controls that are so vital to the living system.

The concept of protein shape as a control mechanism arose from studies of enzymes, a development that is hardly surprising, since enzymes are the easiest to study of all the regulating proteins. It has been estimated that an average living cell contains some 3,000 different enzymes. Each of them catalyzes a distinct chemical reaction in which compounds called substrates are converted into other compounds called products. Fortunately for our understanding of biological systems many enzymes are quite sturdy molecules and can be extracted from a physiological system without destroying their biological properties. Hence they can be studied in the test tube and made to perform the same catalytic role there that they perform in the living organism. Moreover, one can subject them to the same environmental influences in the test tube that they experience in the living cell, thereby getting clues to their role in biological regulation. Finally, enzymes can be obtained in large enough amounts for their physical properties to be studied. However, even though our understanding of the role of shape in protein regulation began with enzymes, the principles of regulation worked out for enzymes appear to be universal and can be applied to other proteins that are more difficult to obtain in bulk.

Shape Changes in Enzyme Catalysis

It has long been known that the basic mechanism by which enzymes catalyze chemical reactions begins with the binding of the substrate (or substrates) to the surface of the enzyme. The enzyme then polarizes the chemical bonds in the substrate, causing a reaction that leads to the formation of products on the surface. The release of these products from the surface regenerates the free enzyme and allows the cycle to be repeated.

Unlike catalysts made in the laboratory, enzymes have the special property called specificity, which means that only one chemical compound or a very few can react with a particular enzyme. This property can be explained by the template, or lock-and-key, hypothesis put forward in 1894 by Emil Fischer, which postulates that the enzyme is designed to allow only special compounds to fit on its surface, just as a key fits a lock or as two pieces of a jigsaw puzzle fit together. The complementary shapes allow one compound to fit and exclude other compounds that lack the correct size, shape or charge distribution [*see illustration on pages 130 and 131*]. Modern X-ray crystallography has revealed in detail precisely such a fit between enzyme and substrate [*see illustration on opposite page*].

Although this concept could explain much of the specificity data, some glaring discrepancies were found. For instance, certain oversized and undersized compounds were found to bind to the surface of the enzyme even though they failed to form products. Furthermore, it was difficult on the basis of the rigid-template theory to explain how sugars can compete with water in enzymatic reactions. It was also difficult to explain why substrates bind in a specific order in many enzymatic reactions. These facts and others like them led to the hypothesis that the enzyme does not exist initially in a shape complementary to that of the substrate but rather is induced to take the complementary shape in much the same way that a hand induces a change in the shape of a glove [*see illustration on pages 132 and 133*]. This "induced fit" theory assumes that the substrate plays a role in determining the final shape of the enzyme and that the enzyme is therefore flexible. Proof that proteins do in fact change their shape

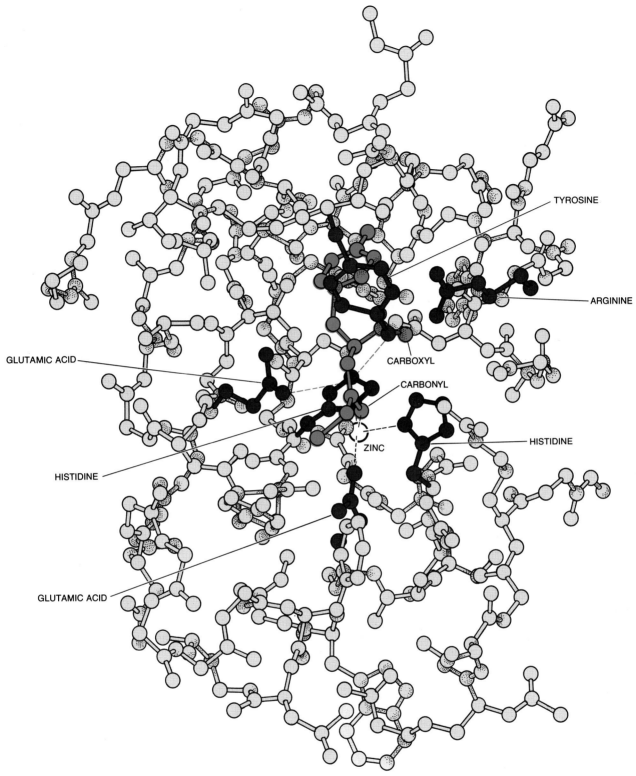

TYROSINE

ARGININE

CARBOXYL

CARBONYL

GLUTAMIC ACID

HISTIDINE

ZINC

HISTIDINE

GLUTAMIC ACID

PRECISE FIT between the active site on the surface of a large protein molecule and the specific substrate molecule with which the protein reacts is evident in this simplified three-dimensional drawing, based on X-ray-crystallographic data obtained by William N. Lipscomb, Jr., and his colleagues at Harvard University. The protein, rendered in shades of gray, is carboxypeptidase *A*, a digestive enzyme that (as its name implies) works by cutting the polypeptide chain of the substrate near its carboxyl end. The substrate, rendered in color, is carbobenzoxyalanylalanyl tyrosine. Approximately a fourth of the total number of atoms in the polypeptide chains that comprise the two molecules are represented in this view. The atoms shown are mostly carbon, with a small admixture of nitrogen and oxygen; all hydrogen atoms have been omitted. The six active-site side chains that specifically interact with the substrate are the darker gray. For example, a positively charged arginine side chain is shown attracting the negatively charged carboxyl group of the substrate. In addition certain hydrophobic, or oily, regions on the substrate are attracted to similar regions on the enzyme, strengthening the attraction between the two molecules. A zinc atom (*white*) forms an additional "coordination bond" involving a carbonyl group on the substrate and three other amino acid side chains (glutamic acid and two histidines) extending from inside the enzyme's bowl-shaped active site. The tyrosine side chain and the second glutamic acid side chain of the active site are catalytic groups that polarize the electrons in one of the substrate's chemical bonds, splitting that bond (*broken colored line*) and thereby dividing the substrate into the two parts that are the reaction products of this particular enzyme-substrate combination.

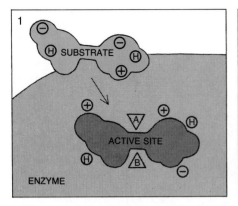
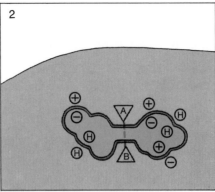
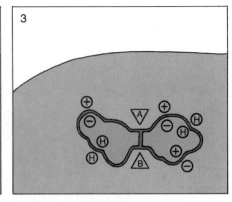

LOCK-AND-KEY MODEL of the mechanism by which enzymes catalyze chemical reactions, put forward in 1894 by Emil Fischer, postulates that the active site of the enzyme is a rigid, templatelike structure that allows only special compounds to fit, just as a key fits a lock. Such a compound, called a substrate, is attracted to the enzyme's active site by mutually attractive groups, such as the electrostatic charges on certain amino acid side chains $(+, -)$, and by the coalescing tendency of adjacent hydrophobic side chains (H). In the lock-and-key model the catalytic groups (A, B) are poised in advance (1) to cause the reaction that ruptures the chemical bond

under the influence of small molecules was initially obtained by chemical studies showing differences in the reactivity to protein reagents of the amino acid side chains that are arrayed along the spine of the protein molecule. The hypothesis has since been verified with the aid of advanced physical techniques, most notably X-ray crystallography.

Once the concept of a flexible enzyme was entertained, the puzzling behavior of the special classes of compounds that were found to bind to the enzyme without forming products could be explained. One type of oversized molecule, for example, binds to the surface of the enzyme, but in doing so it distorts the protein into a shape that does not allow the catalytic groups to be properly aligned. Other compounds can bind to the enzyme too, but they do not have either sufficient size or the correct chemical characteristics to induce the proper alignment. Hence even though both types of compound bind, neither reacts with the enzyme. (Some molecules are of course too big to be bound even to a flexible enzyme; the examples cited here are chosen to show the differences between the rigid-template theory and the induced-fit theory.)

Shape Changes in Regulation

This finding of flexibility does not mean that all proteins must be flexible. Some may indeed be quite rigid, and these are explained very well by the lock-and-key hypothesis. Nor does it mean that enzymes that exhibit flexibility must do so with all chemical compounds. The finding does mean, however, that protein flexibility is a key feature of enzyme action. Indeed, the capacity to induce a change in shape has

been found to be a widespread and vital feature of most, if not all, enzymes.

The concept of protein flexibility led to the deduction that small molecules not themselves involved in the chemical reaction could help to make a deficient molecule act as a substrate by altering the shape of the enzyme [see illustration on page 134]. In the case of a molecule that is too small to induce the proper alignment of catalytic groups, for example, certain molecules that are not consumed in the reaction can be added to produce a stable shape with the right alignment of catalytic groups. One way for this realignment to occur is for a second molecule to bind immediately adjacent to the deficient molecule, thus inducing the proper shape at the active site. This prediction of the flexible-enzyme theory has been confirmed in many cases (for example in the case of the carbohydrate-splitting enzyme hexokinase by Alberto Sols and his co-workers in Spain and in the case of the digestive enzyme trypsin by T. Inagami and T. Murachi in Japan). Molecules that bind far away from the active catalytic site can also induce a proper shape. In that case the induced change is transmitted through the protein like a row of falling dominoes until the active site is altered appropriately.

The reverse process can also occur. Flexible enzymes can be distorted out of the active shape by molecules called inhibitors. These molecules can cause a disruption of either the catalytic function or the binding function of the enzyme, in either case giving rise to an inactive shape [see illustration on page 135].

In short, a regulatory molecule that is not itself involved in the chemical reaction can control the activity of an en-

zyme by changing its shape. It can turn the enzyme on by inducing the correct shape or turn it off by inducing an incorrect shape. In biological systems one of the most important groups of such molecules is the hormones. Although hormones are secreted in small amounts, much too small to be important directly as foodstuffs or sources of energy, they have a tremendous influence on the regulatory processes of the cell. The manner in which hormones exert control is easily explained by the flexible-protein hypothesis. Since these molecules are not consumed, they can be used again and again to activate the enzyme molecules, unlike the substrate molecules, which are consumed. Therefore such regulators need be present only in very small amounts.

Sometimes, as in the case of adrenalin, the initial hormone induces the formation of a second molecule, cyclic AMP, which acts as a regulator for many enzymes by changing their shape [see "Cyclic AMP," by Ira Pastan; SCIENTIFIC AMERICAN Offprint 1256]. In general the shape changes induced by the regulator molecules are similar to the shape changes induced by the substrate.

The Regulation of Pathways

In discussing the shapes of proteins it is difficult, tedious and often unnecessary to show the detailed parts of a large protein molecule. For convenience the different molecular shapes can be symbolized by geometric figures: squares, triangles, circles and so on [see illustration on page 138]. It is nonetheless important to remember that even though such shapes can be expressed by simple line drawings, a change from a circle to a square, say, designates subtle and com-

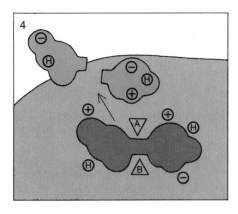

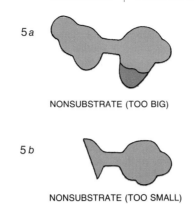

5a

NONSUBSTRATE (TOO BIG)

5b

NONSUBSTRATE (TOO SMALL)

in the neck of the substrate (2), converting the substrate into the compounds called products (3). The release of these products from the surface (4) regenerates the free enzyme and allows the cycle to be repeated. According to this view, the precisely complementary shapes of the substrate and the enzyme that acts on it exclude reactions with compounds that are either too large to fit into the active site (5a) or too small to be attracted to the active site (5b).

plex changes in the orientation of the many amino acid side chains that constitute a protein. This procedure is analogous to using the symbol C for a carbon atom rather than writing out the complete quantum-mechanical description of the electrons, protons and neutrons that constitute the atom. The different shapes of proteins are referred to as conformations, because the protein changes fulfill a chemist's definition of a conformational change, that is, a change in the shape of a molecule caused by rotation around a single chemical bond. The terms "conformation" and "conformational change" are technical synonyms for the terms "shape" and "shape change" employed so far in this article.

The site on the surface of an enzyme at which the catalytic action takes place is called the active site. The binding site for the regulatory molecule is called the regulatory site or the allosteric site to distinguish it from the catalytic site. (The term "allosteric," meaning "the other site," was coined by the French biochemist Jacques Monod, a leader in this area of investigation, and it has gradually come to be used as a general term for regulatory proteins.) Regulatory molecules are also called effectors, modifiers or allosteric effectors.

Let us now consider how these ideas of protein shape help to explain some types of regulatory control. One of the most important decisions a living system must make involves how to process a food substance. We need energy in the form of molecules of adenosine triphosphate, or ATP, for all our bodily processes: to contract our muscles, to see with our eyes, to activate our nerves and to generate our structural materials. This energy comes from the combustion of food. A certain amount of ATP is con-

stantly being used to maintain the system even in a resting state; the heart, for example, continues pumping even when we are asleep.

It is obviously desirable during periods of low energy demand to store energy for future needs. One of the ways of storing energy is in the form of glycogen, a chain of sugar molecules. Thus when a molecule of glucose, say, is ingested by a living system, it can be directed along alternate pathways: either it can be oxidized immediately to form carbon dioxide, giving off large amounts of ATP, or it can be stored in the form of glycogen to be released on future demand [see illustration on page 136]. The enzyme phosphofructokinase, which is involved in the first of these pathways, is turned on and off by variations in the level of ATP in the system. Another enzyme, glycogen synthetase, is involved as part of the second pathway in the regulatory control of the synthetic process; it can be turned on and off by the presence of regulatory compounds in the cell. A third enzyme, glycogen phosphorylase, catalyzes the reaction from glycogen to glucose-1-phosphate in the third pathway and is the enzyme used when the need arises for the retrieval of glucose from storage.

The way these enzymes work under the influence of supply and demand is simple and ingenious. When ATP levels are high, the phosphofructokinase enzyme of pathway 1 and the phosphorylase enzyme of pathway 3 are turned off and the glycogen synthetase of pathway 2 is activated. Glucose is therefore diverted to glycogen for storage. When an animal is frightened, however, it secretes adrenalin, which ultimately activates phosphorylase (pathway 3) and phosphofructokinase (pathway 1) and deac-

tivates glycogen synthetase (pathway 2). In that case glycogen is removed from storage and converted into ATP energy to help the animal escape from the danger. In other words, when we are frightened, the hormone demands that our reserves of glycogen be made available as a source of energy. More than one demand can release glycogen from storage. Hunger and certain stimulators of muscular activity (such as calcium ions) activate phosphorylase and thus can also generate energy. In principle such influences act in the same way that adrenalin does.

This particular regulatory system can be used to illustrate two important points. First, enzymes must be available in every living system and yet they must not be equally active at all times. The alteration in shape under the influence of metabolites and hormones therefore provides a mechanism for turning these enzymes on and off under different external conditions (ranging, say, from starvation to satiety).

The second principle is more subtle. If all the enzymes involved in carbohydrate metabolism were to be activated simultaneously, one would simply have a short circuit going around and around pathways 1, 2 and 3, storing and burning glucose to no avail. It has usually been found for alternate pathways of this kind that a molecule that activates one pathway inhibits the other. In the case of glycogen storage, for example, the regulatory molecule inhibits the enzyme in pathway 2 and activates enzymes in pathways 1 and 3; the pathways for the oxidation of glucose and the removal of glucose from storage are accelerated and the pathway for the storage of glucose is blocked. In the absence of this regulatory molecule, equilibrium favors the inactive form of the enzyme in pathways 1 and 3 and the active form of the enzyme in pathway 2; in other words, the system favors glucose storage. Thus a short circuit is avoided by changing the shape of the enzymes reciprocally so that synthesis to glycogen is favored when ATP levels are high and degradation to carbon dioxide and ATP is favored when energy is needed.

Protein Structure

At this point it is worth considering in a little more detail how the design of proteins enables this regulation to proceed. The proteins that act as enzymes range in weight from approximately 10,000 daltons to many millions of daltons. (One dalton is roughly equal to the weight of one hydrogen atom.) The higher figure is deceptive, however, because all such large proteins are made up of peptide subunits, which usually range between 15,000 and 100,000 daltons. For example, the enzyme phosphorylase, which is important in glycogen storage and degradation, is a dimer (a two-peptide polymer) composed of two identical subunits each with a molecular weight of 96,000 daltons. Similarly, aspartyl transcarbamylase, the first enzyme in the pathway leading to the synthesis of cytidine triphosphate, or CTP, is a dodecamer (a 12-peptide polymer) composed of six subunits of one kind (each with a molecular weight of 35,000 daltons) and six subunits of another kind (each with a molecular weight of 17,000 daltons). These subunits are attracted to one another by noncovalent forces: largely electrostatic attractions or hydrophobic bonds (a name used to describe the tendency of oily regions of a structure to be forced together in the same way that oil droplets tend to coalesce in water). Fortunately qualitative features do not change with the size of the protein, so that by studying the simpler proteins it is possible to understand the properties of proteins in general.

A peptide chain with a molecular weight of only 25,000 daltons is still large compared with the molecular weight of most substrates, which are usually compounds in the molecular-weight range of 100 to 1,000 daltons. Occasionally enzymes act on very large molecules such as DNA (deoxyribonucleic acid), cellulose or other proteins, but when they do, they usually bind only a small portion of these large molecules, so that the effective substrate size is still only about 1,000 daltons or less. This difference in relative size means that only a small portion of the enzyme's sur-

face is actually involved in catalysis. The rest of the surface is available for binding the molecules that are involved in regulation and for the association of subunits with one another.

Cooperativity

The concept of protein flexibility provided an explanation for a long-known phenomenon that had been originally discovered by the Danish physiologist Christian Bohr in the 19th century. Bohr noted that the binding of oxygen to hemoglobin could be described by a sigmoid, or S-shaped, curve instead of the normal hyperbolic curve observed for the binding pattern of most enzymes [see illustration on page 137]. He correctly deduced that this unusual type of binding curve would result if the first molecule bound made it easier for the next molecule to bind, and so forth; hence he called the process cooperative. Since then hemoglobin has been the subject of intensive study by many prominent investigators, and much of our knowledge of the phenomenon of cooperativity results from examination of this vital protein. Cooperativity is not limited to hemoglobin, however. Sigmoid binding curves are also common and important features of the regulatory proteins.

The appearance of a cooperative binding pattern in the case of a regulatory protein can be explained by means of the flexible-protein hypothesis, using the simplest multisubunit protein, a dimer made up of two identical subunits [see illustration on page 139]. A number of such dimer proteins exist in nature; the types of interaction they are involved in are similar to those of the more complex proteins. The binding of the substrate induces conformational chang-

es that depend on the structure of both the substrate and the protein.

Three general types of conformational change are known. In one type the first molecule of substrate alters the subunit to which it is bound but does not alter the interactions between the subunits. The second subunit therefore binds substrate in just the same way and with the same affinity as the first. There is no cooperation between the subunit sites.

The next type of conformational change is quite different. Here the first molecule to be bound induces a conformational change in the first subunit, which induces shape changes in the second subunit. These changes, which are transmitted through the protein structure, change the active site in the second subunit so that it becomes more receptive to the substrate; hence the second molecule of substrate is bound more readily than the first. This phenomenon, called positive cooperativity, explains the sigmoid binding curve discovered by Bohr.

In the third case the first molecule induces a conformational change that makes the binding site in the second subunit less attractive to the substrate because of their incompatible geometries. The binding of the second molecule is therefore discriminated against in favor of the first and proceeds much less readily. This phenomenon is called negative cooperativity (cooperativity because of the subunit interactions, negative because the first molecule has a negative effect on the second).

The induced-fit hypothesis can readily explain the sigmoid curve of multisubunit proteins; a mathematical adaptation of the induced-fit approach was devised by George Némethy, David Filmer and me to do just that. Our solution is not the only possible explanation, but

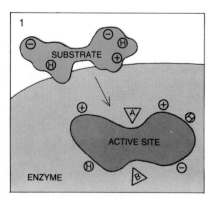
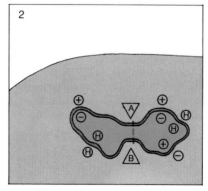
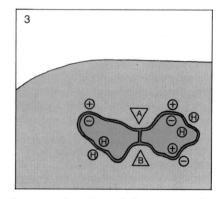

INDUCED-FIT MODEL of enzyme action, developed recently by the author and his colleagues at the University of California at Berkeley, assumes that the enzyme does not exist initially in a shape complementary to that of the substrate (1) but rather is induced to take the complementary shape in much the same way that a hand induces a change in the shape of a glove (2). Once the substrate is bound, the catalytic groups of the enzyme are in position to cut the chemical bond in the substrate's neck, forming the reac-

there is substantial evidence that it is valid for a large number of enzymes exhibiting such sigmoid curves. Besides being able to explain a puzzling phenomenon in terms of the protein structure, the induced-fit approach predicted that a different type of interaction—negative cooperativity—should exist (in other words, that the induced conformational changes would make the second molecule bind less readily than the first). Such a phenomenon was not known in nature at the time, but its prediction from theory led to a determined search for it. In 1968 Abby Conway and I discovered an example of this strange phenomenon in an enzyme that participates in carbohydrate metabolism: glyceraldehyde 3-phosphate dehydrogenase. Negative cooperativity has since been found in many other enzymes.

The mathematical analysis of such a negative-cooperativity pattern explains in part why the phenomenon escaped detection. On superficial inspection a negative-cooperativity binding curve looks like a hyperbola. A more careful analysis shows that in reality the curve is not a true hyperbola. Such a curve can be explained only by assuming that the second molecule binds to the protein less readily than the first. Several examples of each type of cooperativity have now been established. Induced conformational changes can be caused by activators and inhibitors as well as by substrates, and therefore all three types of molecule can give rise to cooperative interactions.

All these concepts apply to molecules with more than two subunits in the same way. In positive interactions the first molecule makes it easier for the second to bind, the second makes it easier for the third and so on. In negative interactions each interacting molecule makes it more difficult for the next molecule to interact. The small molecule (a substrate, an inhibitor or an activator) induces the change and the effect is then transmitted to the neighboring subunits. The greater the number of subunits is, the more dramatic the cooperativity can be, but the actual cooperativity pattern observed depends on the details of the individual protein structure.

Why should such cooperative phenomena exist in nature? Would it not be better for the protein to be designed correctly in the first place, so that these induced conformational changes need not alter the interactions between the subunits or the shape of the protein? One of the reasons protein flexibility is so important, of course, has already been mentioned: flexibility makes it possible for an enzyme to be regulated by molecules that are not themselves consumed in the enzymatic reaction.

The concept of cooperativity, however, apparently provides another reason for protein flexibility. If one examines the binding curves for the three types of cooperativity and compares the change in concentration of a compound to the change in the activity, one finds that approximately an 81-fold change in the concentration of the substrate is needed to go from an activity level of 10 percent to one of 90 percent, assuming that the protein follows the hyperbolic binding curve of a normal noninteractive protein. In sharp contrast, only a ninefold change in concentration is needed if one assumes that the protein is designed with rather mild positive cooperativity. (In hemoglobin a threefold change in concentration will do the job, and in CTP synthetase a 1.5-fold change is sufficient.) In other words, a protein with positive cooperativity is much more sensitive to small fluctuations in the environment than a protein with a normal binding pattern. The sensitivity is increased for inhibitors and activators as well as for substrates. Positive cooperativity is thus an amplification device to make a small signal have a much larger regulatory effect. This increased sensitivity is extraordinarily important for the regulatory function of these proteins.

One example of the physiological importance of the phenomenon of cooperativity can be observed in patients who have a mutant hemoglobin that lacks the positive cooperativity of normal hemoglobin. Positive cooperativity enables hemoglobin molecules to absorb large amounts of oxygen in the lungs and to deposit large amounts of oxygen in the tissues, even though the pressure of oxygen does not vary much from one location to the other. This situation follows from the steepness of the sigmoid curve associated with positive cooperativity. The noncooperative protein is far less efficient in transporting oxygen, and patients possessing it are very sick. In a crude sense it might be said that the positively cooperative protein resembles a truck that takes on a full load of dirt at one location and empties all of it at another, whereas a protein without cooperativity cannot take on a full load to begin with and can only get rid of part of the load.

What, then, is the role of negative cooperativity? A look at its binding curve provides the answer to that question too. Here the protein is less sensitive to fluctuations in the environment than a noncooperative protein; in other words, a much greater change in the concentration of the substrate, the inhibitor or the activator is needed to go from an activity level of 10 percent to one of 90 percent. Some proteins should not be subject to fluctuations in the environ-

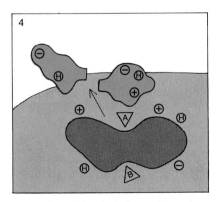

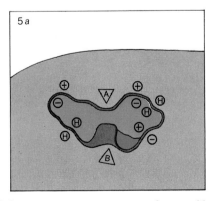

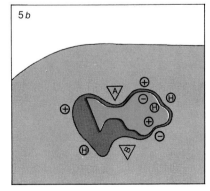

tion products (3), which leave the surface of the enzyme, returning the active site to its original noncomplementary shape (4). This concept of a flexible enzyme made it possible to explain the previously puzzling observation that certain oversized and undersized compounds were able to bind to the surface of the enzyme without forming products (5a, 5b). Even though both types of nonsubstrate compound bind, neither succeeds in inducing the proper alignment of the catalytic groups and hence neither reacts with the enzyme.

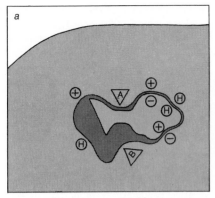
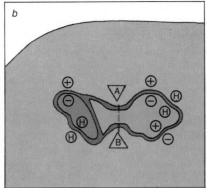
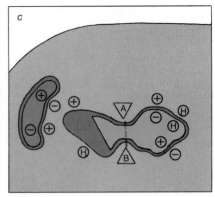

ACTIVATOR MOLECULES can, according to the flexible model of enzyme action, help to make a deficient molecule act as a substrate by altering the shape of the enzyme. For example, in the case of a molecule that is too small to induce the proper alignment of catalytic groups (a) a second molecule can bind immediately adjacent to the deficient molecule inside the active site, thereby inducing a stable shape with the proper alignment of catalytic groups (b). A molecule that binds at a site outside the active site can also induce the proper shape (c). In neither case is the activator molecule itself consumed in the ensuing chemical reaction.

ment, and the proteins that regulate such processes may use negative cooperativity to damp their sensitivity and make them less subject to environmental changes.

The Nature of the Shape Changes

How are conformational changes propagated through the protein molecule and how extensive are they? An understanding of these processes has flowed from two sources: chemical studies of the composition of proteins and X-ray studies of protein structure.

Some of the important structural data were obtained as a result of the pioneering work of M. F. Perutz and his coworkers on hemoglobin [see the article "The Hemoglobin Molecule," by M. F. Perutz, beginning on page 40]. Hemoglobin consists of four similar but not identical subunits arranged in a tetrahedral array. When oxygen is bound to the heme, or iron-containing, binding group of one of these subunits, there is a small shift in the position of the iron atom, which in turn causes a small shift of an adjacent histidine side chain. The histidine shift then compresses several amino acid side chains in a helical portion of the protein and squeezes a tyrosine side chain out of a small pocket. The movement of the tyrosine group dislocates the subunit to which it is attached and breaks the salt linkage between the subunits.

These are not the only regions that are affected by the binding of the oxygen atom, but they are illustrative of the type of change that can occur in a typical protein molecule. The point is that the protein is a tightly structured molecule with intimate close-packing relations between the atoms of the amino acid side chains. A shift of one such group by even a fraction of an angstrom causes realignments of other side chains, generating a "domino" effect that can extend through the entire protein molecule. The changes within the hemoglobin subunit cause a realignment of the subunits with respect to one another.

The importance of these protein shifts can be demonstrated with mutant hemoglobins isolated from sick patients. When the mutant hemoglobins were examined by Perutz and his colleagues, the amino acid side chains involved in the shape changes were found to be altered.

A further understanding of these processes has been obtained from experiments conducted in my laboratory at the University of California at Berkeley by Alexander Levitzki and William Stallcup. They worked with CTP synthetase, a key enzyme that converts uridine triphosphate, or UTP, to CTP in the metabolism of nucleic acids. This regulatory protein usually exists in the form of four identical subunits, but it can be studied as a dimer. Each subunit has binding sites for its three substrates: UTP, ATP and glutamine, and also for its regulatory effector, guanosine triphosphate, or GTP [see illustration on page 140]. The glutamine reacts at the glutamic site on the enzyme, forming what is called the glutamyl enzyme. In doing so it liberates ammonia, which binds to the ammonia site on the enzyme. The ammonia in turn reacts with UTP, which reacts with ATP. The close coupling of the chemical steps indicates that the sites are immediately adjacent to one another.

When the glutamyl enzyme is formed, the reaction produces a covalent bond between glutamine and a cysteine side chain on the surface of the protein. At the beginning the two cysteine groups of the two active sites are equally reactive. When glutamine reacts with the same side chain, however, a strange thing happens. The reaction of the substrate with one subunit turns off the cysteine side chain on the other subunit. The effect is only temporary; the glutamyl enzyme reacts further to regenerate free enzyme.

As it happens, this change in the shape of the enzyme can be "frozen" (somewhat like stopping a motion-picture film at a single frame) with the aid of an "affinity label," dioxoazonorleucine. The molecule of dioxoazonorleucine is enough like the molecule of glutamine so that it forms a covalent bond with the cysteine side chain but enough different so that it cannot react further. The result is that only one of the two initially identical subunits reacts with the dioxoazonorleucine. A reaction at one subunit turns off its neighbor, so that only one of the two potential subunits reacts at any one time, giving rise to what we have termed the "half of the sites" phenomenon.

This phenomenon is observed in a number of enzymes and indicates that in many cases the shape change can be transmitted over long molecular distances. On the basis of the size of the enzyme subunits the cysteine side chains in CTP synthetase are probably between 20 and 60 angstroms apart.

The most surprising part of the finding is that the immediately adjacent sites—the ammonia, UTP and ATP sites—are not altered at all when the cysteine group is modified. All their properties remain the same. Hence the formation of a bond apparently transmits a signal that has dramatic consequences as much as 60 angstroms away without perturbing structures within four or five angstroms! When the reactivity of amino acid side chains to protein reagents is examined, many side chains are found to change position and many others to remain unchanged.

A similar pattern is evoked by GTP, the regulatory molecule. GTP activates the enzyme by altering the reactivity of the glutamine site on the same subunit, so that the glutamyl enzyme is formed more rapidly. The same shape change that activates the glutamyl site deactivates the GTP site of the neighboring subunit. Moreover, the shape changes induced by GTP do not alter the properties of the ammonia, ATP or UTP sites. Thus the GTP merely by binding to the surface of the protein can direct the alteration of some side chains and leave others unchanged. The same is true of many other activators, inhibitors and substrates.

The Scale of the Shape Change

Hence it appears that induced conformational changes in proteins are not like the ever widening concentric ripples produced by throwing a pebble into a pond. They are more like a spider web in which the strands are devised to transmit a perturbation occurring in one corner of the web to another corner. The perturbation can be transmitted over long distances and can alter the positions of many strands, but a clever design can ensure that some strands will remain unchanged at the same time that others are shifting appreciably. The protein, like the spider web, is designed to transmit information in a focused manner to some regions and to leave others unchanged.

The schematic drawings used to illustrate the alteration of protein shapes in this article tend to exaggerate the relative movements necessary to achieve this control. Actually we do not know precisely how big the movement of catalytic or binding side chains must be to achieve on-off switching. There are strong suggestions, however, that the movements do not have to be very large. The length of a carbon-oxygen bond, for example, is about 1.3 angstroms. A catalytic group that needs to be positioned close enough to an oxygen atom to pull electrons out of it would therefore be ineffective as a catalyst if it were positioned next to the carbon atom. Hence a movement on the order of an angstrom or two would appear to be sufficient to make the difference between an effective catalyst and an ineffective one. The effective movement is probably somewhat less than that, particularly because most regulators do not completely turn off the enzyme but rather reduce its function in the direction of either catalysis or binding.

Movements of side chains by 10 or 12 angstroms have been observed in carboxypeptidase molecules by William N. Lipscomb, Jr., and his colleagues at Harvard University, using the X-ray-crystallographic approach. It seems likely that the large movements observed in this type of protein are more than are necessary for regulation in general and that many of the atoms move much less or not at all. It has already been found that the movement of certain atoms such as the iron atom in hemoglobin and the sulfur atom in CTP synthetase may be less than an angstrom, and yet such a movement can trigger large conformational changes. Model experiments with simpler compounds suggest that an alteration of less than half an angstrom can also alter reactivity greatly. In short, the schematic drawings shown here should be taken to indicate that the movements are large enough to alter the function of a protein but that they do not represent the actual scale of conformational changes. Overall observations of protein molecules also indicate that no tremendously large shape changes occur in the subunit as a whole even though a few of the atoms move by several angstroms. By focusing on a few groups and using schematic pictures an erroneous impression could be obtained if one does not remember that one is looking at only a very small portion of the spider web. The gross anatomy of the web is unchanged; only a few strands shift with respect to one another, and yet these shifts are highly significant in terms of function.

Advantages of Induced Changes

One of the advantages of having enzymes that can be induced to change their shape in response to the proper stimulus can be seen in the example we have been considering: CTP synthetase. I have mentioned that the substrates ATP and UTP show strong positive cooperativity. In the case of ATP, for example, the first molecule of ATP alters the shape of the enzyme so that the subsequent molecules of ATP bind rapidly. Actually CTP synthetase is one of the most cooperative enzymes known (even more so than hemoglobin). The binding of the first molecule of substrate has a triggering effect that allows the rapid and complete binding of the subsequent molecules of ATP, so that only the free enzyme and a form of the enzyme with four ATP molecules bound are found in appreciable amounts. This positive cooperativity means that the enzyme is

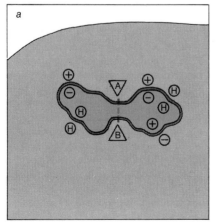 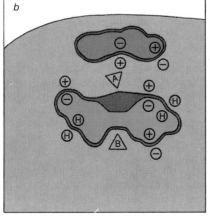 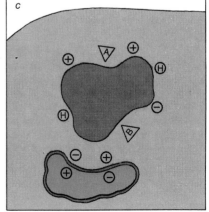

INHIBITOR MOLECULES can, in the reverse process, distort a flexible enzyme out of the active shape by binding at sites outside the active site. In the case of a substrate that would otherwise react with the enzyme (a) such inhibitors can cause a disruption of either the catalytic function (b) or the binding function (c) of the enzyme, in effect "turning off" the enzyme-substrate reaction.

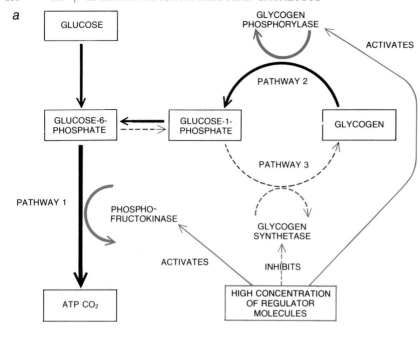

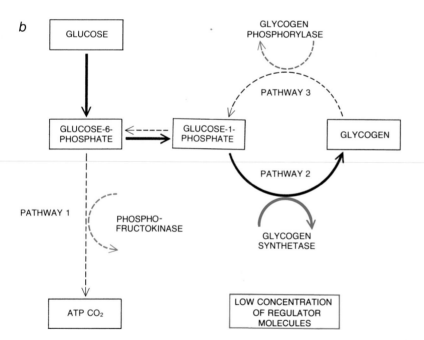

ALTERNATE BIOLOGICAL PATHWAYS are controlled by regulatory molecules through the alteration of enzyme shapes. In the human body, for example, ingested glucose is normally converted to glucose-6-phosphate, which can either be oxidized immediately to provide energy in the form of molecules of ATP (adenosine triphosphate) or stored in the form of glycogen (a chain of sugar molecules) to be released on future demand. Which pathways are chosen depends on the concentration of regulatory molecules. In times of energy demand high concentrations of these molecules convert the enzyme of pathway 1 (phosphofructokinase) and the enzyme of pathway 3 (glycogen phosphorylase) to active forms while at the same time converting the enzyme of pathway 2 (glycogen synthetase) to an inactive form, thereby channeling both the ingested glucose and the stored glycogen into the production of ATP (*top*). Low concentrations of the regulatory molecules enable the enzymes of pathways 1 and 3 to return to their inactive forms while enabling the enzyme of pathway 2 to become active again, thereby diverting the glucose into glycogen for storage (*bottom*). The reciprocal effect of the regulatory molecules on pathways 2 and 3 prevents their both being active at the same time and therefore rules out a futile short circuit in the system.

very susceptible to small fluctuations in the concentration of ATP and hence is highly sensitive to environmental influences on this molecule. In contrast, the regulatory molecule, GTP, induces conformational changes leading to negative cooperativity, which desensitizes the enzyme toward fluctuations of GTP in the environment.

What does all this mean in terms of enzyme action? It means that the enzyme has been programmed by its design through evolutionary time so that it can be responsive to small changes in ATP levels and at the same time be desensitized to rather wide fluctuations of GTP levels. A conceivable reason for this ability is that ATP is highly controlled in the biological system, since it is such a central compound for so many pathways. As a result enzymes such as CTP synthetase must respond readily to even small changes in the levels of ATP when they are regulated by the ATP level itself. On the other hand, GTP may fluctuate greatly from time to time in the organism's life cycle because it plays a role in different but less vital pathways. If a fairly constant level of CTP-synthetase activity is needed throughout these fluctuations, the negative cooperativity of GTP will ensure this desensitization to fluctuations in GTP levels. Of course, complete desensitization would allow the production of CTP in the absence of GTP, which would be a wasteful operation because both are needed for the synthesis of RNA. The protein ensures GTP control by requiring it to serve as the activator but eliminates excessive sensitivity through the device of negative cooperativity.

Another advantage of these induced conformational changes arises from the sequence of steps on an enzyme surface. High-energy intermediates are frequently formed in chemical syntheses, but if they are not isolated from water or other reactive substances, they decompose in side reactions that lower the yield of the reaction. Induced conformational changes can enable one step in the reaction to trigger the next step, which in turn triggers the third step and so on. In this way the high-energy intermediates exist for only brief intervals and are nestled in the protective harbor of the active site during the chemical changes. Wasteful side reactions are prevented, and the characteristically high yields of enzymatic reactions are achieved. In short, the induced-fit conformational change both explains the anomaly of a required order in the binding of substrates and provides a reason for it.

It therefore seems likely that flexible enzymes arose early in evolutionary time because of catalytic needs and certain specificity requirements. One of the greatest of these needs was the exclusion of water from some reactions. The cooperative property probably developed later because of the multisubunit structure of the enzyme; cooperativity survived because it had a useful function of its own: the amplification of some responses and the damping of other responses.

It is important to note that the parallelism between activators and inhibitors (positive effectors and negative effectors) and positive and negative cooperativity should not obscure their different and complementary roles. An activator is designed to turn an enzyme on and an inhibitor is designed to turn it off. Cooperativity is designed to increase or decrease the sensitivity of the enzyme to the environmental fluctuations of these regulators. An activator may show either positive cooperativity, negative cooperativity or noncooperativity, and the same is true of a substrate or an inhibitor. GTP, for example, is an activator of enzyme action but shows negative cooperativity in its binding pattern. In a simple sense one might say that activation or inhibition defines the key role of the regulatory molecules, whereas cooperativity provides the fine tuning of the system.

Finally, it is important to emphasize that although this account of the development of the induced-fit theory of enzyme action concentrates on the particular approach taken in our laboratory, many other workers in both the U.S. and Europe have made outstanding contributions to our understanding of the role of protein shape in biological control. Moreover, several possible alternative explanations for some of the regulatory properties of proteins have been put forward.

In particular Monod, Jeffries Wyman and Jean-Pierre Changeux at the Pasteur Institute have proposed an explanation of cooperativity that has quite different features from the induced-fit mechanism proposed above. It can be shown mathematically that their model can account for positive cooperativity and can also explain many features of activation and inhibition by effector molecules. Many features of their model appear to be present in the cooperative binding of oxygen to hemoglobin and in the properties of some enzymes; moreover, it has been of great value in clarifying the properties of regulatory proteins. Their approach cannot apply to all enzymes, however, because negative cooperativity requires an induced-fit model. In addition, Sidney A. Bernhard and his co-workers at the University of Oregon have postulated that in some cases an asymmetry of the type observed in insulin crystals by Dorothy Crowfoot Hodgkin and her colleagues at the University of Oxford may be important in regulatory proteins. In spite of the usefulness of these alternatives in describing some properties of proteins, it would appear that the main mechanism used throughout nature in the control of biological processes is the induced alteration of protein structure I have described.

Triggering Events

I have so far mentioned two types of event that can trigger conformational changes: (1) the reaction of a substrate to form a new covalent bond with an amino acid side chain and (2) the binding of a regulatory molecule (a substrate, an inhibitor or an activator) to the surface of a protein without the formation of a covalent bond. Both of these events occur in the reaction of substrates and in regulatory processes. In the preceding examples I used only the binding of effectors such as GTP to illustrate regulation, but regulation by the formation of covalent bonds can also occur at regulatory sites. Phosphorylation of the enzyme phosphorylase by ATP, for example, can lock the enzyme into a new structure that has different properties of reactivity and sensitivity to regulatory compounds. Earl W. Sutherland, Jr., and his colleagues at the Vanderbilt University School of Medicine showed that this covalent change is controlled by the hormone adrenalin through its mes-

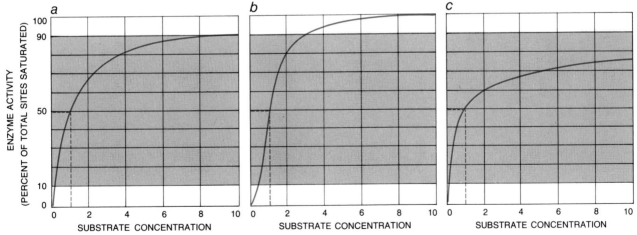

THREE TYPES OF COOPERATIVITY observed in the binding patterns of different enzyme-substrate combinations are represented by these curves, which relate the change in the concentration of a substrate to the change in the level of enzyme activity induced by the substrate. Curve A depicts the normal hyperbolic binding pattern exhibited by most enzymes; in this case an 81-fold increase in concentration of the substrate is needed to go from an activity level of 10 percent to one of 90 percent. Curve B shows the sigmoid, or S-shaped, binding pattern associated with an enzyme that exhibits positive cooperativity (the tendency for the first molecule bound to make it easier for the next molecule to bind, and so forth); in this case only a ninefold change in substrate concentration is needed to go from an activity level of 10 percent to one of 90 percent. Curve C shows the binding pattern associated with an enzyme that exhibits negative cooperativity (the tendency for the first molecule bound to make it more difficult for the second molecule to be bound, and so forth); in this case the curve looks somewhat like a hyperbola, but in reality it is not. The curve approaches the final saturation state so slowly that it does not even reach the 90 percent activity level in this graph. Actually an increase in substrate concentration of 6,541-fold would be needed for such an enzyme to go from an activity level of 10 percent to one of 90 percent. To clarify relations between types of cooperativity all three curves are presented so that the situation at which half of the sites are occupied corresponds to a substrate concentration of 1. Similar curves can be obtained for activator and inhibitor molecules.

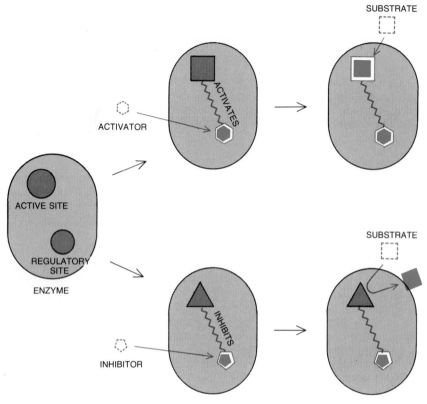

SIMPLIFIED SCHEME is presented here for illustrating the mechanism by which enzyme activity is controlled by the induced conformational changes brought about by small regulatory molecules that are not themselves involved in the primary enzymatic reaction. The complex polypeptide chain that comprises the actual enzyme is represented in this convention by the light gray area; the darker gray circles symbolize the shapes of the enzyme's active site and regulatory site in the absence of any bound molecules. The binding of the activator molecule (*colored hexagon*) induces a conformational change at the regulatory site (*symbolized by the change from a circle to a hexagon*) and also alters the enzyme's structure (*symbolized by the colored zigzag line*), which in turn changes the shape of the active site (*symbolized by the change from a circle to a square*). As a result the substrate (*colored square*) can now bind more easily than it can in the absence of the activator, since no further induced conformational change is needed for binding. In the reverse process the inhibitor molecule (*colored pentagon*) induces a different type of conformational change in the regulatory site (*symbolized by the change from a circle to a pentagon*), which is transmitted through the enzyme's structure, so that the resulting alteration at the active site (*symbolized by the change from a circle to a triangle*) makes the active site repel the substrate.

senger, cyclic AMP. Similar covalent changes are caused by the adenylation of the enzyme glutamine synthetase, as shown by Earl R. Stadtman, Helmut Holzer and their co-workers at the National Heart and Lung Institute. These covalent changes can be used to activate or to inhibit the enzyme. In such instances the reversal to the original enzyme is achieved by a second enzyme, which breaks the covalent bond of the regulatory molecule.

Why should nature use two devices, covalent changes and noncovalent binding, to induce the same kind of shape change for regulation? The answer is not known with certainty, but the suspicion arises that time is one factor that is involved. If an animal is frightened and needs large amounts of energy for a short period of time, it would be de-

sirable to have an override mechanism that would ensure the high activity of some crucial enzymes until the crisis is past. The phosphorylation of phosphorylase induced by adrenalin would seem to provide such a mechanism. The formation of the covalent phosphoryl-enzyme bond converts the enzyme to a more active shape. When the crisis is over, the phosphate group can be removed by a second enzymatic process to regenerate the original, more placid enzymes, but for a time the normal instantaneous controls are eliminated to mobilize glucose for the crisis.

In a similar fashion, conformational changes that occur by the aggregation of subunits may also play an important role in regulation. In many instances it is found that a protein is active as a monomer but inactive as a dimer, or inactive

as a dimer and active as a tetramer. The association of the polypeptide chains with one another causes a shape change with a resulting alteration in the activity of the protein. In essence, then, one subunit becomes the regulator of the other. If that is true, compounds that cause shape changes directly should also induce indirect changes in the polymeric structure of proteins. Indeed, some compounds induce changes in shape that cause them to dissociate. Such association-dissociation reactions appear to be important for many enzymes in metabolic pathways. The types of shape change in these associations and dissociations are exactly analogous to those induced by small molecules. It is also quite possible that certain hormones act as polypeptide regulators in much the same way that small molecules or subunits act in other instances.

Shape changes in proteins can occur very rapidly, in some cases in only a billionth of a second. Other shape changes occur in the millisecond range and have been shown to limit the rate of some enzymatic reactions. Still others take minutes to be effected; they are largely involved in association-dissociation reactions. This range of speeds serves the purposes of regulation well because some processes require an instantaneous correction in response to a stimulus, whereas a slower response is required to prevent an overshoot or to "lock in" a response until a crisis is passed. Furthermore, a slow conformational change has recently been suggested as a memory device in bacterial chemotaxis, in which a bacterium moves in response to a gradient in the concentration of a chemical substance. Such a change may have its counterpart in the neural systems of animals. Shape changes, in short, may occur over many time intervals from very fast to very slow, and each may have its usefulness.

Application to Other Systems

Up to this point I have emphasized the regulatory control of enzymes because they are the regulatory proteins that have been studied the most intensively and they are readily available in the laboratory. As biochemistry has progressed, however, other molecules that have key regulatory roles in biological systems have been isolated. Receptor molecules involved in sensory systems have been shown to be similar to enzymes in terms of structure and binding properties. These molecules have the specificity characteristic of enzymes, and it is generally believed, but not absolutely proved, that induced conformational changes are the signals that trigger the sensory impulse. When we smell or taste a compound, the compound induces a change in the shape of the receptor molecule, which triggers a response in our nervous system. Light induces a change in the shape of a protein in our eyes. Sensory phenomena are the highest form of regulation, and the brain is the ultimate regulatory apparatus of the most complex biological system: man.

Similarly, molecules that control protein synthesis have been isolated. Repressor molecules, for example, bind to DNA and prevent the reading of the genetic message unless they are removed by inducers. Again it is generally believed, although not yet absolutely proved, that the inducer causes a conformational change that peels the repressor molecule off the DNA, thereby initiating protein synthesis. Conformational changes are also believed to be crucial in promoter molecules, such as the cyclic-AMP-binding protein, which binds to DNA and aids in the initiation of protein synthesis. Such initiation and control of the reading of selected portions of the DNA allow different proteins to be made according to circumstances. A control of this type has been postulated to be a mechanism of differentiation, the reason nerve cells have one mixture of proteins and muscle cells have another. Differentiation, the key process that allows multicellular organisms to have specialized functions, is thus also dependent on changes in protein shape.

Antibody molecules, which protect us against invasion by foreign substances, induce a series of reactions in the apparatus called the complement-fixation system. That system is activated to destroy harmful cells and proteins by digesting them. Moreover, the presence of a foreign protein, such as a diphtheria toxin, can induce antibody-producing cells to multiply rapidly so that antibody to that specific harmful agent can be generated. This selective inducing of certain types of protective behavior is another regulatory device of living systems and is again believed to be caused by induced conformational changes of antibody molecules or antibody-receptor molecules. The transport of foodstuffs across cell walls is also thought to be triggered by induced changes in the

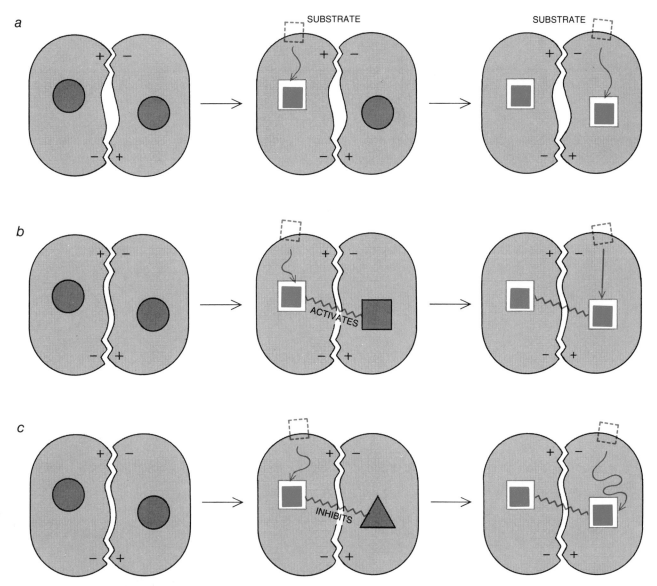

COOPERATIVITY IS EXPLAINED according to the induced-fit model of enzyme action in these diagrams, which use the simplified schematic convention introduced in the illustration on the opposite page. The type of protein represented here is the simplest known multisubunit enzyme: a dimer (that is, a two-peptide polymer) made up of two identical subunits. The three different enzymes shown exhibit respectively noncooperativity (*top row*), positive cooperativity (*middle row*) and negative cooperativity (*bottom row*).

shapes of proteins, as is the contraction of muscle.

There is no single regulatory process in the living system any more than there is a single type of control in a big city. The controls in the city are mediated by traffic lights, judges, payrolls, telephones, foremen, mayors and so on. The controls in biological systems are mediated by enzymes, antibodies, receptors, repressors and so on. Each of these protein molecules is designed differently and therefore each has a special history, giving rise to enormous diversity. Nevertheless, out of all this complexity comes a great simplicity. The flexibility of pro-

teins and the shape changes they can undergo are the key features of regulatory control. Because the protein can exist in more than one shape and because the shape can be altered by external agents, the living system can respond to external stimuli and protect itself against environmental changes.

Nature thus places in the living cell the most powerful catalysts known and tames them to obey commands by subtle shifts in their structure, shifts measured in angstroms. These changes in shape can be imagined in a gross way as the alteration of a glove by the insertion of a hand, but on closer examination they

seem much more like the delicate responses of a spider web designed with exquisite balance and interweaving. The delicate web we call a protein can be altered in shape by subtle perturbations, and through these perturbations functions can be turned on and off. As a result we feel like eating when we need food and lose our appetite when we are full. We can control the use of energy and the growth of specialized tissues, protect ourselves against hostile invading substances and develop a brain. All these essential regulations rest on the ability of a protein molecule to bend flexibly from one shape to another.

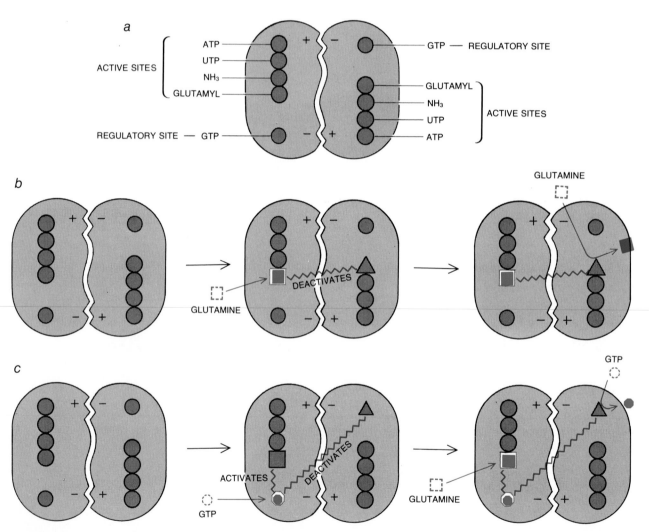

HIGHLY SPECIFIC NATURE of the changes induced in the structure and action of proteins as a result of the binding of selected substrate and regulator molecules is demonstrated in these schematic diagrams, which portray the behavior of the dimer of CTP synthetase, a key enzyme in the metabolism of nucleic acids. Experiments conducted in the author's laboratory have demonstrated the existence of specific binding sites on the surface of this enzyme for four substrate molecules and one regulator molecule (a). When the amino acid glutamine binds to the glutamyl site on one of the enzyme's subunits (b), it forms a glutamyl-enzyme intermediate. The conformational changes induced in this subunit in turn cause changes in the shape of the glutamyl site on the other subunit, mak-

ing it less reactive. Since only one subunit of every dimer can engage in this reaction, the process is sometimes referred to as the "half of the sites" phenomenon. In contrast the binding of the regulatory molecule, GTP (guanosine triphosphate), to one of the subunits induces a conformational change that makes the glutamyl site of the subunit react more readily with glutamine at the same time that it causes the GTP site in the other subunit to assume a new shape that has less affinity for the second GTP molecule (c). In this case the design of the protein enables GTP to act as an activator of a catalytic reaction and an inhibitor of further GTP binding. The binding of both glutamine and GTP causes negligible changes in the shapes of the ATP, UTP or NH₃ (ammonia) sites.

The Three-dimensional Structure of an Enzyme Molecule

by David C. Phillips
November 1966

*The arrangement of atoms in an enzyme molecule
has been worked out for the first time. The enzyme
is lysozyme, which breaks open cells of bacteria.
The study has shown how lysozyme performs its task.*

One day in 1922 Alexander Fleming was suffering from a cold. This is not unusual in London, but Fleming was a most unusual man and he took advantage of the cold in a characteristic way. He allowed a few drops of his nasal mucus to fall on a culture of bacteria he was working with and then put the plate to one side to see what would happen. Imagine his excitement when he discovered some time later that the bacteria near the mucus had dissolved away. For a while he thought his ambition of finding a universal antibiotic had been realized. In a burst of activity he quickly established that the antibacterial action of the mucus was due to the presence in it of an enzyme; he called this substance lysozyme because of its capacity to lyse, or dissolve, the bacterial cells. Lysozyme was soon discovered in many tissues and secretions of the human body, in plants and most plentifully of all in the white of egg. Unfortunately Fleming found that it is not effective against the most harmful bacteria. He had to wait seven years before a strangely similar experiment revealed the existence of a genuinely effective antibiotic: penicillin.

Nevertheless, Fleming's lysozyme has proved a more valuable discovery than he can have expected when its properties were first established. With it, for example, bacterial anatomists have been able to study many details of bacterial structure [see "Fleming's Lysozyme," by Robert F. Acker and S. E. Hartsell; SCIENTIFIC AMERICAN, June, 1960]. It has now turned out that lysozyme is the first enzyme whose three-dimensional structure has been determined and whose properties are understood in atomic detail. Among these properties is the way in which the enzyme combines with the substance on which it acts—a complex sugar in the wall of the bacterial cell.

Like all enzymes, lysozyme is a protein. Its chemical makeup has been established by Pierre Jollès and his colleagues at the University of Paris and by Robert E. Canfield of the Columbia University College of Physicians and Surgeons. They have found that each molecule of lysozyme obtained from egg white consists of a single polypeptide chain of 129 amino acid subunits of 20 different kinds. A peptide bond is formed when two amino acids are joined following the removal of a molecule of water. It is customary to call the portion of the amino acid incorporated into a polypeptide chain a residue, and each residue has its own characteristic side chain. The 129-residue lysozyme molecule is cross-linked in four places by disulfide bridges formed by the combination of sulfur-containing side chains in different parts of the molecule [see illustration on page 143].

The properties of the molecule cannot be understood from its chemical constitution alone; they depend most critically on what parts of the molecule are brought close together in the folded three-dimensional structure. Some form of microscope is needed to examine the structure of the molecule. Fortunately one is effectively provided by the techniques of X-ray crystal-structure analysis pioneered by Sir Lawrence Bragg and his father Sir William Bragg.

The difficulties of examining molecules in atomic detail arise, of course, from the fact that molecules are very small. Within a molecule each atom is usually separated from its neighbor by about 1.5 angstrom units (1.5×10^{-8} centimeter). The lysozyme molecule, which contains some 1,950 atoms, is about 40 angstroms in its largest dimension. The first problem is to find a microscope in which the atoms can be resolved from one another, or seen separately.

The resolving power of a microscope depends fundamentally on the wavelength of the radiation it employs. In general no two objects can be seen separately if they are closer together than about half this wavelength. The shortest wavelength transmitted by optical microscopes (those working in the ultraviolet end of the spectrum) is about 2,000 times longer than the distance between atoms. In order to "see" atoms one must use radiation with a much shorter wavelength: X rays, which have a wavelength closely comparable to interatomic distances. The employment of X rays, however, creates other difficulties: no satisfactory way has yet been found to make lenses or mirrors that will focus them into an image. The problem, then, is the apparently impossible one of designing an X-ray microscope without lenses or mirrors.

Consideration of the diffraction theory of microscope optics, as developed by Ernst Abbe in the latter part of the 19th century, shows that the problem can be solved. Abbe taught us that the formation of an image in the microscope can be regarded as a two-stage process. First, the object under examination scatters the light or other radiation falling on it in all directions, forming a diffraction pattern. This pattern arises because the light waves scattered from different parts of the object combine so as to produce a wave of large or small amplitude in any direction according to whether the waves are in or out of phase—in or out of step— with one another. (This effect is seen most easily in light waves scattered by a regularly repeating structure, such as a diffraction grating made of lines scribed at regular intervals on a glass plate.) In the second stage of image formation, according to Abbe, the objective lens of the microscope collects the diffracted waves and recombines them to form an image of the object. Most important, the nature of the image depends critically on how much of the diffraction pattern is used in its formation.

X-Ray Structure Analysis

In essence X-ray structure analysis makes use of a microscope in which the two stages of image formation have been separated. Since the X rays cannot be focused to form an image directly, the diffraction pattern is recorded and the image is obtained from it by calculation. Historically the method was not developed on the basis of this reasoning, but this way of regarding it (which was first suggested by Lawrence Bragg) brings out its essential features and also introduces the main difficulty of applying it. In recording the intensities of the diffracted waves, instead of focusing them to form an image, some crucial information is lost, namely the phase relations among the various diffracted waves. Without this information the image cannot be formed, and some means of recovering it has to be found. This is the well-known phase problem of X-ray crystallography. It is on the solution of the problem that the utility of the method depends.

The term "X-ray crystallography" reminds us that in practice the method was developed (and is still applied) in the study of single crystals. Crystals suitable for study may contain some 10^{15} identical molecules in a regular array; in effect the molecules in such a crystal diffract the X radiation as though they were a single giant molecule. The crystal acts as a three-dimensional diffraction grating, so that the waves scattered by them are confined to a number of discrete directions. In order to obtain a three-dimensional image of the structure the intensity of the X rays scattered

in these different directions must be measured, the phase problem must be solved somehow and the measurements must be combined by a computer.

The recent successes of this method in the study of protein structures have depended a great deal on the development of electronic computers capable of performing the calculations. They are due most of all, however, to the discovery in 1953, by M. F. Perutz of the Medical Research Council Laboratory of Molecular Biology in Cambridge, that the method of "isomorphous replacement" can be used to solve the phase problem in the study of protein crystals. The method depends on the preparation and study of a series of protein crystals into which additional heavy atoms, such as atoms of uranium, have been introduced without otherwise affecting the crystal structure. The first successes of this method were in the study of sperm-whale myoglobin by John C. Kendrew of the Medical Research Council Laboratory and in Perutz' own study of horse hemoglobin. For their work the two men received the Nobel prize for chemistry in 1962 [see the articles "The Three-dimensional Structure of a Protein Molecule," by John C. Kendrew, beginning on page 25, and "The Hemoglobin Molecule," by M. F. Perutz, which begins immediately following on page 40].

Because the X rays are scattered by the electrons within the molecules, the image calculated from the diffraction pattern reveals the distribution of electrons within the crystal. The electron density is usually calculated at a regular array of points, and the image is made visible by drawing contour lines

through points of equal electron density. If these contour maps are drawn on clear plastic sheets, one can obtain a three-dimensional image by assembling the maps one above the other in a stack. The amount of detail that can be seen in such an image depends on the resolving power of the effective microscope, that is, on its "aperture," or the extent of the diffraction pattern that has been included in the formation of the image. If the waves diffracted through sufficiently high angles are included (corresponding to a large aperture), the atoms appear as individual peaks in the image map. At lower resolution groups of unresolved atoms appear with characteristic shapes by which they can be recognized.

The three-dimensional structure of lysozyme crystallized from the white of hen's egg has been determined in atomic detail with the X-ray method by our group at the Royal Institution in London. This is the laboratory in which Humphry Davy and Michael Faraday made their fundamental discoveries during the 19th century, and in which the X-ray method of structure analysis was developed between the two world wars by the brilliant group of workers led by William Bragg, including J. D. Bernal, Kathleen Lonsdale, W. T. Astbury, J. M. Robertson and many others. Our work on lysozyme was begun in 1960 when Roberto J. Poljak, a visiting worker from Argentina, demonstrated that suitable crystals containing heavy atoms could be prepared. Since then C. C. F. Blake, A. C. T. North, V. R. Sarma, Ruth Fenn, D. F. Koenig, Louise N. Johnson and G. A. Mair have played important roles in the work.

ALA	ALANINE	GLY	GLYCINE	PRO	PROLINE
ARG	ARGININE	HIS	HISTIDINE	SER	SERINE
ASN	ASPARAGINE	ILEU	ISOLEUCINE	THR	THREONINE
ASP	ASPARTIC ACID	LEU	LEUCINE	TRY	TRYPTOPHAN
CYS	CYSTEINE	LYS	LYSINE	TYR	TYROSINE
GLN	GLUTAMINE	MET	METHIONINE	VAL	VALINE
GLU	GLUTAMIC ACID	PHE	PHENYLALANINE		

TWO-DIMENSIONAL MODEL of the lysozyme molecule is shown on the opposite page. Lysozyme is a protein containing 129 amino acid subunits, commonly called residues (*see key to abbreviations above*). These residues from a polypeptide chain that is cross-linked at four places by disulfide (—S—S—) bonds. The amino acid sequence of lysozyme was determined independently by Pierre Jollès and his co-workers at the University of Paris and by Robert E. Canfield of the Columbia University College of Physicians and Surgeons. The three-dimensional structure of the lysozyme molecule has now been established with the help of X-ray crystallography by the author and his colleagues at the Royal Institution in London. A painting of the molecule's three-dimensional structure appears on pages 172 and 173. The function of lysozyme is to split a particular long-chain molecule, a complex sugar, found in the outer membrane of many living cells. Molecules that are acted on by enzymes are known as substrates. The substrate of lysozyme fits into a cleft, or pocket, formed by the three-dimensional structure of the lysozyme molecule. In the two-dimensional model on the opposite page the amino acid residues that line the pocket are shown in dark green.

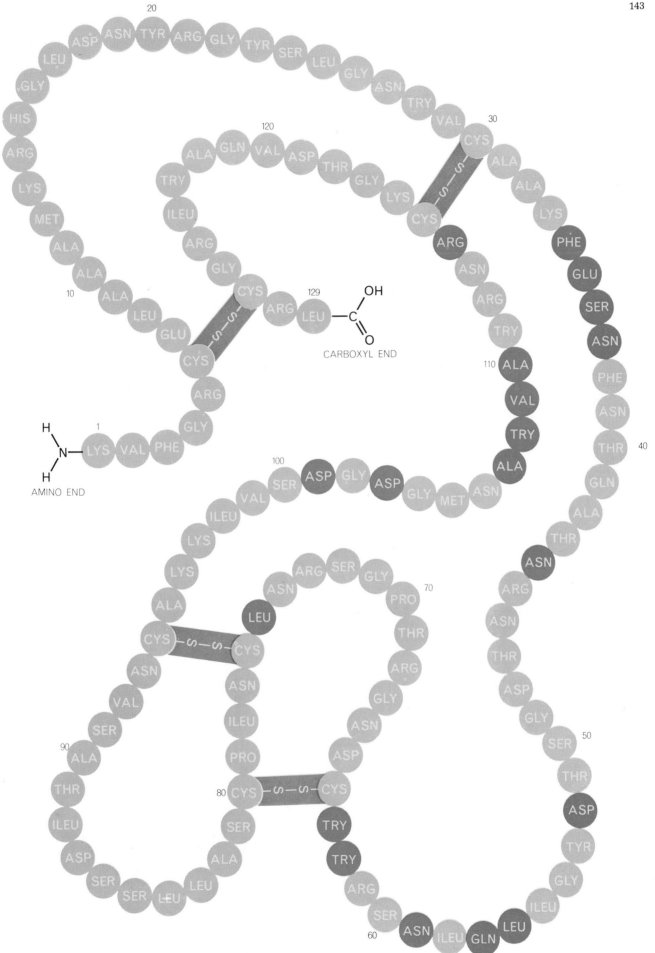

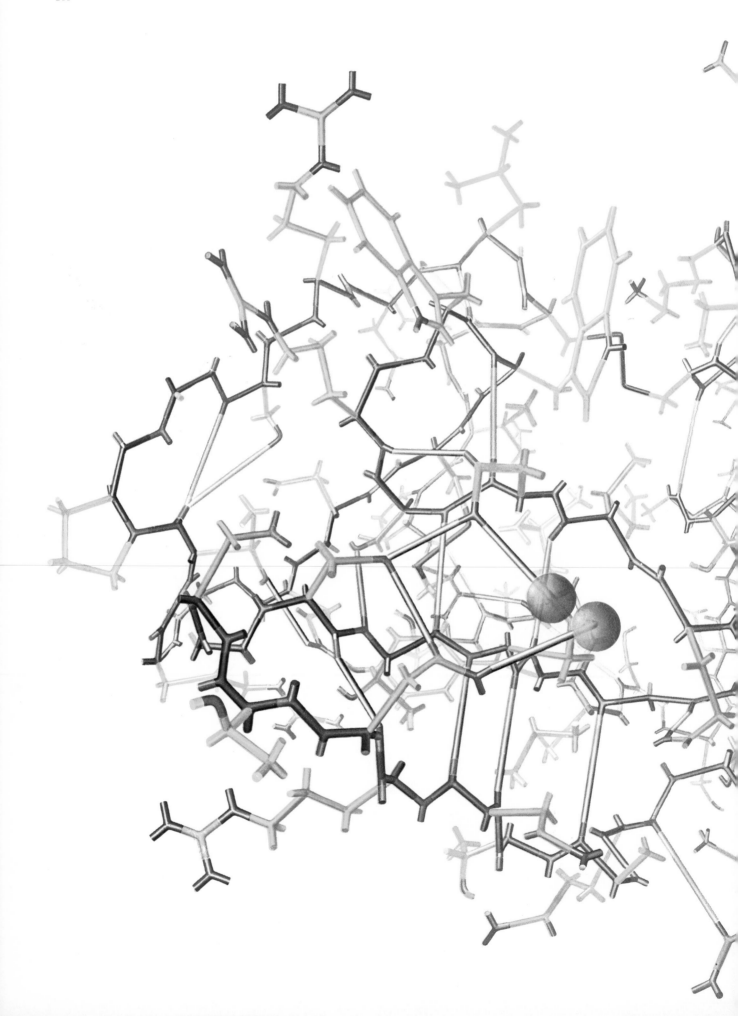

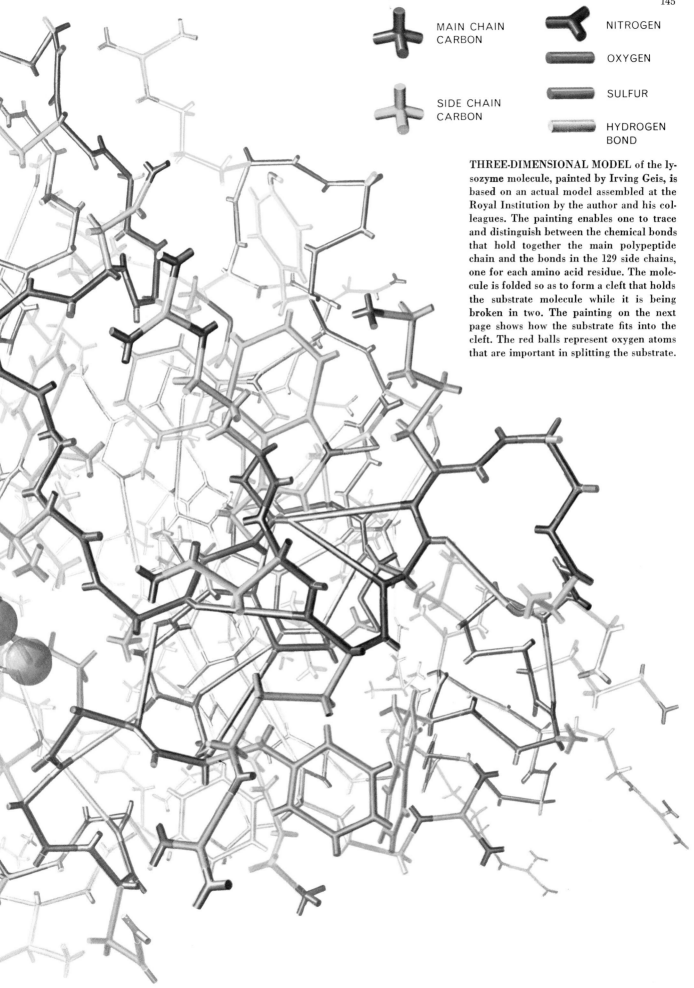

MAIN CHAIN CARBON

SIDE CHAIN CARBON

NITROGEN

OXYGEN

SULFUR

HYDROGEN BOND

THREE-DIMENSIONAL MODEL of the lysozyme molecule, painted by Irving Geis, is based on an actual model assembled at the Royal Institution by the author and his colleagues. The painting enables one to trace and distinguish between the chemical bonds that hold together the main polypeptide chain and the bonds in the 129 side chains, one for each amino acid residue. The molecule is folded so as to form a cleft that holds the substrate molecule while it is being broken in two. The painting on the next page shows how the substrate fits into the cleft. The red balls represent oxygen atoms that are important in splitting the substrate.

In 1962 a low-resolution image of the structure was obtained that revealed the general shape of the molecule and showed that the arrangement of the polypeptide chain is even more complex than it is in myoglobin. This low-resolution image was calculated from the amplitudes of about 400 diffraction maxima measured from native protein crystals and from crystals containing each of three different heavy atoms. In 1965, after the development of more efficient methods of measurement and computation, an image was calculated on the basis of nearly 10,000 diffraction maxima, which resolved features separated by two angstroms. Apart from showing a few well-separated chloride ions, which are present because the lysozyme is crystallized from a solution containing sodium chloride, the two-angstrom image still does not show individual atoms as separate maxima in the electron-density map. The level of resolution is high enough, however, for many of the groups of atoms to be clearly recognizable.

The Lysozyme Molecule

The main polypeptide chain appears as a continuous ribbon of electron density running through the image with regularly spaced promontories on it that are characteristic of the carbonyl groups (CO) that mark each peptide bond. In some regions the chain is folded in ways that are familiar from theoretical studies of polypeptide configurations and from the structure analyses of myoglobin and fibrous proteins such as the keratin of hair. The amino acid residues in lysozyme have now been designated by number; the residues numbered 5 through 15, 24 through 34 and 88 through 96 form three lengths of "alpha helix," the conformation that was proposed by Linus Pauling and Robert B. Corey in 1951 and that was found by Kendrew and his colleagues to be the most common arrangement of the chain in myoglobin. The helixes in lysozyme, however, appear to be somewhat distorted from the "classical" form, in which four atoms (carbon, oxygen, nitrogen and hydrogen) of each peptide group lie in a plane that is parallel to the axis of the alpha helix. In the lysozyme molecule the peptide groups in the helical sections tend to be rotated slightly in such a way that their CO groups point outward from the helix axes and their imino groups (NH) inward.

The amount of rotation varies, being slight in the helix formed by residues 5 through 15 and considerable in the one formed by residues 24 through 34. The effect of the rotation is that each NH group does not point directly at the CO group four residues back along the chain but points instead between the CO groups of the residues three and four back. When the NH group points directly at the CO group four residues back, as it does in the classical alpha helix, it forms with the CO group a hydrogen bond (the weak chemical bond in which a hydrogen atom acts as a bridge). In the lysozyme helixes the hydrogen bond is formed somewhere between two CO groups, giving rise to a structure intermediate between that of an alpha helix and that of a more symmetrical helix with a three-fold symmetry axis that was discussed by Lawrence Bragg, Kendrew and Perutz in 1950. There is a further short length of helix (residues 80 through 85) in which the hydrogen-bonding arrangement is quite close to that in the three-fold helix, and also an isolated turn (residues 119 through 122) of three-fold helix. Furthermore, the peptide at the far end of helix 5 through 15 is in the conformation of the three-fold helix, and the hydrogen bond from its NH group is made to the CO three residues back rather than four.

Partly because of these irregularities in the structure of lysozyme, the proportion of its polypeptide chain in the alpha-helix conformation is difficult to calculate in a meaningful way for comparison with the estimates obtained by other methods, but it is clearly less than half the proportion observed in myoglobin, in which helical regions make up about 75 percent of the chain. The lysozyme molecule does include, however, an example of another regular conformation predicted by Pauling and Corey. This is the "antiparallel pleated sheet," which is believed to be the basic structure of the fibrous protein silk and in which, as the name suggests, two lengths of polypeptide chain run parallel to each other in opposite directions. This structure again is stabilized by hydrogen bonds between the NH and CO groups of the main chain. Residues 41 through 45 and 50 through 54 in the lysozyme molecule form such a structure, with the connecting residues 46 through 49 folded into a hairpin bend between the two lengths of comparatively extended chain. The remainder of the polypeptide chain is folded in irregular ways that have no simple short description.

Even though the level of resolution achieved in our present image was not enough to resolve individual atoms, many of the side chains characteristic of the amino acid residues were readily identifiable from their general shape. The four disulfide bridges, for example, are marked by short rods of high electron density corresponding to the two relatively dense sulfur atoms within them. The six tryptophan residues also were easily recognized by the extended electron density produced by the large double-ring structures in their side chains. Many of the other residues also were easily identifiable, but it was nevertheless most important for the rapid and reliable interpretation of the image that the results of the chemical analysis were already available. With their help more than 95 percent of the atoms in the molecule were readily identified and located within about .25 angstrom.

Further efforts at improving the accuracy with which the atoms have been located is in progress, but an almost complete description of the lysozyme molecule now exists [see illustration on pages 144 and 145]. By studying it and the

MODEL OF SUBSTRATE shows how it fits into the cleft in the lysozyme molecule. All the carbon atoms in the substrate are shown in purple. The portion of the substrate in intimate contact with the underlying enzyme is a polysaccharide chain consisting of six ringlike structures, each a residue of an amino-sugar molecule. The substrate in the model is made up of six identical residues of the amino sugar called N-acetylglucosamine (NAG). In the actual substrate every other residue is an amino sugar known as N-acetylmuramic acid (NAM). The illustration is based on X-ray studies of the way the enzyme is bound to a trisaccharide made of three NAG units, which fills the top of the cleft; the arrangement of NAG units in the bottom of the cleft was worked out with the aid of three-dimensional models. The substrate is held to the enzyme by a complex network of hydrogen bonds. In this style of model-making each straight section of chain represents a bond between atoms. The atoms themselves lie at the intersections and elbows of the structure. Except for the four red balls representing oxygen atoms that are active in splitting the polysaccharide substrate, no attempt is made to represent the electron shells of atoms because they would merge into a solid mass.

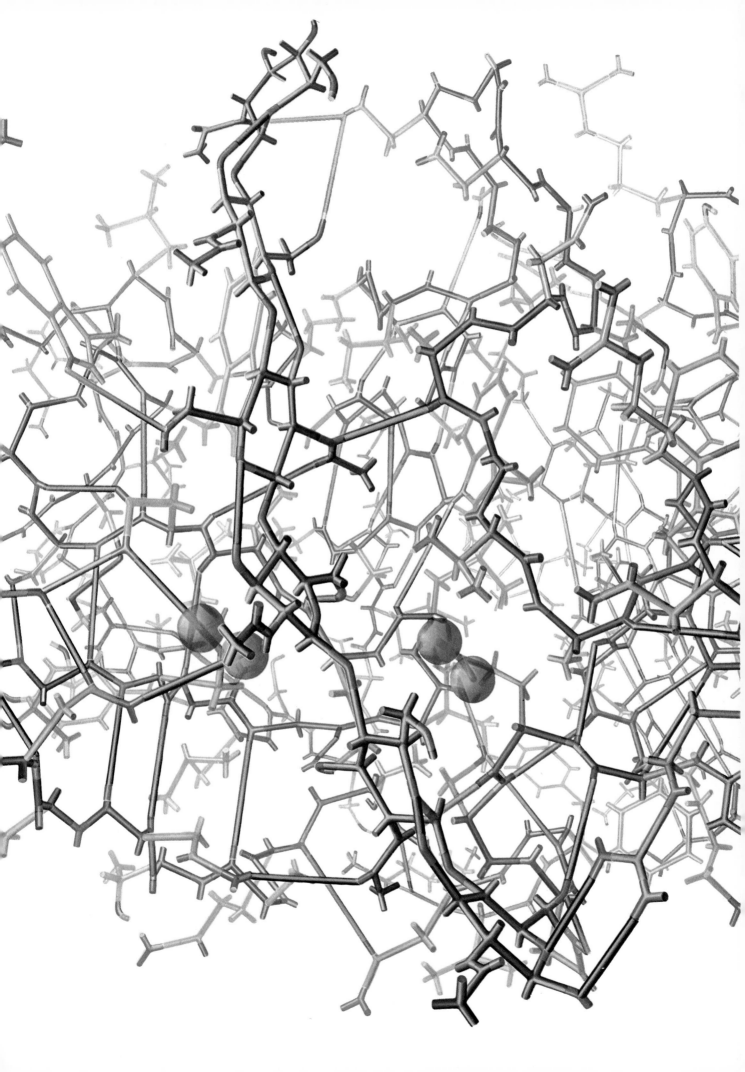

results of some further experiments we can begin to suggest answers to two important questions: How does a molecule such as this one attain its observed conformation? How does it function as an enzyme, or biological catalyst?

Inspection of the lysozyme molecule immediately suggests two generalizations about its conformation that agree

well with those arrived at earlier in the study of myoglobin. It is obvious that certain residues with acidic and basic side chains that ionize, or dissociate, on contact with water are all on the surface of the molecule more or less readily accessible to the surrounding liquid. Such "polar" side chains are hydrophilic—attracted to water; they

are found in aspartic acid and glutamic acid residues and in lysine, arginine and histidine residues, which have basic side groups. On the other hand, most of the markedly nonpolar and hydrophobic side chains (for example those found in leucine and isoleucine residues) are shielded from the surrounding liquid by more polar parts of the mole-

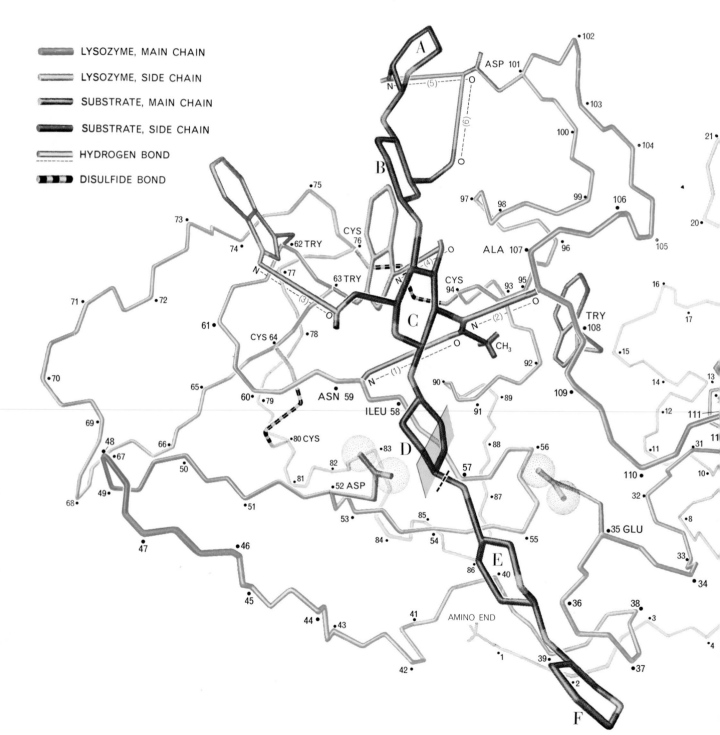

MAP OF LYSOZYME AND SUBSTRATE depicts in color the central chain of each molecule. Side chains have been omitted except for those that produce the four disulfide bonds clipping the lysozyme molecule together and those that supply the terminal con-

nections for hydrogen bonds holding the substrate to the lysozyme. The top three rings of the substrate (A, B, C) are held to the underlying enzyme by six principal hydrogen bonds, which are identified by number to key with the description in the text. The lyso-

cule. In fact, as was predicted by Sir Eric Rideal (who was at one time director of the Royal Institution) and Irving Langmuir, lysozyme, like myoglobin, is quite well described as an oil drop with a polar coat. Here it is important to note that the environment of each molecule in the crystalline state is not significantly different from its natural environment in the living cell. The crystals themselves include a large proportion (some 35 percent by weight) of mostly watery liquid of crystallization. The effect of the surrounding liquid on the protein conformation thus is likely to be much the same in the crystals as it is in solution.

It appears, then, that the observed conformation is preferred because in it

FIRST 56 RESIDUES in lysozyme molecule contain a higher proportion of symmetrically organized regions than does all the rest of the molecule. Residues 5 through 15 and 24 through 34 (*right*) form two regions in which hydrogen bonds (*gray*) hold the residues in a helical configuration close to that of the "classical" alpha helix. Residues 41 through 45 and 50 through 54 (*left*) fold back against each other to form a "pleated sheet," also held together by hydrogen bonds. In addition the hydrogen bond between residues 1 and 40 ties the first 40 residues into a compact structure that may have been folded in this way before the molecule was fully synthesized (*see illustration at the bottom of the next two pages*).

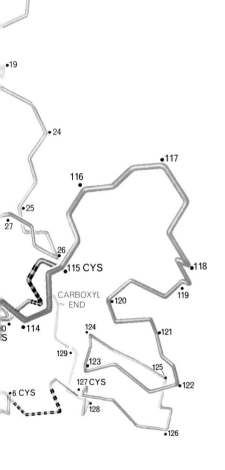

zyme molecule fulfills its function when it cleaves the substrate between the *D* and the *E* ring. Note the distortion of the *D* ring, which pushes four of its atoms into a plane.

the hydrophobic side chains are kept out of contact with the surrounding liquid whereas the polar side chains are generally exposed to it. In this way the system consisting of the protein and the solvent attains a minimum free energy, partly because of the large number of favorable interactions of like groups within the protein molecule and between it and the surrounding liquid, and partly because of the relatively high disorder of the water molecules that are in contact only with other polar groups of atoms.

Guided by these generalizations, many workers are now interested in the possibility of predicting the conformation of a protein molecule from its chemical formula alone [see "Molecular Model-building by Computer," by Cyrus Levinthal; SCIENTIFIC AMERICAN Offprint 1043]. The task of exploring all possible conformations in the search for the one of lowest free energy seems likely, however, to remain beyond the power of any imaginable computer. On a conservative estimate it would be necessary to consider some 10^{129} different conformations for the lysozyme molecule in any general search for the one with minimum free energy. Since this number is far greater than the number of particles in the observable universe,

it is clear that simplifying assumptions will have to be made if calculations of this kind are to succeed.

The Folding of Lysozyme

For some time Peter Dunnill and I have been trying to develop a model of protein-folding that promises to make practicable calculations of the minimum energy conformation and that is, at the same time, qualitatively consistent with the observed structure of myoglobin and lysozyme. This model makes use of our present knowledge of the way in which proteins are synthesized in the living cell. For example, it is well known, from experiments by Howard M. Dintzis and by Christian B. Anfinsen and Robert Canfield, that protein molecules are synthesized from the terminal amino end of their polypeptide chain. The nature of the synthetic mechanism, which involves the intracellular particles called ribosomes working in collaboration with two forms of ribonucleic acid ("messenger" RNA and "transfer" RNA), is increasingly well understood in principle, although the detailed environment of the growing protein chain remains unknown. Nevertheless, it seems a reasonable assumption that, as the synthesis proceeds, the amino end of the chain becomes separated by an increasing distance from the point of attachment to the ribosome, and that the folding of the protein chain to its native conformation begins at this end even before the synthesis is complete. According to our present ideas, parts of the polypeptide chain, particularly those near the terminal amino end, may fold into stable conformations that can still be recognized in the finished molecule and that act as "internal templates," or centers, around which the rest of the chain is folded [see illustration at bottom of these two pages]. It may therefore be useful to look for the stable conformations of parts of the polypeptide chain and to avoid studying all the possible conformations of the whole molecule.

Inspection of the lysozyme molecule provides qualitative support for these ideas [see top illustration on page 149]. The first 40 residues from the terminal amino end form a compact structure (residues 1 and 40 are linked by a hydrogen bond) with a hydrophobic interior and a relatively hydrophilic surface that seems likely to have been folded in this way, or in a simply related way, before the molecule was fully synthesized. It may also be important to observe that this part of the molecule includes more alpha helix than the remainder does.

These first 40 residues include a mixture of hydrophobic and hydrophilic side chains, but the next 14 residues in the sequence are all hydrophilic; it is interesting, and possibly significant, that these are the residues in the antiparallel pleated sheet, which lies out of contact with the globular submolecule formed by the earlier residues. In the light of our model of protein folding the obvious speculation is that there is no incentive to fold these hydrophilic residues in contact with the first part of the chain until the hydrophobic residues 55 (isoleucine) and 56 (leucine) have to be shielded from contact with the surrounding liquid. It seems reasonable to suppose that at this stage residues 41 through 54 fold back on themselves, forming the pleated-sheet structure and burying the hydrophobic side chains in the initial hydrophobic pocket.

Similar considerations appear to govern the folding of the rest of the molecule. In brief, residues 57 through 86 are folded in contact with the pleated-sheet structure so that at this stage of the process—if indeed it follows this course—the folded chain forms a structure with two wings lying at an angle to each other. Residues 86 through 96 form a length of alpha helix, one side of which is predominantly hydrophobic, because of an appropriate alternation of polar and nonpolar residues in that part of the sequence. This helix lies in the gap between the two wings formed by the earlier residues, with its hydrophobic side buried within the molecule. The gap between the two wings is not completely filled by the helix, however; it is transformed into a deep cleft running up one side of the molecule. As we shall see, this cleft forms the active site of the enzyme. The remaining residues are folded around the globular unit formed by the terminal amino end of the polypeptide chain.

This model of protein-folding can be tested in a number of ways, for example by studying the conformation of the first 40 residues in isolation both directly (after removal of the rest of the molecule) and by computation. Ultimately, of course, the model will be regarded as satisfactory only if it helps us to predict how other protein molecules are folded from a knowledge of their chemical structure alone.

The Activity of Lysozyme

In order to understand how lysozyme brings about the dissolution of bacteria we must consider the structure of the bacterial cell wall in some detail. Through the pioneer and independent studies of Karl Meyer and E. B. Chain, followed up by M. R. J. Salton of the University of Manchester and many others, the structures of bacterial cell walls and the effect of lysozyme on them are now quite well known. The important part of the cell wall, as far as lysozyme is concerned, is made up of glucose-like amino-sugar molecules linked together into long polysaccharide chains, which are themselves cross-connected by short lengths of polypeptide chain. This part of each cell wall probably forms one enormous molecule—a "bag-shaped macromolecule," as W. Weidel and H. Pelzer have called it.

The amino-sugar molecules concerned in these polysaccharide structures are of two kinds; each contains an acetamido ($-NH \cdot CO \cdot CH_3$) side group, but one of them contains an additional major group, a lactyl side chain [see illustration at top of opposite page]. One of these amino sugars is known as N-acetylglucosamine (NAG) and the other as N-acetylmuramic acid (NAM). They occur alternately in the polysaccharide chains, being connected by bridges that include an oxygen atom (glycosidic linkages) between carbon atoms 1 and 4 of consecutive sugar rings; this is the same linkage that joins glucose residues in

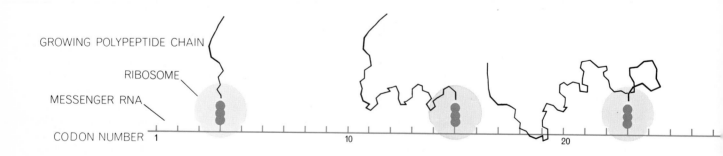

GROWING POLYPEPTIDE CHAIN

RIBOSOME

MESSENGER RNA

CODON NUMBER   1   10   20

FOLDING OF PROTEIN MOLECULE may take place as the growing polypeptide chain is being synthesized by the intracellular particles called ribosomes. The genetic message specifying the amino acid sequence of each protein is coded in "messenger" ribonucleic acid (RNA). It is believed several ribosomes travel simultaneously along this long-chain molecule, reading the message as they go.

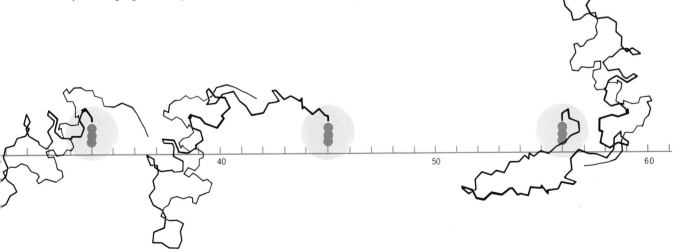

POLYSACCHARIDE MOLECULE found in the walls of certain bacterial cells is the substrate broken by the lysozyme molecule. The polysaccharide consists of alternating residues of two kinds of amino sugar: N-acetylglucosamine (NAG) and N-acetylmuramic acid (NAM). In the length of polysaccharide chain shown here

A, C and E are NAG residues; B, D and F are NAM residues. The inset at left shows the numbering scheme for identifying the principal atoms in each sugar ring. Six rings of the polysaccharide fit into the cleft of the lysozyme molecule, which effects a cleavage between rings D and E (see illustration on pages 148 and 149).

cellulose. The polypeptide chains that cross-connect these polysaccharides are attached to the NAM residues through the lactyl side chain attached to carbon atom 3 in each NAM ring.

Lysozyme has been shown to break the linkages in which carbon 1 in NAM is linked to carbon 4 in NAG but not the other linkages. It has also been shown to break down chitin, another common natural polysaccharide that is found in lobster shell and that contains only NAG.

Ever since the work of Svante Arrhenius of Sweden in the late 19th century enzymes have been thought to work by forming intermediate compounds with their substrates: the substances whose chemical reactions they catalyze. A proper theory of the en-

zyme-substrate complex, which underlies all present thinking about enzyme activity, was clearly propounded by Leonor Michaelis and Maude Menton in a remarkable paper published in 1913. The idea, in its simplest form, is that an enzyme molecule provides a site on its surface to which its substrate molecule can bind in a quite precise way. Reactive groups of atoms in the enzyme then promote the required chemical reaction in the substrate. Our immediate objective, therefore, was to find the structure of a reactive complex between lysozyme and its polysaccharide substrate, in the hope that we would then be able to recognize the active groups of atoms in the enzyme and understand how they function.

Our studies began with the observa-

tion by Martin Wenzel and his colleagues at the Free University of Berlin that the enzyme is prevented from functioning by the presence of NAG itself. This small molecule acts as a competitive inhibitor of the enzyme's activity and, since it is a part of the large substrate molecule normally acted on by the enzyme, it seems likely to do this by binding to the enzyme in the way that part of the substrate does. It prevents the enzyme from working by preventing the substrate from binding to the enzyme. Other simple amino-sugar molecules, including the trisa charide made of three NAG units, behave in the same way. We therefore decided to study the

Presumably the messenger RNA for lysozyme contains 129 "codons," one for each amino acid. Amino acids are delivered to the site of synthesis by molecules of "transfer" RNA (dark color). The

illustration shows how the lysozyme chain would lengthen as a ribosome travels along the messenger RNA molecule. Here, hypothetically, the polypeptide is shown folding directly into its final shape.

binding of these sugar molecules to the lysozyme molecules in our crystals in the hope of learning something about the structure of the enzyme-substrate complex itself.

My colleague Louise Johnson soon found that crystals containing the sugar molecules bound to lysozyme can be prepared very simply by adding the sugar to the solution from which the lysozyme crystals have been grown and in which they are kept suspended. The small molecules diffuse into the protein crystals along the channels filled with water that run through the crystals. Fortunately the resulting change in the crystal structure can be studied quite simply. A useful image of the electron-density changes can be calculated from measurements of the changes in amplitude of the diffracted waves, on the assumption that their phase relations have not changed from those determined for the pure protein crystals. The image shows the difference in electron density between crystals that contain the added sugar molecules and those that do not.

In this way the binding to lysozyme of eight different amino sugars was studied at low resolution (that is, through the measurement of changes in the amplitude of 400 diffracted waves). The results showed that the sugars bind to lysozyme at a number of different places in the cleft of the enzyme. The investigation was hurried on to higher resolution in an attempt to discover the exact nature of the binding. Happily these studies at two-angstrom resolution (which required the measurement of 10,000 diffracted waves) have now

shown in detail how the trisaccharide made of three NAG units is bound to the enzyme.

The trisaccharide fills the top half of the cleft and is bound to the enzyme by a number of interactions, which can be followed with the help of the illustration on pages 148 and 149. In this illustration six important hydrogen bonds, to be described presently, are identified by number. The most critical of these interactions appear to involve the acetamido group of sugar residue C [third from top], whose carbon atom 1 is not linked to another sugar residue. There are hydrogen bonds from the CO group of this side chain to the main-chain NH group of amino acid residue 59 in the enzyme molecule [bond No. 1] and from its NH group to the main-chain CO group of residue 107 (alanine) in the enzyme molecule [bond No. 2]. Its terminal CH$_3$ group makes contact with the side chain of residue 108 (tryptophan). Hydrogen bonds [No. 3 and No. 4] are also formed between two oxygen atoms adjacent to carbon atoms 6 and 3 of sugar residue C and the side chains of residues 62 and 63 (both tryptophan) respectively. Another hydrogen bond [No. 5] is formed between the acetamido side chain of sugar residue A and residue 101 (aspartic acid) in the enzyme molecule. From residue 101 there is a hydrogen bond [No. 6] to the oxygen adjacent to carbon atom 6 of sugar residue B. These polar interactions are supplemented by a large number of nonpolar interactions that are more difficult to summarize briefly. Among the more important nonpolar interactions, however, are those between sugar residue B and the ring system of residue 62; these deserve special mention because they are affected by a small change in the conformation of the enzyme molecule that occurs when the trisaccharide is bound to it. The electron-density map showing the change in electron density when tri-NAG is bound in the protein crystal reveals clearly that parts of the enzyme molecule have moved with respect to one another. These changes in conformation are largely restricted to the part of the enzyme structure to the left of the cleft, which appears to tilt more or less as a whole in such a way as to close the cleft slightly. As a result the side chain of residue 62 moves about .75 angstrom toward the position of sugar residue B. Such changes in enzyme conformation have been discussed for some time, notably by Daniel E. Koshland, Jr., of the University of California at Berkeley,

whose "induced fit" theory of the enzyme-substrate interaction is supported in some degree by this observation in lysozyme.

The Enzyme-Substrate Complex

At this stage in the investigation excitement grew high. Could we tell how the enzyme works? I believe we can. Unfortunately, however, we cannot see this dynamic process in our X-ray images. We have to work out what must happen from our static pictures. First of all it is clear that the complex formed by tri-NAG and the enzyme is not the enzyme-substrate complex involved in catalysis because it is stable. At low concentrations tri-NAG is known to behave as an inhibitor rather than as a substrate that is broken down; clearly we have been looking at the way in which it binds as an inhibitor. It is noticeable, however, that tri-NAG fills only half of the cleft. The possibility emerges that more sugar residues, filling the remainder of the cleft, are required for the formation of a reactive enzyme-substrate complex. The assumption here is that the observed binding of tri-NAG as an inhibitor involves interactions with the enzyme molecule that also play a part in the formation of the functioning enzyme-substrate complex.

Accordingly we have built a model that shows that another three sugar residues can be added to the tri-NAG in such a way that there are satisfactory interactions of the atoms in the proposed substrate and the enzyme. There is only one difficulty: carbon atom 6 and its adjacent oxygen atom in sugar residue D make uncomfortably close contacts with atoms in the enzyme molecule, unless this sugar residue is distorted a little out of its most stable "chair" conformation into a conformation in which carbon atoms 1, 2 and 5 and oxygen atom 5 all lie in a plane [see illustration at left]. Otherwise satisfactory interactions immediately suggest themselves, and the model falls into place.

At this point it seemed reasonable to assume that the model shows the structure of the functioning complex between the enzyme and a hexasaccharide. The next problem was to decide which of the five glycosidic linkages would be broken under the influence of the enzyme. Fortunately evidence was at hand to suggest the answer. As we have seen, the cell-wall polysaccharide includes alternate sugar residues of two kinds, NAG and NAM, and the bond broken is between NAM and NAG. It was

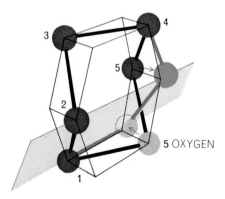

"CHAIR" CONFIGURATION (gray) is that normally assumed by the rings of amino sugar in the polysaccharide substrate. When bound against the lysozyme, however, the D ring is distorted (color) so that carbon atoms 1, 2 and 5 and oxygen atom 5 lie in a plane. The distortion evidently assists in breaking the substrate below the D ring.

therefore important to decide which of the six sugar residues in our model could be NAM, which is the same as NAG except for the lactyl side chain appended to carbon atom 3. The answer was clear-cut. Sugar residue C cannot be NAM because there is no room for this additional group of atoms. Therefore the bond broken must be between sugar residues B and C or D and E. We already knew that the glycosidic linkage between residues B and C is stable when tri-NAG is bound. The conclusion was inescapable: the linkage that must be broken is the one between sugar residues D and E.

Now it was possible to search for the origin of the catalytic activity in the neighborhood of this linkage. Our task was made easier by the fact that John A. Rupley of the University of Arizona had shown that the chemical bond broken under the influence of lysozyme is the one between carbon atom 1 and oxygen in the glycosidic link rather than the link between oxygen and carbon atom 4. The most reactive-looking group of atoms in the vicinity of this bond are the side chains of residue 52 (aspartic acid) and residue 35 (glutamic acid). One of the oxygen atoms of residue 52 is about three angstroms from carbon atom 1 of sugar residue D as well as from the ring oxygen atom 5 of that residue. Residue 35, on the other hand, is about three angstroms from the oxygen in the glycosidic linkage. Furthermore, these two amino acid residues have markedly different environments. Residue 52 has a number of polar neighbors and appears to be involved in a network of hydrogen bonds linking it with residues 46 and 59 (both asparagine) and, through them, with residue 50 (serine). In this environment residue 52 seems likely to give up a terminal hydrogen atom and thus be negatively charged under most conditions, even when it is in a markedly acid solution, whereas residue 35, situated in a nonpolar environment, is likely to retain its terminal hydrogen atom.

A little reflection suggests that the concerted influence of these two amino acid residues, together with a contribution from the distortion to sugar residue D that has already been mentioned, is enough to explain the catalytic activity of lysozyme. The events leading to the rupture of a bacterial cell wall probably take the following course [see illustration on this page].

First, a lysozyme molecule attaches itself to the bacterial cell wall by interacting with six exposed amino-sugar

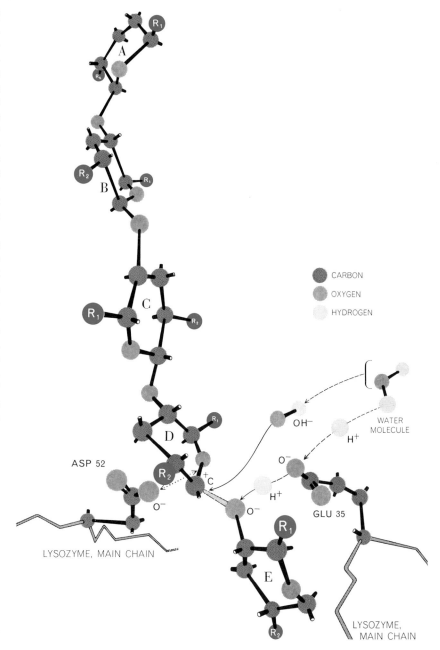

residues. In the process sugar residue D is somewhat distorted from its usual conformation.

Second, residue 35 transfers its terminal hydrogen atom in the form of a hydrogen ion to the glycosidic oxygen, thus bringing about cleavage of the bond between that oxygen and carbon atom 1 of sugar residue D. This creates a positively charged carbonium ion (C^+)

SPLITTING OF SUBSTRATE BY LYSOZYME is believed to involve the proximity and activity of two side chains, residue 35 (glutamic acid) and residue 52 (aspartic acid). It is proposed that a hydrogen ion (H^+) becomes detached from the OH group of residue 35 and attaches itself to the oxygen atom that joins rings D and E, thus breaking the bond between the two rings. This leaves carbon atom 1 of the D ring with a positive charge, in which form it is known as a carbonium ion. It is stabilized in this condition by the negatively charged side chain of residue 52. The surrounding water supplies an OH^- ion to combine with the carbonium ion and an H^+ ion to replace the one lost by residue 35. The two parts of the substrate then fall away, leaving the enzyme free to cleave another polysaccharide chain.

where the oxygen has been severed from carbon atom 1.

Third, this carbonium ion is stabilized by its interaction with the negatively charged aspartic acid side chain of residue 52 until it can combine with a hydroxyl ion (OH^-) that happens to diffuse into position from the surrounding water, thereby completing the reaction. The lysozyme molecule then falls away, leaving behind a punctured bacterial cell wall.

It is not clear from this description that the distortion of sugar residue D plays any part in the reaction, but in fact it probably does so for a very interesting reason. R. H. Lemieux and G. Huber of the National Research Council of Canada showed in 1955 that when a sugar molecule such as NAG incorporates a carbonium ion at the carbon-1 position, it tends to take up the same conformation that is forced on ring D by its interaction with the enzyme molecule. This seems to be an example, therefore, of activation of the substrate by distortion, which has long been a favorite idea of enzymologists. The binding of the substrate to the enzyme itself favors the formation of the carbonium ion in ring D that seems to play an important part in the reaction.

It will be clear from this account that although lysozyme has not been seen in action, we have succeeded in building up a detailed picture of how it may work. There is already a great deal of chemical evidence in agreement with this picture, and as the result of all the work now in progress we can be sure that the activity of Fleming's lysozyme will soon be fully understood. Best of all, it is clear that methods now exist for uncovering the secrets of enzyme action.

15

Enzymes Bound to Artificial Matrixes

by Klaus Mosbach
March 1971

*A new technique imitates the way enzymes are held
in place in the living cell. Besides clarifying cell
mechanisms, enzymes bound to matrixes act as
biocatalysts in industry and offer a new medical tool*

The complex chemistry of the living cell is engineered by thousands of different enzymes, each of which is a catalyst for a particular chemical reaction. It has gradually become clear that the cell could not function if it were simply a little bag filled with enzymes in solution. There is now much evidence that the great majority, and perhaps all, of the intracellular enzymes function either in an environment resembling a gel, or while adsorbed at interfaces, or in solid-state assemblages such as seem to exist in mitochondria and other organelles of the cell. The architectural distribution of enzymes within the cell must be extremely precise; otherwise the different enzymes, the substrates on which they act, the thousands of reaction products and the wide variety of substances that inhibit specific reactions would become chaotically mixed.

Until recently most laboratory investigations of enzymes were conducted with purified extracts in dilute aqueous solution and therefore under conditions far removed from those existing in the living cell. Probably only a few enzymes, the true extracellular enzymes, actually perform their primary biological function under such conditions. Ideally one would like to study the intracellular enzymes in their natural environment by recombining isolated enzymes and the gel-like or matrix environment with which they are normally associated inside the cell, but progress in this direction is beset with many difficulties.

An alternative and much more practical approach is to attach isolated enzymes to mechanically stable artificial matrixes such as hydrophilic (water-loving) polymers. Such systems not only can provide valuable models of how enzymes behave in their natural milieu but also can serve as efficient biological cata-lysts with many practical applications, including medical ones.

It is now 74 years since Eduard Buechner made the discovery, startling in its day, that a cell-free extract from yeast can ferment sugar to alcohol. Since Buechner's time it has been recognized that enzymes, which were subsequently identified as proteins, are the biocatalysts responsible for the myriad reactions that enable living cells to survive and reproduce. In 1926 James B. Sumner succeeded in crystallizing the enzyme urease, thereby demonstrating that enzymes are distinct chemical compounds and hence are amenable to detailed characterization and analysis. Subsequent progress was so rapid that by 1964 the number of enzymes listed by the International Enzyme Commission was close to 900, and by 1968 it was some 1,300. A complete three-dimensional structure has been worked out for about 20 enzymes, and one enzyme (ribonuclease) has been synthesized in the laboratory. With the recognition that enzymes in the living cell are normally attached to surfaces, the term "allotopy" (from the Greek for "other" and "position") has been introduced to describe the differences between the properties of membrane-bound enzymes and the properties of the same enzymes in solution.

In retrospect it is hard to say whether the interest in enzymes affixed to matrixes was originally stimulated by the hope that such systems would provide insight into cellular mechanisms or whether early investigators were attracted primarily by practical applications. In either case the new approach to enzyme technology has in barely half a decade initiated a remarkable volume of work. In one of the earliest references to enzyme-binding, published in 1954,

Nikolaus Grubhofer and Lotte Schleith of the Institute for Virus Research in Heidelberg describe the fixation of pepsin and other enzymes to polyaminostyrene, and they appear interested only in the technological potential of their new technique. Since then workers in many laboratories have developed a variety of artificial matrixes and enzyme-binding methods. Outstanding contributions have been made by the group under Ephraim Katchalsky at the Weizmann Institute of Science, by Jerker Porath, Rolf Axen and their co-workers at the University of Uppsala, and by a large group under Garth Kay and Malcolm Lilly at University College London.

Among the successful matrixes are cross-linked dextran gels (Sephadex), cross-linked acrylic polymers (Biogel), polyamino acids, various kinds of cellulose and even ordinary filter paper and glass. There are three principal methods for binding enzymes to matrixes: ordinary (covalent) chemical linkage, adsorption (which involves the attraction of opposite electric charges) and entrapment of the enzyme within a gel lattice whose pores are large enough to allow the molecules of substrate and product to pass freely but small enough to retain the enzyme [*see illustrations on page 158*]. A less common method is to convert the enzyme molecules themselves into insoluble matrixes by using bifunctional compounds to cross-link them into large aggregates.

Let us look first at some investigations in which enzymes bound to matrixes have been used as model systems for in vivo enzyme reactions taking place in the living cell. One property usually studied in characterizing an enzyme is the dependence of the enzymic reaction on the degree of acidity or alkalinity of

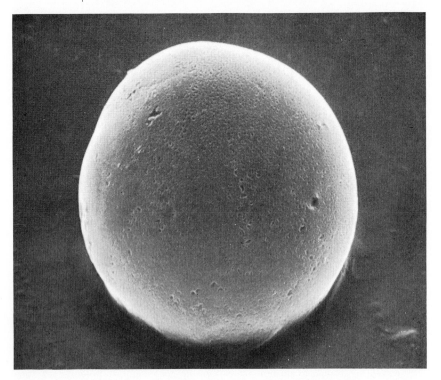

POLYMER BEAD CARRYING BOUND ENZYME looks like this under the scanning electron microscope at a magnification of 700 diameters. The bead is a copolymer of acrylamide and acrylic acid. At this low magnification the enzyme, lactate dehydrogenase, is invisible.

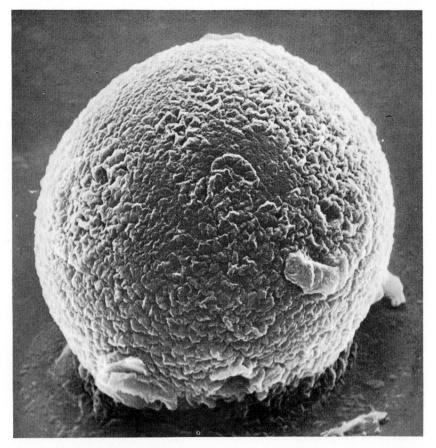

WRINKLED MATERIAL ON BEAD consists of molecules of formazan, the substance produced by the catalytic action of lactate dehydrogenase on lactate, the substrate compound. This micrograph, like the one at the top of the page, is at a magnification of 700 diameters. Both pictures were made by the author's colleague Folke Larsson of the University of Lund.

the solution containing the enzyme and its substrate. Acidity or alkalinity is measured in terms of the solution's hydrogen-ion concentration, expressed on a logarithmic scale as pH. (Values of pH below 7 indicate the high hydrogen-ion concentrations typical of acid solutions; values above 7 indicate the low concentrations typical of alkaline solutions.) This is usually done with the enzyme dissolved in an aqueous solution.

One can now compare the activity at various pH levels of an enzyme bound to a matrix with the activity of a freely dissolved enzyme. Such studies on pH-activity profiles for enzymes have been carried out in particular by Leon Goldstein of Katchalsky's group. Goldstein finds, for example, that when the enzyme trypsin is attached to a negatively charged matrix, the pH-activity profile is markedly shifted toward the alkaline side; the bound enzyme reaches 100 percent of maximum activity at apparently higher alkalinity than the unbound enzyme [see illustration on page 159].

Evidently the negatively charged groups on the polymer matrix attract a thin "film" of positive hydrogen ions, thereby creating a microenvironment for the bound enzyme that has a higher hydrogen-ion concentration (lower pH) than the concentration in the surrounding solution where the pH is actually measured. In other words, the enzyme, bound or unbound, "prefers" to work in the same slightly alkaline environment. When the enzyme is bound to a negatively charged matrix, however, the pH of the macroenvironment of the solution is even more alkaline than the microenvironment of the "working" enzyme.

Another way to demonstrate the effect of the enzyme's microenvironment on the functioning of the enzyme is to measure the dependence of the reaction rate on the concentration of the substrate. For example, one can observe the effect of an enzyme matrix whose charge is opposite to the charge of the substrate. One can predict that the microenvironment next to the matrix will have a higher concentration of substrate than the external solution, in analogy with the hydrogen ions in the preceding example. That is indeed what happens; the enzyme bound to such a matrix works efficiently at what seems to be a low concentration of substrate. (The enzymologist would say that the apparent Michaelis constant, $K_m^{app.}$, is low.)

Such studies emphasize the importance of the microenvironment when one is considering the enzymic processes actually operating in the living cell. As an example, Israel H. Silman of the Weizmann Institute and Arthur Karlin of the Columbia University College of Physicians and Surgeons have found that the optimum *p*H for the activity of acetylcholine esterase in its natural state, that is, membrane-bound, is different from the optimum *p*H observed when the free enzyme is placed in solution. With a knowledge of the work described above they were able to attribute their unexpected findings to local *p*H effects within the cell membrane.

Another area in which model studies can be valuable concerns the investigation of reactions responsible for cell metabolism, for example the sequential steps by which glucose is broken down into carbon dioxide and water with the release of energy. Most of the hundreds of different enzymic reactions that take place within the cell are organized in sequences—often cyclic sequences—in which the product of one reaction serves as the substrate for the next one. Either the participating enzymes must be arranged in proper sequence inside the cell or they must be at least concentrated in certain areas. Two closely related questions immediately arise: What is the effect of the microenvironment on such systems, and how does the distance between individual enzymes in a sequence influence the efficiency of the system?

In an attempt to answer these questions my colleague Bo Mattiasson and I at the University of Lund have recently bound two enzymes, hexokinase and glucose-6-phosphate-dehydrogenase, by covalent linkages to the same polymer-bead matrix [*see bottom illustration on page 159*]. The product from the first enzymic reaction, glucose-6-phosphate, serves as the substrate for the second enzyme. When we compared the efficiency of this matrix-bound two-enzyme system with the efficiency of the same two enzymes unbound but present in a homogeneous solution, we found that in the initial stage the matrix-bound system was twice as effective as the solution system, and that at no stage was it less efficient than the solution system. Our interpretation is that in the matrix system, because of the close proximity of the two enzymes, the product of the first reaction, glucose-6-phosphate, is available in higher concentration for the second enzyme. The proximity of the two enzymes may not be the only contributing factor; when the matrix particles are stirred, they are surrounded by a "cage" of water molecules (the "diffusion layer") that may impede the diffusion of the product of the first reaction into the sur-

rounding medium. Thus the concentration of substrate for the second reaction may be higher in the microenvironment of the matrix than one would infer from measuring the concentration of substrate even a short distance away. A reasonable conclusion is that the rate of enzymic reactions within the living cell is deter-

mined not by the concentration of a substrate in the cell as a whole, or even in a small region of the cell, but by the concentration in the immediate vicinity of the operative enzyme and by other conditions in the microenvironment.

Let us turn our attention now from theoretical matters to some of the practi-

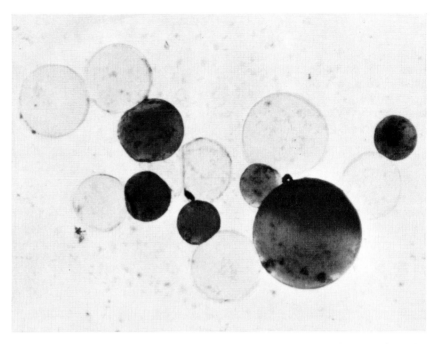

MIXTURE OF COPOLYMER BEADS appears in this micrograph with and without the attached enzyme. The dark beads are copolymers of acrylamide and acrylic acid to which the enzyme lactate dehydrogenase has been bound using a carbodiimide. The light beads lack the enzyme. The enzyme-bound beads are dark (actually blue) owing to the presence of formazan, a blue precipitate formed from the dye tetrazolium blue during the reaction.

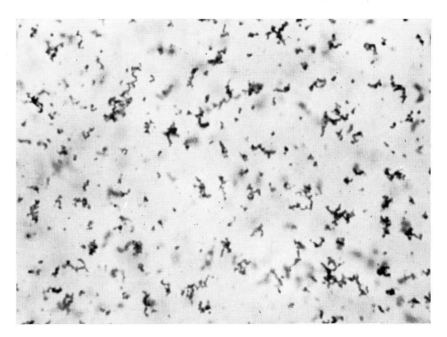

SOLUTION OF UNBOUND ENZYME, lactate dehydrogenase, when incubated with lactate, yields a product (formazan) that is evenly distributed throughout the area of the micrograph. The magnification of 70 diameters is twice that in the micrograph at the top of the page. Both micrographs were made by E. Carlemalm of the University of Lund.

cal applications that have been found for matrix-bound enzymes. In industry a growing number of processes depend on the catalytic activity provided by natural enzymes. Compared with ordinary chemical catalysts, biocatalysts offer the advantage of allowing mild reaction conditions and often the advantage of high specificity. The industrial use of biocatalysts has heretofore been restricted, however, by the high cost of the enzymes themselves and by the difficulty of separating them from the end product. By binding the enzymes to an insoluble matrix these limitations can be circumvented. A column packed with matrix-bound enzymes can be used repeatedly, and the product that emerges is uncontaminated. A further advantage is that enzymes have been reported in a number of cases to be more stable when they are bound to a matrix.

As an illustration of this approach I shall describe a project that Per-Olof Larsson and I have been working on for

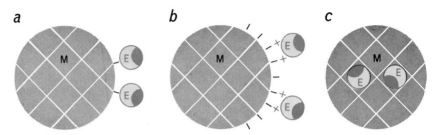

THREE METHODS OF BINDING ENZYMES to matrixes are in general use. The enzyme (E) can be bound to the matrix (M) by direct chemical linkages, known as covalent chemical bonds (a). The enzyme can be held in place by the attraction of unlike electric charges, the phenomenon known as adsorption (b), or the enzyme can be trapped within a gel lattice (c) whose pores are large enough to permit the substrate and the product into which it is converted to enter and leave freely. The colored regions of the enzyme represent active sites of the enzyme molecule, where the enzyme performs its catalytic work on the substrate.

some time in our laboratory in collaboration with the chemical and pharmaceutical firm of CIBA Limited. The problem had its origin some 20 years ago, well before my own involvement, when cortisol was found to be a potent drug in the treatment of rheumatoid arthritis. Cortisol belongs to the family of steroid hormones produced by the cortex of the adrenal gland. In the original process of its synthetic manufacture it was necessary to add to the starting material an oxygen or hydroxyl (OH) group at a particular molecular position designated 11-beta; this had been found essential for physiological activity. The introduction

PREPARATION METHODS are shown for binding enzymes to matrixes or trapping them in a gel. If the matrix for a covalent linkage (a) is a cross-linked dextran gel (Sephadex), adjacent hydroxyl (OH) groups of the gel can react with cyanogen bromide and combine with amino (H_2N-) groups of the enzyme. In b the enzyme has been bound to carboxyl (COOH) groups of the matrix (a cross-linked acrylic copolymer) after treatment with a carbodiimide. (The one here is dicyclohexyl carbodiimide.) In c the enzyme is shown in solution with monomers (1) and after the monomers have been polymerized to form polyacrylamide gel (2).

of functional groups at specific positions in complex molecules such as steroids by conventional chemical means often requires several steps, and it can give rise to undesirable side reactions and poor yields. Later it was discovered that 11-beta hydroxylation yielding cortisol from the cheap starting material Compound S could be achieved biologically in a single step with the aid of a hydroxylase enzyme present in certain fungi.

When two hydrogen atoms are removed from the 1–2 position on the A ring of cortisol, the resulting compound is prednisolone, which is superior to cortisol in relieving rheumatoid arthritis. Again it was found that the most efficient way to carry out the desired reactions, in this case the dehydrogenation of cortisol, is to use another biocatalyst: a dehydrogenase found in bacteria. In the original process the reaction step involved exposure of the substrate (cortisol) to intact microorganisms. There was then the problem of separating the end product from a large mass of bacteria. Our contribution was to isolate the desired enzyme, a steroid dehydrogenase, and trap it in a matrix consisting of a hydrophilic gel. The reaction can now be conducted in a column packed with the trapped enzyme; the flow rate is adjusted so that all the cortisol that enters is transformed to prednisolone.

As an alternative in working with enzymes that are either highly unstable or difficult to isolate we can also use intact cells trapped in a gel. This approach was chosen for handling the fungal cells that carry out the hydroxylation of Compound S to cortisol. In this way we obtained a two-step continuous transformation unit that biocatalytically converts Compound S to cortisol and cortisol to prednisolone [see illustration on opposite page].

One can imagine almost any molecule, including for instance complex antibiotics, being synthesized at least in part by passing a suitable starting material through a battery of enzyme columns, each effecting a single transformation with high efficiency. An example of what can be achieved in this way has recently been provided by Harry D. Brown and his colleagues at Columbia P&S. They arranged within one column in proper order the four different enzymes that participate in the natural breakdown of sugar: hexokinase, phosphogluco-isomerase, phosphofructokinase and aldolase. When glucose is poured into the top of the column, it is transformed in four successive steps into the expected end product: glyceraldehyde-3-phosphate.

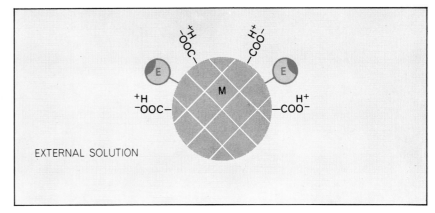

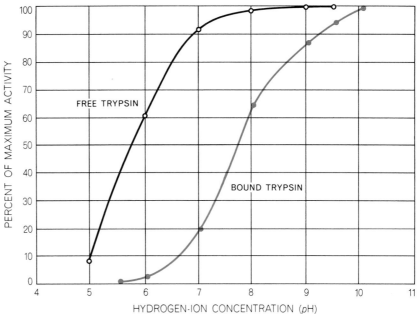

SHIFT IN ACTIVITY PROFILE results when the enzyme trypsin (E) is bound to a negatively charged matrix. As the diagram at the top shows, the matrix raises the local concentration of hydrogen ions (H^+) to a higher level than in the external solution where the concentration is measured. Thus (curves at bottom) the maximum activity of bound trypsin apparently takes place in more alkaline solution than when the trypsin is unbound.

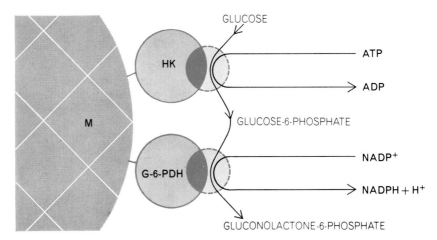

TWO ENZYMES BOUND TO ONE MATRIX prove twice as effective in the initial phase as a system in which the same enzymes are present unbound. The first enzyme, hexokinase (HK), converts glucose to glucose-6-phosphate, consuming energy and a phosphate supplied by adenosine triphosphate (ATP). The second enzyme, glucose-6-phosphate dehydrogenase (G-6-PDH), converts the product of the first reaction to gluconolactone-6-phosphate with the help of the coenzyme nicotinamide adenine dinucleotide phosphate ($NADP^+$).

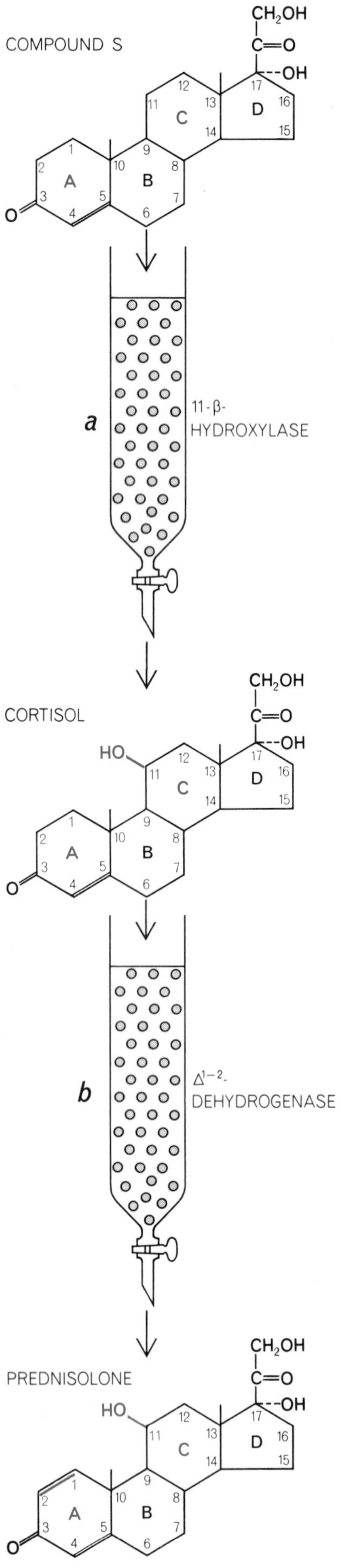

COMPOUND S

CORTISOL

PREDNISOLONE

There are now commercially available several matrix-bound enzymes, most of which break down proteins. It is fair to say that enzyme technology has received an enormous boost through the introduction of techniques for binding enzymes to carriers. The potential of the new technology will be increased still further if the stability of polymer-bound enzymes can be improved, if cheaper ways can be developed to isolate enzymes from biological materials on a large scale and if more progress can be made toward the ultimate goal of routinely synthesizing enzymes on a commercial scale using, for instance, the solid-phase synthesis technique developed by R. B. Merrifield of Rockefeller University.

Biochemical analysis is another area with great inherent potential for matrix-bound enzymes. Enzymic methods are being used increasingly as analytical tools, particularly in medicine. The use of immobilized enzyme systems has the same advantages here as in the applications mentioned above. In addition, interference from proteins present in the solution to be analyzed, a state of affairs that often creates problems in conventional enzyme analysis, can be avoided when the enzyme is securely held within a gel matrix.

Two recent developments will suggest what is being done in this area. The first involves a sensitive test for hydrogen peroxide (H_2O_2) in solution that has been developed by Howard H. Weetall and Norman Weliky of the Jet Propulsion Laboratory. In the presence of the enzyme peroxidase and hydrogen peroxide, a colorless dye is oxidized and immediately turns blue. To provide a convenient test system the enzyme is bound to the cellulose in strips of paper. The test is performed by spotting the strip with the colorless dye and with a small amount of solution suspected of contain-

ing hydrogen peroxide. Depending on the amount of hydrogen peroxide present, more or less of the dye is oxidized. The intensity of color provides a rapid and semiquantitative determination of hydrogen peroxide down to concentrations as low as .03 microgram per milliliter.

The second example is the development of enzyme electrodes by Stuart J. Updike and John W. Hicks of the University of Wisconsin. The electrode is simply covered with a thin polymer film in which the enzyme is trapped. The electrode thus represents a miniaturized chemical transducer that functions by combining an electrochemical procedure with the activity of an immobilized enzyme. George G. Guilbault and Joseph H. Montalvo of Louisiana State University have applied such an enzyme electrode to the measurement of urea in body fluids. In their device the enzyme urease is embedded in a polyacrylamide membrane coated in a layer about .1 millimeter thick on an electrode sensitive to ammonium ions (NH_4^+). In the presence of urease, urea and water react to form ammonium ions and bicarbonate ions (HCO_3^-). The concentration of ammonium ions that builds up at the surface of the electrode yields a direct measure of the urea present in the sample [see illustration at left on next page]. This has a direct analogy in the conventional determination of hydrogen ions with a glass electrode. Enzyme electrodes of this type have operated continuously at room temperature for three weeks without loss of activity.

The clinical potentials of enzyme technology using matrix-bound enzymes are also substantial. More than 120 anomalies and diseases are known that can be regarded as inborn errors of metabolism. Of these a large number represent enzyme-deficiency diseases, in which certain enzymes normally found in the body are either lacking or inactive. One of the best-known enzyme-deficiency diseases is phenylketonuria, which leads to mental retardation. The disease is believed to be caused by the lack of an enzyme that introduces a hydroxyl group into the amino acid phenylalanine, thereby converting it to tyrosine. At present children with the disease are given a costly diet free of phenylalanine.

A more direct approach would be to supply the patient with the missing enzyme. Unfortunately the foreign protein would immediately produce an adverse immunological reaction in the patient receiving it. If the enzyme could be

STEROID TRANSFORMATIONS can be carried out in two simple steps by enzymes bound to matrixes. With conventional chemical procedures the reactions would require many steps and the yields would be low. The starting material, known as Compound S, is a simple steroid obtainable from natural sources. In Step *a* the enzyme 11-beta-hydroxylase bound to a gel matrix inserts a hydroxyl (OH) group at the No. 11 position in the steroid's *C* ring. The product, cortisol, is a close relative of cortisone. In Step *b* the enzyme △1-2-dehydrogenase, also in a gel matrix, removes two hydrogen atoms from cortisol and creates a double bond (*color*) in the *A* ring. The product, prednisolone, is more potent than cortisone or cortisol in treatment of rheumatoid arthritis.

trapped in a gel, however, the reaction could be prevented. The enzymes could be enclosed in tiny semipermeable polymer beads that would allow the substrate to diffuse in and the product to diffuse out. The beads could be introduced directly into the bloodstream of the patient, where they might remain active for a considerable time. Alternatively they might be packed in a shunt chamber connected to the circulatory system. The problems to be solved, however, are not simple. For instance, enzymic processes such as the conversion of phenylalanine to tyrosine often involve the participation of coenzymes, which should be bound together with the enzyme. Recently we have been successful in binding such a coenzyme, nicotinamide adenine dinucleotide, to a polymeric matrix in such a way that its coenzymic activity is retained. I feel certain that this type of approach will be the most promising avenue for treating enzyme-deficiency diseases until such time as genetic engineering—the direct modification of the organism's genetic inheritance—can offer a solution.

Matrix-bound enzymes have been utilized for the construction of a new type of artificial kidney that was tested for the first time last year on a patient with kidney failure. As the agent for removing toxic substances dissolved in the blood, Thomas M. S. Chang of McGill University employed microcapsules consisting of tiny pellets of activated charcoal coated with a thin film of collodion. The microcapsules were packed in a chamber connected to the patient's bloodstream. The system represents a valuable alternative to the bulky and costly dialysis mechanism normally employed as an artificial kidney. The capacity of charcoal to remove toxic substances is, however, rather unspecific. The next step will be to replace the charcoal pellets with encapsulated enzymes selected for their ability to remove specific toxic substances and to leave desirable components of the bloodstream unaffected. As an example, if the enzyme urease were encapsulated, it would convert urea into ammonium ions and bicarbonate ions [*see illustration at right below*]. The rather toxic ammonium ions could then be removed by an ammonia absorbent or by an additional enzyme (such as glutamic dehydrogenase) that incorporates ammonia into organic nitrogen compounds.

Related techniques are being investigated with the general goal of preparing "artificial cells" or important parts of cells. Last year, for instance, Grazia L. Sessa and Charles Weissmann of the New York University School of Medicine incorporated the enzyme lysozyme in small phospholipid spheres, thereby producing a liposome: an artificial organelle with some of the properties of the organelle called the lysosome.

The general concept of binding biological material to matrixes has also had a major impact on the development of a purification technique known as affinity chromatography. Let me illustrate this technique with the following example. It is known that for most enzymes there exist specific inhibitors that function by blocking the active site of the enzyme. As a rule the complex of enzyme plus inhibitor is readily formed and just as readily dissociated.

In the course of a certain investigation Hans von Fritz and his co-workers at the University of Munich were faced with the problem of separating two enzyme inhibitors of about the same molecular weight present in the pancreas. One was nonspecific in that it inhibited both alpha-chymotrypsin and trypsin,

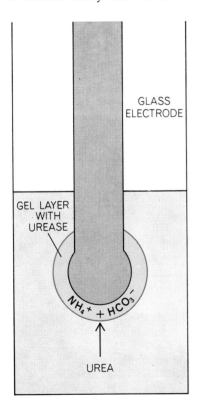

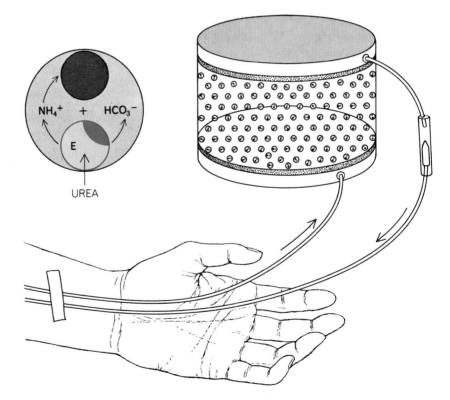

MEASUREMENT OF UREA in body fluids can be accomplished by coating the end of a glass electrode with a thin gel layer, about .1 millimeter thick, to which the enzyme urease is kept bound. The enzyme catalyzes the reaction of urea and water into ammonium ions (NH_4^+) and bicarbonate ions (HCO_3^-). The potential of the electrode is altered by the buildup of ammonium ions, providing a direct measure of the amount of urea present in the sample.

REMOVAL OF UREA from body fluids could be achieved by a new kind of artificial kidney consisting of a vessel filled with microcapsules containing the enzyme urease. The enzyme would convert urea and water into ammonium and bicarbonate ions. The microcapsules would also have to contain either an ammonia absorbent or an additional enzyme for the removal of ammonium ions. Charcoal has been used as an absorbent in clinical trials.

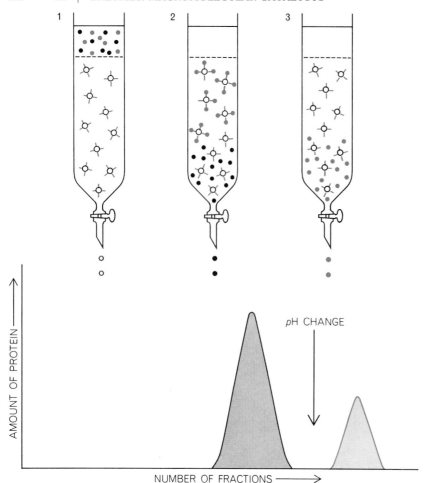

AFFINITY CHROMATOGRAPHY uses enzymes or other biological materials bound to polymer beads to separate one or more substances from a complex mixture. In this example the polymer beads carry an inhibitor that forms a complex with the enzyme alpha-chymotrypsin. In Step *1* a solution containing a small amount of the enzyme (*color*) together with several other enzymes (*black*) is poured into the column. In Step *2* alpha-chymotrypsin is retained on the beads while the other enzymes pass through freely. The liquid drawn from the column at this step will contain all the enzymes except alpha-chymotrypsin (*curves at bottom*). In Step *3* a slightly acid solution releases the alpha-chymotrypsin.

two enzymes that cleave proteins. The other inhibitor was specific for trypsin. To separate the two the Munich workers employed a column packed with alpha-chymotrypsin bound to a matrix. When a solution containing the two inhibitors was passed through the column, the non-specific inhibitor was retained as the inhibitor specific for trypsin passed through. Later the nonspecific inhibitor could be freed from the column with an agent that dissociated the enzyme-inhibitor complex.

One can readily see that with a reverse procedure it should be possible to pack a column with a matrix-bound inhibitor capable of removing a specific enzyme from a complex mixture of enzymes. This was first accomplished in 1953 by Leonard S. Lerman of the University of Chicago. Since then affinity chromatography has been extensively developed in several laboratories, particularly by Christian B. Anfinsen and his co-workers Pedro Cuatrecasas and Meir Wilchek at the National Institute of Arthritis and Metabolic Diseases. In one example of interest they used a covalently bound inhibitor to form a complex with alpha-chymotrypsin and thus remove it from a mixture. They discovered, however, that when the inhibitor was directly attached to the matrix (a Sepharose gel), it was not effective. In subsequent experiments they found that the inhibitor would function if the distance between the matrix and the inhibitor was extended by the insertion of a short carbon chain about seven angstroms long. Apparently this made the active site on the enzyme, the molecular diameter of which is about 40 angstroms, more accessible to the inhibitor [*see top illustration at left*].

Affinity chromatography has also shown wide usefulness in the field of immunology, where it is frequently important to select one particular type of antibody molecule out of a highly complex mixture of these key molecules of the immune system [see "The Structure and Function of Antibodies," by Gerald M. Edelman; SCIENTIFIC AMERICAN Offprint 1185]. This is accomplished quite simply by using matrixes carrying the antigen that corresponds specifically to the desired antibody. Conversely, matrix-bound antibodies can be employed to isolate their specific antigens [*see bottom illustration at left*]. Affinity chromatography thus adds to the arsenal of different purification methods available to the biologist a new and sophisticated procedure allowing the tailor-made design of materials for the separation of biological molecules.

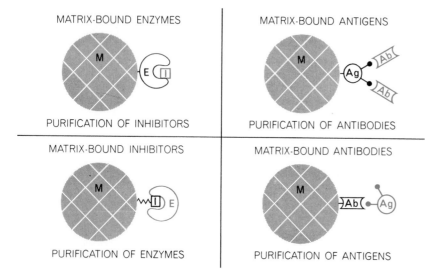

VERSATILITY OF AFFINITY CHROMATOGRAPHY is depicted schematically. In principle one can extract any substance from a complex mixture by attaching the right "grappling hook" to a suitable matrix. In these diagrams an enzyme, an antigen, an enzyme inhibitor and an antibody are used to capture substances for which they have a natural affinity.

IV

METHODS TO
CHARACTERIZE
MACROMOLECULES

IV

METHODS TO CHARACTERIZE MACROMOLECULES

INTRODUCTION

B iophysical methods have developed at the same time that fundamental relationships between the physical and biological sciences have been demonstrated, and in considerable measure these methods have helped bridge the two sciences. Numerous techniques that were once virtually the exclusive province of the analytical chemist, the physical chemist, and even the experimental physicist have been "borrowed" by the biophysical scientist to study biological problems. At the same time, techniques developed by the biophysical scientist have been used by workers in organic chemistry and biochemistry to complement their own traditional methods.

Because of the extraordinary complexity of many of the significant problems in biophysical chemistry, only rarely can a single experimental method satisfactorily solve a given problem. Often, confidence in a proposed solution to a problem depends not only on the choice of experimental techniques, but also the variety of techniques used. The most compelling arguments are usually those based upon the largest and most diverse base of experimental data; the prodigious breakthrough based upon "one crucial experiment" is more often than not an idealized fiction.

As is evident in articles in the preceding sections, much of biophysical chemistry is concerned with quite complicated macromolecules. Inevitably, therefore, much experimental methodology in biophysical chemistry is drawn from research in polymers. One of the early giants of polymer physical chemistry, as well as many other areas of physical science, was the late Peter J. W. Debye, who played a large role in developing many of the physical characterization methods described in his article "How Giant Molecules are Measured." He provides valuable insights into some fundamental differences between the physical chemistry of high polymers and that of ordinary small molecules in solution. Among the more important differences is imprecision in the definition of molecular weight and the inapplicability of the Law of Definite Proportions to many polymer systems. The heterogeneity of molecular size and weight in a polymer system requires that a single molecule be selected as a statistical representative or "average" of the entire sample. Different physical methods may lead to different definitions for the average of a measured property in a polymer system. It is extremely important, therefore, that the type of average be clearly specified in order to avoid ambiguity in the comparison and interpretation of results obtained for various samples and for various types of polymers. In general, different averages are the result of differences in the way the various physical characterization methods "weight" the counting of molecules.

In his article "How Giant Molecules are Measured," Debye describes two of the so-called "hydrodynamic" methods of characterizing polymers: viscosity and the velocity of sedimentation of macromolecules in the high centrifugal field of the ultracentrifuge. (For some amplification of the latter phenomenon, see the articles by George W. Gray, "The Ultracentrifuge," p. 176, and Gerald Oster, "Density Gradients," p. 186.) These methods are based upon the motion of macromolecules through a fluid, and owing to the considerable uncertainties regarding the friction of the fluid past the polymer particles, they are *relative* methods. That is, the theoretical aspects of the flow problem are not in general sufficiently understood to permit an absolute calculation of molecular size, weight, or conformation from the observed property; instead, measurement is calibrated in terms of a known and well-characterized standard. Nevertheless, because of their comparative simplicity and the wealth of information that can frequently be obtained from them, these methods are widely used to characterize macromolecules in solution.

Two other widely used physical techniques for polymer characterization described by Debye are the *absolute* methods of osmotic pressure and ordinary scattering of light without absorption by polymer molecules, or Rayleigh scattering—the latter is one to which Debye himself has made major contributions. The first of these is based upon the laws of thermodynamics, which predict that the osmotic pressure is simply proportional to the number of molecules present; this method is therefore a direct way of counting the number of molecules in a sample. The second method, light scattering, is somewhat less simple than osmotic pressure in interpretation, but has a wider range of application and yields more diverse information on a polymer system in a single experiment. These two absolute characterization methods are also widely used by biophysical chemists to characterize and investigate complicated biological macromolecules in solution.

Out of the vast array of techniques available to the biophysical scientist in his attack upon problems of biopolymer structure and function, almost certainly the one most widely used throughout the broad range of biophysics, biochemistry, and molecular biology is ultracentrifugation. If you were to ask biophysical researchers, "If you were stranded on a desert island with access to but a single technique, which one would it be?" the answer would almost invariably be the ultracentrifuge.

The reasons for the widespread use of the ultracentrifuge are many and diverse: sensitive and highly engineered instrumentation is commercially available and accessible to virtually every laboratory and investigator; the technique has a basic simplicity that lends itself well to molecular interpretation of observed phenomena; it can characterize dissolved particles either in terms of their sedimentation velocity or in terms of their distribution at equilibrium between the forces of sedimentation and diffusion. The ultracentrifuge is especially well adapted to the study of size-, shape-, or even species-heterogeneous systems. But perhaps most important is the extraordinary versatility of the method and thus its applicability to a rich variety of systems, ranging from cell cultures, bacteria, and virus particles, to polymeric and even nonpolymeric molecules of biological interest.

In "The Ultracentrifuge" George W. Gray traces the development of the theory and practice of centrifugation and describes some of its basic principles. From the earliest practical ultracentrifuges of the Swedish chemist Thé Svedberg to the elegant and sophisticated instruments designed by Jesse W. Beams at the University of Virginia, the evolution of the instrumentation, technique, and theory of centrifugation has proceeded at a relatively steady pace under the basic guidance of a few extraordinary individuals. At the same time, applications of the ultracentrifuge have multiplied rapidly in number and variety; at present ultracentrifugation contributes significantly to many areas of science and technology. In the article "Density Gradients," Gerald Oster describes the special technique of density-gradient centrifugation in considerable detail, as well as additional methods of preparation and uses of density gradients. The technique of density-gradient centrifugation was first developed by Meselson, Stahl, and Vinograd for their well-known experiment to test the replication hypothesis advanced for DNA by James Watson and Francis Crick. It now constitutes one of the most widespread uses of the ultracentrifuge in biology and biological chemistry. In the future, ultracentrifugation and ultracentrifugation-related techniques are likely to have an even greater impact than they have had so far. Furthermore, future advances are likely to be increasingly tied to the development and availability of more advanced instrumentation for equilibrium and density-gradient centrifugation, so that studies can be performed more readily on very large macromolecular and particulate systems.

The next two articles in this section describe separation methods of great generality that are widely used throughout chemistry and biochemistry; in innumerable instances these methods have played quite central roles in the solutions of important problems. In the first of these articles, "Chromatography," William H. Stein and Stanford Moore narrate the history and outline the vast collection of analytical and preparative separation techniques that are now traditionally grouped under chromatography. As the name implies, chromatography originated with the separation of pigments or color. Today, however, the "chromatographic" method includes an enormous variety of gaseous or liquid separation processes based on physical and chemical interactions with substrates of different kinds, including permeation through paper and through a number of gels. Chromatographic methods have played a significant role in many of the great recent advances in biochemistry and molecular biology. A well-known example is the separation and analysis of amino acids and polypeptides, leading to the determination of amino acid sequences in proteins and the ultimate resolution of the genetic code.

A second separatory-analytical method of enormous importance to biology and biochemistry is quite closely related to chromatographic processes, but unlike these, uses an applied electric field to provide the "driving force" for the physical separation process. The technique is called "electrophoresis," and in the article of the same name George W. Gray discusses the history and development, the conceptual framework, and the many uses of this powerful tool. Like chromatography, electrophoresis has been widely used in both preparative and analytical modes, and has used many of the same substrates. Because many of the molecular species of paramount interest in biology and medicine are polyions, electrophoresis has been particularly useful in these areas.

In "Relaxation Methods in Chemistry," Larry Faller describes a group of techniques developed comparatively recently in connection with the study of rates and mechanisms of extremely rapid chemical and physical processes. All of these techniques involve observing the return to equilibrium of a system that has been perturbed by an abrupt change in its chemical or physical state. The development of measurement techniques to accommodate the short time scales typically involved in relaxation processes has opened an entirely new world of rapid-rate processes to chemists and biochemists.

The relaxation methods most commonly used in chemistry and biochemistry involve the rapid mixing of chemical reactants—the so-called "stopped-flow" techniques—and rapid perturbations in temperature and pressure. As is well known to the physical chemist, changes in temperature and pressure usually cause a shift in chemical equilibrium—that is, a change in the relative balance of reactants and products in a chemical reaction. The rates, and often the mechanisms, of chemical reactions can be determined from the time required for the reestablishment of an equilibrium state following rapid perturbation.

The usefulness of relaxation methods lies not only in the wide range of rates that can be investigated, but also in the fact that they greatly facilitate separation of complex reacting systems into component "elementary" reactions.

These advantages have been illustrated and utilized perhaps most dramatically in studies of such biochemical processes as enzyme-catalyzed reactions (see the articles in Section III on enzyme catalysis). Enzymatic processes are enormously complicated, and in order to understand enzyme function in relation to structure, a detailed mechanistic understanding of the reaction is required. Clearly, the ability of relaxation methods to follow short-time processes and resolve complexities of mechanism in enzyme systems is important, but of even greater significance is the ability of these methods to focus directly upon reaction intermediates and to help illuminate so-called "cooperative" enzyme-substrate binding.

The variety and complexity of the experimental methods used in biophysical chemistry nearly rival the variety and complexity of the problems themselves. Although the articles that follow describe only some of these methods, the methods described here have an unusual range of applicability and potential for use in unsolved problems in biochemistry and molecular biology. For those who wish to read beyond the articles included here, we recommend the following *Scientific American* articles: "A High-Resolution Scanning Electron Microscope" by Albert V. Crewe, April, 1971; "Magnetic Resonance" by George E. Pake, August, 1958 (Offprint 233); "The Mass Spectrometer" by Alfred O. C. Nier, March, 1953 (Offprint 256); "X-Ray Crystallography" by Sir Lawrence Bragg, July, 1968 (Offprint 325); "Ion Exchange" by Harold F. Walton, November, 1950; "Immuno-Electrophoresis" by Curtis A. Williams, Jr., March, 1960 (Offprint 84); "Gas Chromatography" by Roy A. Keller, October, 1961 (Offprint 276); "Molecular Sieves" by D. W. Breck and J. V. Smith, January, 1959; "The Scanning Electron Microscope" by Thomas E. Everhart and Thomas L. Hayes, January, 1972; "Flash Photolysis" by Leonard A. Grossweiner, May, 1960; "Chemical Analysis by Infrared" by Bryce Crawford, Jr., October, 1953 (Offprint 257); "Light Scattered by Particles" by Victor K. LaMer and Milton Kerker, February, 1953; "Mössbauer Spectroscopy" by R. H. Herber, October, 1971; "Ultrahigh-Speed Rotation" by Jesse W. Beams, April, 1961; "Neutron-Activation Analysis" by Henry W. Kramer and Werner H. Wahl, April, 1967 (Offprint 310); "Laser Spectroscopy" by M. S. Feld and V. S. Letokhov, December, 1973.

16

How Giant Molecules
Are Measured

by Peter J. W. Debye
September 1957

When light passes through such molecules in solution, some of it is scattered. From this and other effects it is possible to determine the weight of the molecules and the degree to which they are folded

The first requirement in attempting to understand the properties of giant molecules is to determine their size and form. Indeed it is the great size of these molecules, resulting in very complicated structures, that is primarily responsible for their extraordinary be-

havior. A chemist seeking to describe or identify a high polymer therefore begins by measuring its molecular weight.

It is not a simple task. Most of the customary chemical or physical procedures for weighing small molecules do not avail with the macromolecules. They re-

quire special methods. We shall consider in this article the principal techniques that have been employed.

First, what do we mean by the molecular weight of a substance? Consider glucose, a sugar molecule. Its formula is $C_6H_{12}O_6$, meaning that it has six carbon

SCATTERED LIGHT makes a vertical beam of light visible when viewed from the side. Even pure liquids scatter a certain amount of light, but the scattering is greatly increased when they contain dissolved polymer molecules, as in the experiment shown here.

atoms, 12 hydrogen and six oxygen. The atomic weight of a hydrogen atom is taken as 1, of carbon 12 and of oxygen 16. Hence the atoms composing glucose add up to a weight of 180 units for the molecule. The molecular weight of glucose, as of most other monomers, is known precisely. It is also known that the giant molecule of cellulose is made up of many glucose molecules linked together. But how many? The number runs into the thousands. The task of the polymer investigator is to weigh molecules composed of unnumbered thousands of monomers, with total molecular weights running into the hundreds of thousands or the millions.

A method of estimating the weight of giant molecules was worked out many years ago by Hermann Staudinger of Germany, a pioneer in the chemistry of high polymers. It is based upon the effect of the molecules on the viscosity of a liquid in which they are dissolved. Let us suppose that we have measured the viscosity of the solvent by its rate of flow in a capillary tube. If we now dissolve a substance in the liquid, the dissolved molecules, acting like mud particles suspended in a stream, will increase the viscosity of the fluid. Albert Einstein, who first studied this problem theoretically, calculated how much the viscosity would increase if a number of solid little spheres were suspended in a liquid and drifted along with it, turning and tumbling in the stream. Staudinger derived a formula for molecules in solution, relating the percentage of increase in the viscosity of the fluid to the molecular weight of the dissolved molecules. He used this rule to calculate molecular weights.

However, reliable measurements by other methods later showed that the relative increase of viscosity is not directly proportional to molecular weight in the case of large molecules. The actual situation is that, at a rough approximation, when the weight of the dissolved molecules is quadrupled, the viscosity only doubles. We can account for this theoretically by assuming that with increasing molecular weight molecules capable of coiling up do not, on the average, enlarge in diameter in direct proportion to their gain in weight. Furthermore, other complications arise, having to do with interactions between parts of the molecular chain. We cannot say yet that viscosity measurements are a reliable guide to the molecular weight of large molecules of unknown structure. Nevertheless, they are useful for estimat-

ing the weight of a molecule structured like one whose weight has already been determined, and also for indicating whether a molecule is tightly coiled or spread out. Viscometers are standard equipment in every high-polymer laboratory.

A second instrument for weighing large molecules, invented by Thé Svedberg of Sweden, is the ultracentrifuge, which has become an invaluable aid in studying proteins and other biological substances. The weight of the molecules is calculated from the rate at which they move out to the rim of the whirling centrifuge—the larger the molecule, the faster it migrates. However, because of the unknown magnitude of frictional effects and the fact that very large molecules tend to get in one another's way, the ultracentrifuge becomes ineffective as a weighing instrument for giant molecules.

If we wish to obtain a precise measurement of molecular weight, the most straightforward procedure is to take a weighed amount of a substance, count the number of molecules in the sample, and then compute the average weight of the molecules. This can be done in two ways, and these are the principal means employed for the measurement of macromolecules.

As we learned many years ago from J. H. van't Hoff of the Netherlands, if we dissolve a measured amount of a substance in a given volume of a solvent, we can count the number of dissolved molecules by measuring the osmotic pressure of the solution. Recall the familiar demonstration of osmotic pressure in high-school chemistry class. Into a beaker of pure water is lowered a glass tube containing a certain amount of sugar solution. The bottom end of the tube is covered with a membrane which lets water pass through but not sugar molecules. Water from the beaker slowly infiltrates into the tube (as it always does from a less concentrated to a more concentrated solution), and the level of the liquid in the tube rises [see diagram at right]. Eventually it stops rising, and the height of the column of solution in the tube then represents the osmotic pressure—i.e., the pressure necessary to counteract the tendency of water to move into the tube.

In a solution dilute enough to avoid any appreciable interactions among the sugar molecules the osmotic pressure is proportional to the number of sugar molecules per cubic centimeter of solution. Thus the height of the liquid in the

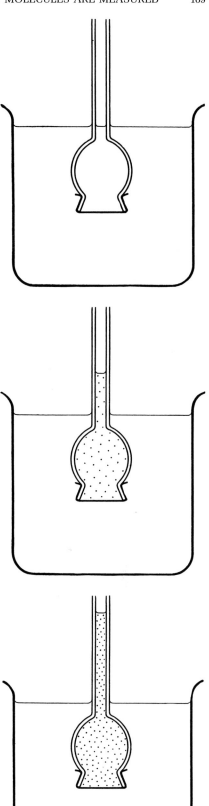

OSMOTIC PRESSURE apparatus is a tube sealed at bottom with semipermeable membrane. At top tube contains pure solvent. Dots in lower drawings are large molecules.

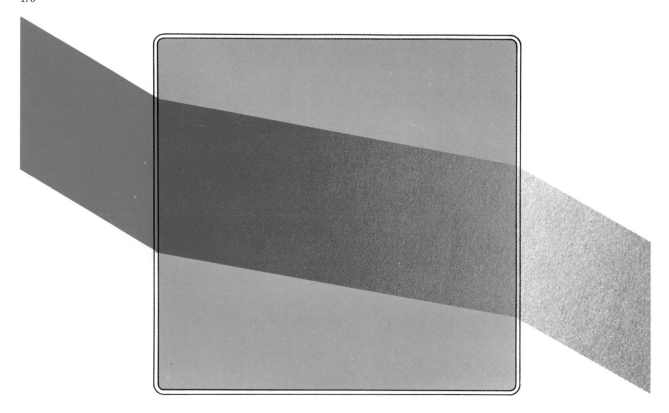

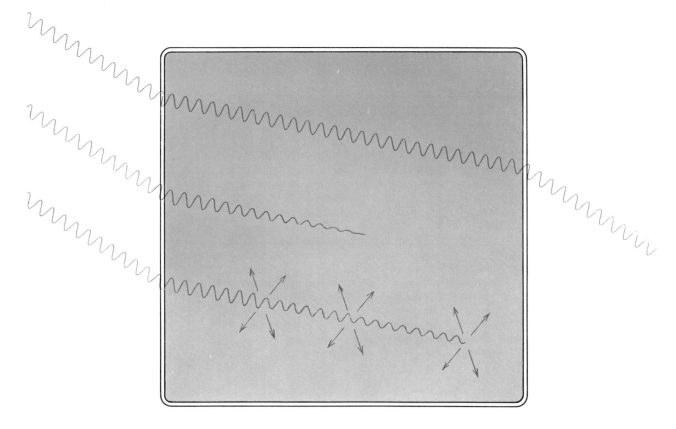

LIGHT BEAM transmitted through a tank of liquid is shown schematically in the top drawing. The tank is seen from above. As the beam passes from left to right it is refracted and attenuated. In an actual experiment some energy would also be reflected each time the light struck a new material, but the reflected beams are not shown here. The lower drawing indicates the fate of the transmitted beam symbolically. The top wave represents energy that is refracted but that gets through the liquid. The center wave shows energy that is absorbed and converted to heat. The bottom wave represents light that is scattered by molecules in the solution.

tube tells us how many sugar molecules it contains. Knowing the weight of the sugar that we put in the tube, we can compute the weight of a sugar molecule.

We have in this case a large effect: even at a dilution of about one ounce of glucose (molecular weight 180) per quart of water in the tube, the osmotic pressure amounts to one atmosphere, equivalent to a column of water 32 feet high. The osmotic pressure falls with increasing size of molecules, because there are fewer molecules in a given concentration of solution by weight. Thus if the molecules are 1,000 times heavier than glucose, a one-ounce-per-quart concentration gives an osmotic pressure of one thousandth of an atmosphere. This still corresponds to a column of water one centimeter high. But by the time we reach molecules with a molecular weight of the order of one million, the effect becomes inconveniently small: a 1 per cent solution of such a substance yields an osmotic pressure amounting only to a water column a few millimeters high. In short, the osmotic method is a simple and effective measure but declines in efficiency as we approach the giant molecules.

The second method of counting molecules improves in efficiency with increasing molecular weight. Developed in the last few years, it has become a useful instrument not only for weighing molecules but also for picturing their size and structure. The method is founded upon the phenomenon that molecules scatter light.

As light passes through any material medium, it loses energy. The distance we can see through the clearest waters is measured in feet, and even air is not perfectly transparent. If it were completely clean of dust particles and water droplets, it would still weaken light. The air molecules themselves (oxygen, nitrogen and so on) extract energy from a light beam. Part of this effect is due simply to absorption and conversion of the light energy into heat. But usually a much larger proportion of the beam's loss of energy is due to scattering of the light by the air molecules. It is the scattering of sunlight by air molecules, as Lord Rayleigh deduced many years ago, that gives the sky its blue color.

The molecules of a gas or liquid give it a certain graininess. Each molecule excited by a beam of light acts like a tiny antenna, picking up the energy and re-radiating secondary light waves of the same frequency in all directions. Suppose, then, we start with a certain liquid,

a pure solvent, and measure its scattering, or weakening, of a light beam per centimeter of distance—this we call the turbidity of the liquid. If we now dissolve some foreign molecules in the liquid, each molecule constitutes an additional antenna which scatters light. Consequently the increase in turbidity of the solution should be a measure of the number of molecules added. However, there is a complication. The amount of light scattered by dissolved molecules depends not only on their number but also on the size or amplitude of the oscillations of their electrons. Different molecules respond differently to the light waves of a given frequency. Hence in order to count the molecules we must find a way to calculate the unknown amplitude factor.

Fortunately this can be done by taking account of another property of the solution which also involves its effect upon light. A liquid refracts (bends) a beam of light entering it from the air. When a substance is dissolved in the liquid, it changes the index of refraction. The amount of change again depends upon the number of dissolved molecules and the amplitude of the electronic vibrations. But this change is directly proportional to the amplitude, whereas the increase in the scattering of light is proportional to the *square* of the amplitude.

The change in refraction represents the number of molecules times the amplitude; the increase in turbidity indicates the number of molecules times the square of the amplitude. Therefore if we make both measurements, we obtain two equations which can be solved to give the number of molecules.

A 1 per cent solution of large molecules is about 100 times more turbid than the solvent itself. This increase in scattering of light can easily be measured with a photocell. Moreover, the efficiency of the turbidity method for counting molecules increases with the size of the molecules. The following thought experiment illustrates why. Let us say we have dissolved a certain number of molecules in a given volume of solvent. If we were now to hook pairs of these molecules together to double the size of the molecules, what would be the effect on the turbidity of the solution? There would only be half as many scattering centers as before. But by doubling the size of the molecules we would have doubled the amplitude of the oscillations; since the scattering of light increases in proportion to the square of the amplitude, we would have increased the scattering by each center fourfold. Thus

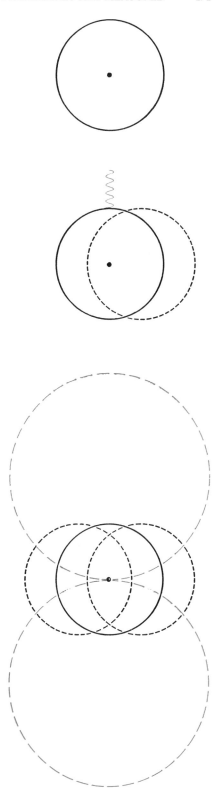

SCATTERING is produced by particles acting as antennas. The simplest scattering center is a hydrogen atom, shown schematically at top. The dot represents the nucleus, and the circle the electron cloud. In the center the electric field of an incoming light wave displaces the electron cloud. The cloud then oscillates (*bottom*), radiating light in the pattern indicated by colored circles.

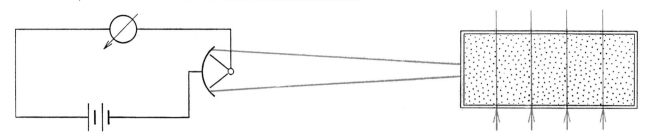

SCATTERING IS MEASURED by photocell. Light, indicated by colored lines, passes up through tank at right. Scattered energy strikes the cell, releasing electrons (*black lines*). The resulting current is then measured by the meter at top of photocell circuit.

the net result is that large molecules in solution have twice as strong a scattering effect as an equal weight of molecules half their size.

So far we have assumed that the molecules we are counting in a given experiment to arrive at the molecular weight are all of the same size. But of course this is not usually the case in a batch of a polymerized substance. The stringlike molecules do not necessarily grow to one uniform length; the molecular chains in the batch may vary widely in weight. Consequently the molec-

ular weight we obtain is only an average, both in the osmotic and the light-scattering methods. The interesting fact is that one average differs from the other!

In the osmosis experiment every dissolved macromolecule contributes equally to the osmotic pressure, whatever its weight. Therefore the count of molecules, divided into the combined weight of the molecules, gives an average molecular weight in the usual sense of the word average: for example, if the sample is a mixture in which half the molecules have a molecular weight of 10,000 and half 30,000, the average will come out to

20,000. But the light-scattering method will give a different result, because the larger molecules scatter light more strongly—in this case in the ratio of 3 to 1. The smaller molecules contribute one fourth to the total effect, the larger molecules three fourths. Therefore the computed average molecular weight will be one fourth of 10,000 plus three fourths of 30,000, or 25,000. It is easy to see that we can obtain extra information by making both measurements on a given sample of material. If both methods yield the same average, we know that all the molecules in the sample have the same

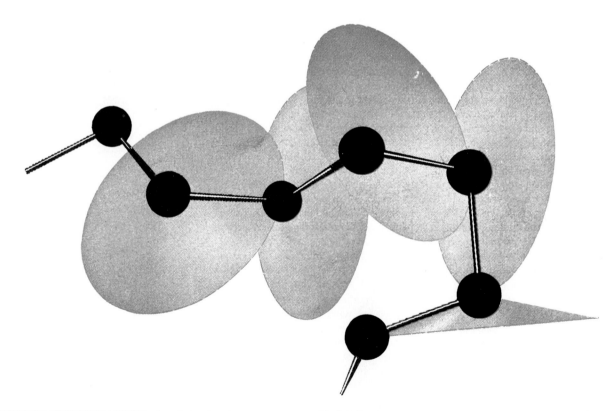

POSSIBLE CONFIGURATIONS of a chain of carbon atoms are illustrated in the drawings at the bottom of these two pages. The black balls represent carbon atoms; the gray lines, the bonds which join them. The colored cones show, for each bond, all the pos-

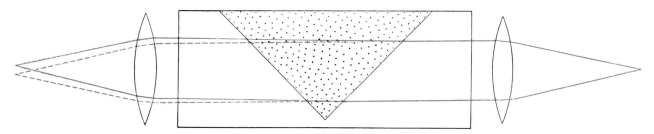

REFRACTOMETER measures changes in refraction. Hollow prism (*triangle*) containing solution of giant molecules (*dots*) is immersed in vessel of pure solvent. Colored lines show light paths, the broken lines giving the path when prism contains pure solvent.

weight. If the averages differ, the difference indicates roughly the spread of their weights.

Let us now give thought to what the scattering of light may tell us about the structure or configuration of giant molecules. We think of a long, flexible molecule which can coil up or stretch out into a great number of different shapes. Considering all its possible convolutions, what is its average size (the volume it occupies) likely to be? We can make a rough estimate by considering the distance between one end of the molecule and the other in all of the molecule's possible configurations.

The problem is reminiscent of the "random walk" problem in mathematics. Suppose a person takes a step in some random direction, forgets what he has done and takes a step in another direction, turns and steps again, and so on. Assuming that the turns are made completely at random, how far will he move from his starting point, on the average, in a given number of steps of the same length? It turns out that the average distance is proportional to the square root of the number of steps. If he takes four times as many steps, he will wind up only twice as far from his starting point.

In a similar way we examine the possible twists and turns of a long chain molecule. Here the distance between one carbon atom and the next corresponds to a step. The molecule can rotate more or less freely around each C-C bond, but the bonds are restricted to an angle of 109.5 degrees to one another. Knowing the length of the step and the angle of the permitted turns, we can calculate the distance from one end of the molecule to the other in all its allowed configurations. Now the square root of the

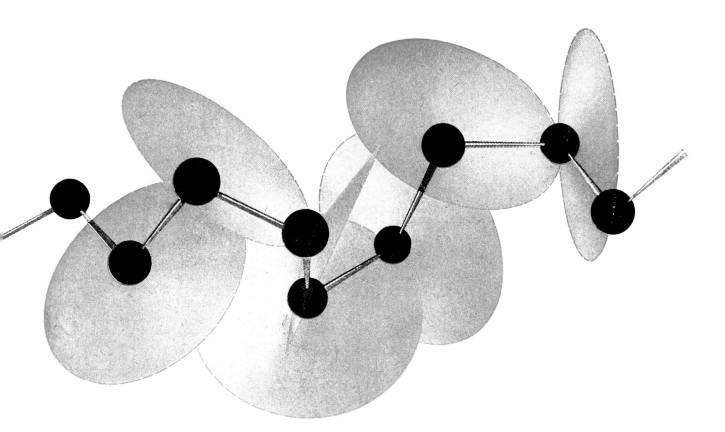

sible positions through which it could move, assuming that the bond to its left is fixed. Any two successive bonds must make an angle of 109.5 degrees, but the bonds are free to move with respect to one another as long as the size of the angle is maintained.

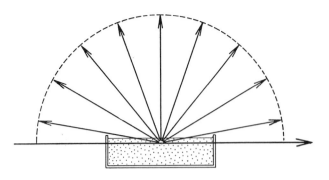 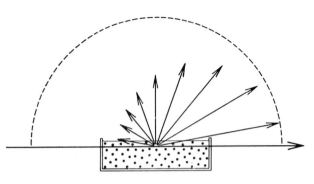

SCATTERING PATTERN for solution of small molecules is at left; for solution of large molecules, at right. Horizontal arrows give direction of beam. Other arrows, by their relative lengths, indicate the relative intensity of light scattered at various angles.

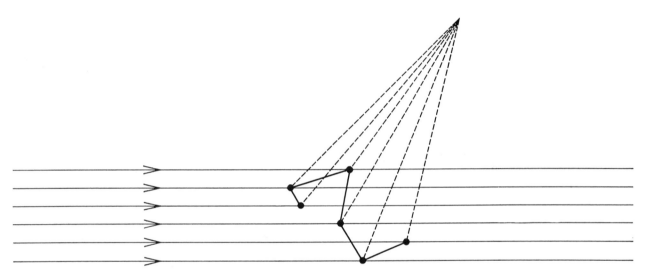

LARGE MOLECULE scatters light asymmetrically because each point on the molecule acts as a separate antenna. Light received at any outside point depends on the amount of interference among trains of energy (*broken lines*) from different radiating centers.

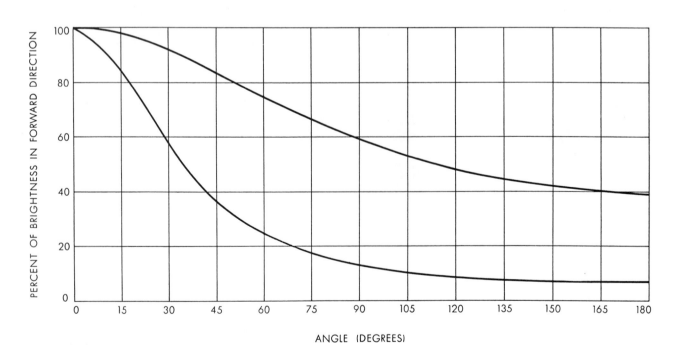

LIGHT CURVES give scattering pattern for polystyrene in benzene. Upper curve is for molecules of weight one million; lower curve, for molecules of weight nine million. Vertical scale compares intensity at any angle with the intensity scattered forward.

average of the squares of all these distances is taken to define the average size of the molecule.

As a specific example let us take the molecule of polystyrene in solution in benzene. A polystyrene molecule of molecular weight one million has about 20,000 C-C links, each 1.5 Angstrom units long (one Angstrom is a hundred millionth of a centimeter). From the information aforementioned, we compute that the average end-to-end distance, or diameter, of the molecule is about 300 Angstroms.

Now this size begins to be comparable with the wavelengths of visible light (which are in the neighborhood of 5,000 Angstroms). A molecule of such a size no longer acts like a single point in intercepting and reradiating light; instead, light comes from various points on the molecule, and as a result we have interference effects because of phase differences in the scattered light waves. The intensity of light-scattering now depends on the angle from which we view the beam [see diagrams on opposite page]. On the average a solution of large molecules scatters more energy forward, in the direction of the beam, than backward. What is most important, the larger the molecule, the higher the proportion of light scattered in the forward direction.

We can consult this scattering pattern, then, as a guide to the size of molecules. Polystyrene has been so studied, and it has confirmed the prediction that the size of a molecule is proportional to the square root of the number of C-C links. However, the size in each case is substantially larger than had been calculated theoretically: polystyrene of molecular weight one million has an average "diameter" (end-to-end distance) of 1,100 Angstroms instead of 300. The chief reason for the discrepancy is that in considering the possible configurations the calculation failed to take account of such factors as the impossibility of two monomers in the chain simultaneously occupying the same position in space. Naturally the molecules cannot be as compact, on the average, as such a calculation predicts.

The mathematical difficulties in dealing with these enormously complex molecules are so formidable that we are far from any fully satisfactory quantitative description of them. But all in all, the combined information from the various measurements—of viscosity, osmotic pressure and scattering of light—is beginning to give us a clear picture of their size and structure.

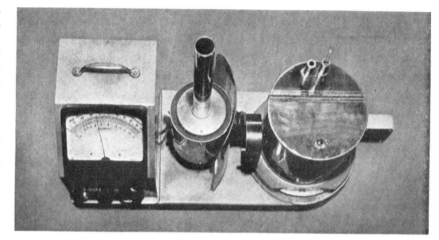

PHOTOMETER measures light scattering. Light source is a mercury arc in the cylinder in the center. The beam passes through a shutter to the dark chamber at right. This contains the scattering solution and a photocell which can be rotated by means of the graduated dial at the bottom of the chamber. The meter at the right indicates the current in the photocell.

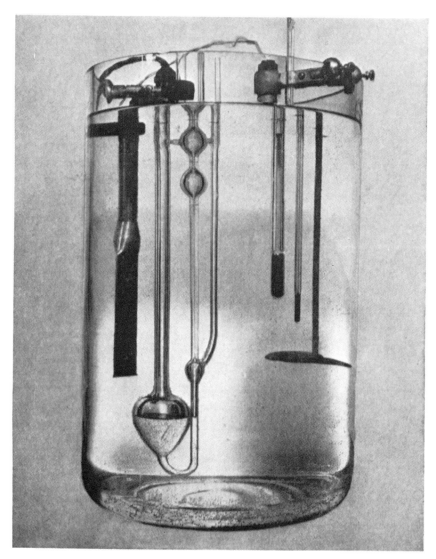

VISCOMETER is essentially a capillary tube. In this photograph the capillary is the center tube in the assembly of glass at left in the large vessel. Liquid drawn up from the bottom bulb fills the two small bulbs above the capillary. It is then timed as it flows out. The rest of the apparatus in the vessel is for the purpose of maintaining constant temperature.

17 The Ultracentrifuge

by George W. Gray
June 1951

*By spinning a rotor at very high speed it can generate
enormous gravitational forces. It has been used notably
to study large molecules by causing their sedimentation*

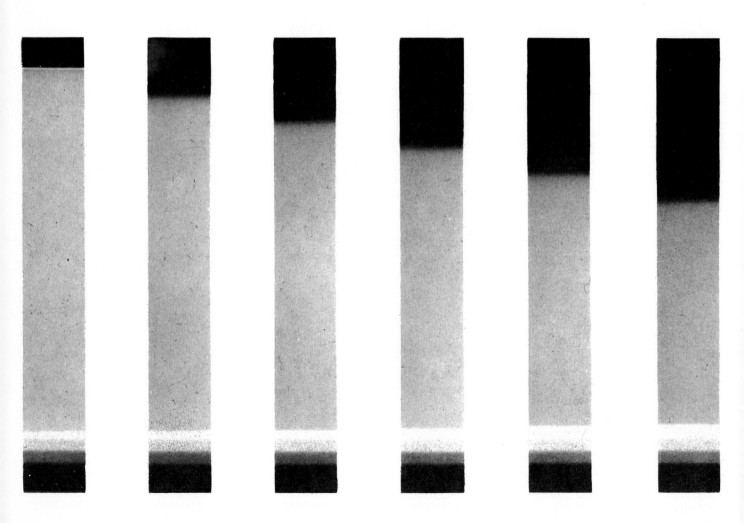

IOLOGICAL research, which began with the forms of plants and animals and later sharpened its focus to examine the tissues and cells of which they are composed, is occupied today with the study of still smaller units—the organic molecules known as proteins.

These structures of carbon and nitrogen interlinked with atoms of other species are the present frontiers of experimental biology. Every essential chemical reaction of life, including such fundamental processes as the fertilization of the egg, growth, nutrition, metabolism, muscular contraction, resistance to infection, nerve communication and response, is governed by one or more kinds of specialized proteins, most particularly enzymes and hormones, which the cells ceaselessly manufacture.

Proteins differ in weight, shape and electrical polarity. These physical properties critically affect molecular behavior. They determine, for example, whether or not a molecule will pass through a cell wall, whether or not it will attract, repel or be indifferent to another molecule. Biochemists long ago recognized that to understand living processes they must appraise physically the molecules that participate in those processes. And first of all, it is important to know how *large* the molecules are.

A common method of determining molecular weight is to dissolve a measured quantity of the substance in water or some other solvent and see how much the added material raises the boiling point or depresses the freezing point of the liquid. When one part of sodium chloride is dissolved in 1,000 parts of water, it lowers the water's freezing point by .061 of a degree Centigrade. Cane sugar in the same proportion lowers the freezing point by .0054 of a degree, acetic acid by .032 of a degree. There is a fixed relationship between the molecular weight of a substance and the freezing point of a solution of it, and this relationship is widely used as a means of identifying many compounds. But as the test is applied to substances of higher and higher molecular weight, the difference in freezing point becomes progressively smaller, and when the proteins are

reached, the difference is so nearly infinitesimal as to be beyond measurement.

Similar limitations apply to other traditional methods of analysis. It is comparatively easy to study small molecules, but when one tackles the larger ones, the measurements grow more and more elusive. It is only within the last two and a half decades that science has acquired any reliable means of appraising the physical properties of proteins—or indeed of knowing for sure that proteins are molecules. A powerful device which reached development during this period and became one of biochemistry's principal tools for protein research is the ultracentrifuge, a machine that weighs particles by measuring their sedimentation.

Sedimentation

The force of gravity has an orderly way of sorting substances out of a mixture. If, for example, you let a suspension of muddy water stand undisturbed, the fragments of rock, being heaviest, will drop first; particles of coarse sand will come out next and form a layer on top of the rock fragments; lighter sand grains will deposit in another layer above this; after several hours particles of clay will follow, and eventually still finer clay will form a top layer. But no matter how long the solution stands, there are some motes that will never settle. They are too light for gravity to overcome the upward recoil from countless collisions with the perpetually moving water particles—the so-called Brownian movement.

The laws of sedimentation and diffusion have been the subject of much mathematical research. One of the basic approaches to the problem was a joint study by Max Mason and Warren Weaver at the University of Wisconsin in the early 1920s; they solved the problem for sedimentation in a uniform gravitational field. Physicists have used gravity to study the behavior of particles in a vacuum. Biochemists, on the other hand, have usually found that the Earth's pull is too weak to assist their molecular studies. Consequently they have turned to centrifugal force, which closely parallels gravitational force in effects and can be magnified to any degree by increasing the speed of rotation.

The separation of substances by centrifugation is an old story: a thousand years ago tung oil was extracted in this way, and the principle is familiar to everyone in the form of the dairyman's cream separator. The principle of a centrifuge of course is that under centrifugal force the heavier particles in a mixture tend to move farther out toward the periphery than the lighter ones, and as rotation continues they pile up against the outer wall and are separated. Sedimentation here means the movement of

particles outward toward the rim of the centrifuge.

In the laboratory chemists began early in this century to use centrifugal force to fractionate colloidal suspensions. Even the best of their centrifuges, however, were of limited velocity and were subject to incalculable fluctuations. They could separate gross particles, like hookworm eggs and blood cells, but were unable to sort out molecules or determine their dimensions.

To separate and measure molecules chemists needed not only centrifuges with faster and steadier rotational speeds but also an optical means of recording the behavior of the mixture during rotation and a mathematical formula by which to interpret the behavior in units of molecular weight. The first to bring about this partnership of mechanics, optics and mathematics was Thé Svedberg, professor of physical chemistry at the University of Uppsala, Sweden.

Svedberg

From his student days Svedberg had been fascinated by the performance of particles in emulsions, the state of dispersion between the solid and the liquid states in which matter is called colloid. He became widely known for these studies, and in 1922 the University of Wisconsin invited him to spend a year on its campus as visiting professor of colloid chemistry. During this visit he began to try out an idea that had long been hatching in his brain. He enlisted the help of a graduate student, J. Burton Nichols, and together they built a rotating machine from the professor's specifications. Because the apparatus included a system of lenses for viewing and photographing the sedimentation through a window in the rotor, Svedberg called it an "optical centrifuge."

This optical centrifuge of 1923 could turn only a few hundred revolutions per minute, but it provided a means of studying the conditions that had to be fulfilled in harnessing centrifugal force to the exacting tasks of molecule measurement. The fundamental problem was to attain rotation without heating, for heat stirred up convection currents in the whirling liquid, and these currents, no matter how slight, distorted the sedimentation.

Returning to Uppsala after his American sabbatical, Svedberg began a comprehensive survey of the conditions for convection-free sedimentation. He and H. Rinde experimented with different rotor sizes and investigated ways of removing heat from bearings and of reducing air friction. Within a few months they had built a machine with a rotor installed in a sealed chamber filled with hydrogen gas at low pressure. This arrangement reduced friction and at the same time insured that whatever heat

THE SIX PHOTOGRAPHS on the opposite page show the sedimentation of a protein in the ultracentrifuge of the International Health Division Laboratories of The Rockefeller Foundation. The protein is hemocyanin from the blood of *Limulus polyphemus*, the horseshoe crab. In the photograph at the far left the protein is suspended throughout the whole solution. The succeeding five photographs, which were made at 15-minute intervals, show the protein settling as it is whirled at a speed of 18,000 revolutions per minute.

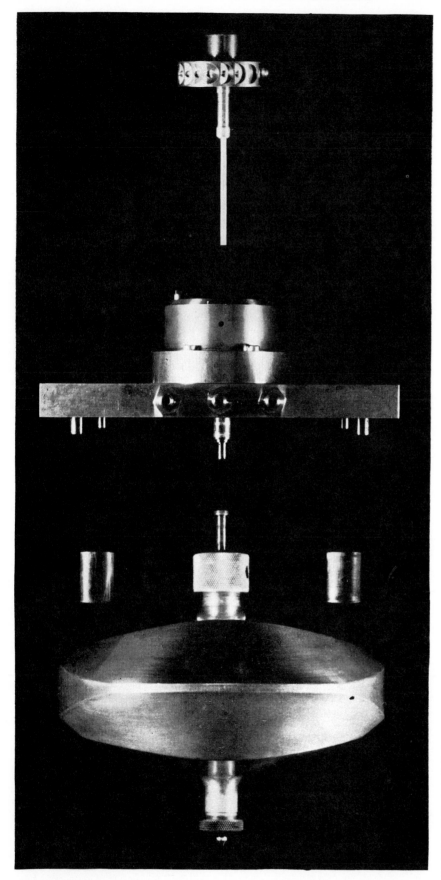

DISSECTED VIEW shows the principal parts of the ultracentrifuge of the International Health Division Laboratories. At the top is the rotor of an air turbine; the stator is below it. At the bottom is the rotor of the ultracentrifuge and its two cells. The two rotors are connected by a shaft.

was generated by the rotation would be quickly conducted away. The motor-driven rotor, 45 millimeters in radius, whirled at 10,000 revolutions per minute and produced a centrifugal force 5,000 times gravity. It was the first centrifuge to give quantitative data of sedimentation. Svedberg named it the "ultracentrifuge."

With their new machine Svedberg and his associates soon determined the molecular weight of hemoglobin and certain other proteins of the blood. They employed a measurement criterion known as sedimentation equilibrium, which depends on two opposing effects: sedimentation and diffusion. As the large molecules move outward toward the periphery of the rotor, they collide repeatedly with smaller molecules, and many are driven back against the urge of centrifugal force. Under certain conditions, after a long period of continuous rotation, the particles streaming outward in sedimentation and those streaming inward by diffusion establish an equilibrium. The greatest concentration of heavy particles will still be at the outer rim of the container, with the concentration progressively less all the way in toward the center, but the concentration at each stage remains constant. The molecular weight of the large molecules can then be determined by comparing their concentration at two distances from the center and calculating the result with a mathematical formula devised by Svedberg.

The equilibrium method gave reliable results for relatively small proteins such as hemoglobin, but Svedberg found it less successful with larger molecules. Moreover, it consumed a lot of time. Days, even weeks, of continuous rotation were required to obtain a single determination. He therefore began to explore the possibilities of a different procedure, which has come to be known as the velocity-of-sedimentation method. The idea here was to rotate the material so fast that sedimentation would predominate over diffusion. Then all, instead of merely a preponderance, of the heaviest molecules would be thrown into the zone nearest the outer rim of the rotor. Theory said that there would be a definite boundary between these heaviest particles and the adjacent inner zone of the next heaviest, and that this boundary would shift progressively outward with continued rotation. By measuring the rate at which the boundary moved toward the periphery, it should be possible to determine the velocity of sedimentation. Using this velocity as one factor in an equation, it should then be possible to calculate the molecular weight. Such was the reasoning back of the new approach in ultracentrifugation which Svedberg and his associates entered upon in 1925.

The first requisite was a means of rotating liquids not at 10,000 but at many tens of thousands of r.p.m. without stirring up convection currents. Such velocities were beyond the powers of the 1924 ultracentrifuge, and Svedberg turned his thought now to the design and construction of a faster instrument.

The Oil Drive

The first high-speed ultracentrifuge was completed in 1926. Its rotor, like that of the older, slower machine, was whirled in a sealed chamber of hydrogen. The engine that turned it, however, was a tiny turbine driven by a stream of oil under pressure, instead of an electric motor. Svedberg has said that he chose the turbine because it was more efficient for high-speed rotation, and he chose oil to drive the turbine because it threatened no serious danger of contaminating the atmosphere around the rotor by evaporation. Moreover, there was an added advantage in the fact that the compressor which pumped the oil stream served also to deliver oil for lubricating and damping the bearings. With this apparatus he turned a 52-millimeter rotor at 45,000 r.p.m., generating centrifugal forces that were 100,000 times gravity.

The greater the radius of the rotor and the faster the velocity, the greater are the centrifugal forces. As these are accelerated to higher and higher magnitudes, even the toughest metal will fly to pieces. Svedberg and his co-workers progressively modified the 1926 design and pushed centrifugation to unprecedented dimensions. By 1931 they were generating forces 200,000 times gravity with an oil-driven rotor of 65-millimeter radius; early in 1932 they obtained 260,000 g, and in the spring of 1933 this whirligig was still holding together at 400,000 g. In later experiments to see how far centrifugation of liquids could be pushed, the group reduced the radius and finally, in 1934, produced 900,000 times gravity. But all except one of the rotors used in the superspeed experiments exploded into fragments after a few runs.

Svedberg finally fixed on the 65-millimeter radius (about seven inches overall diameter) as the most satisfactory, and this dimension has since become standard in machines built for biochemical research. For that size, the maximum safe operating speed is about 67,000 r.p.m., producing 350,000 g. Even so, ultracentrifugers never operate without barricading the instrument with a protective wall of heavy timber, concrete or other shock-absorbing material.

In the midst of his experimenting with whirling rotors, Svedberg was awarded the 1926 Nobel Prize in Chemistry. In its citation the Nobel Committee made no direct reference to ultracentrifuges, merely stating that Svedberg was chosen because of "his writings on dispersive

systems." Increased support for his work now came from many directions. In 1928 the Swedish Government allotted funds to build an Institute of Physical Chemistry at Uppsala. The same year the Rockefeller International Education Board appropriated $50,000 toward equipping the new institute. This initial aid has been followed by further grants from The Rockefeller Foundation totaling over $200,000 and by additional grants from the Nobel Foundation and other Swedish trusts.

Most of these funds went to finance studies of proteins. In the course of the research the Uppsala group thoroughly demonstrated that the protein particles were giant molecules and determined the weights of several hundred of them. The success of this work electrified physical chemists all over the world. From many quarters came requests for this powerful research tool. In the course of the years Svedberg's shop built six high-speed, oil-driven ultracentrifuges for outside clients in addition to the three installed in the Uppsala laboratory. The six went to the Universities of Gothenberg and Copenhagen, the Lister Institute in London, the Biochemical Laboratory of Oxford University, the Du Pont Laboratories at Wilmington, Del., and the Laboratory of Physical Chemistry at the University of Wisconsin. Workers in many other institutions yearned for a molecule weigher, but the oil-driven ultracentrifuge is an intricate and costly machine, and no one outside Svedberg's group has yet succeeded in building one. With a view to simplifying the engineering problems and lessening the cost, American efforts to design ultracentrifuges turned to pressurized air as a motive force.

The Air Drive

Two Parisian scientists, E. Henriot and E. Huguenard, are the pioneers of the air drive. Their first machine, details of which were published about the time that Svedberg was deciding on the oil drive, consisted of a solid cone-shaped rotor which was both supported and turned by a whirling layer of air discharged under pressure from properly directed jets. It was, in effect, a spinning top. In their early experiments Henriot and Huguenard made no attempt to exploit the device beyond demonstrating that it could provide high rotational speeds. Later they harnessed their spinning top to a narrow cylindrical centrifuge and got it up to 65,000 r.p.m. Then they mounted a small mirror on top of the rotor and spun the mirror at 360,000 r.p.m.

Publication of these results caught the attention of a young physicist at the University of Virginia who was working on an experiment which required very rapid reflections of light. Applying the Henriot-Huguenard principle, the Vir-

ginian, Jesse W. Beams, rigged up an air-driven top 30 millimeters in diameter which rotated a mirror 180,000 times a minute. That was the beginning of a lifelong, imaginative interest in whirling mechanisms for Beams. In the years since 1930, when he made his first machine, he has lavished much of his time, thought and inventive skill on studies of high-speed rotation. Practically all current American development of the ultracentrifuge stems from the work of this Virginia professor, his students and his associates.

Beams was interested in the sheer physics of centrifugation—to see how high it could be pushed and to test its effects on materials. He enlisted the interest of a graduate student, Edward G. Pickels, and soon Pickels became captive to the subject that was fascinating his professor. They whirled many a piece of metal to destruction and thereby learned what to do as well as what to avoid. Before the end of 1933 they had mounted a 10-millimeter rotor and driven it up to one million r.p.m. Later, using compressed hydrogen in place of air as the motive power, Beams and Pickels spun a nine-millimeter rotor at better than 1,200,000 r.p.m. These devices were essentially spinning tops. When the time came for Pickels to concentrate on a subject for his doctoral thesis, Beams suggested that he study the application of the spinning-top principle, either to the measurement of the velocity of light or to the ultracentrifugation of biological materials. Pickels chose the latter alternative. He was just beginning to explore the requirements when a visitor arrived who was especially interested in biological ultracentrifugation—Johannes H. Bauer of The Rockefeller Foundation.

Bauer was a yellow-fever researcher, a former member of the Foundation team in West Africa which had identified the virus of yellow fever. Nothing was then known of the weight, shape or other physical characteristics of the yellow-fever virus, and Bauer was interested in any device that might throw light on virus properties. Could the ultracentrifuge be useful? Beams and Pickels demonstrated their spinning tops, pointed out the limitations and suggested the possibilities.

At Bauer's invitation Pickels visited the Foundation's International Health Division Laboratories in New York in the summer of 1934. He spent two months there getting acquainted with the particular requirements of virus research. This convinced him that the spinning-top technique was impractical: the spinning tops were too small to provide sufficient distance for virus sedimentation to take place at a measurable rate. Moreover, since the spinning top whirled in the open air, enlarging its diameter would cause it to heat up substantially and would require a prohibi-

ULTRACENTRIFUGE ROTOR on page 178 is shown from the top. The cell that contains the solution to be centrifuged is at the bottom of the rotor; an enlarged view of it is below. When the centrifuge is in operation, the curved surface of the solution flattens. The rotor is some seven inches in diameter.

tive multiplication of power, due to the increase in air friction.

Would it be possible to rotate the thing in a vacuum? Back at Charlottesville Pickels and Beams took up a serious study of that question. Before the winter was over their answer was yes. In April they published in *Science* an account of the first vacuum ultracentrifuge.

The Vacuum Ultracentrifuge

What Pickels and Beams did was to separate the driving element of the spinning top from its virus-carrying element. The driving element was made into a tiny air turbine which operated in the open air. Suspended from the turbine by a short stretch of piano wire was a larger rotor, hollowed out to accommodate a tiny cell carrying the solution. This rotor hung in a sealed chamber from which air was evacuated; the piano wire passed into the chamber through an airtight oil gland. The wire served both as a support for the rotor and as a vertical driving shaft to whirl it under the torque of the turbine.

In June, after Pickels had received his Ph.D. degree at Virginia, he agreed to join the staff of the Foundation's International Health Division. Thereafter for a full decade this laboratory in New York was the principal developmental center of the vacuum type of air-driven ultracentrifuge. Pickels, whose interests were primarily mechanical, optical and mathematical, became a close collaborator of Bauer, who approached the problem from the biological point of view. Many improvements in design and advances in application came out of the joint efforts of these two men. The Rockefeller Institute for Medical Research also was interested in the possibilities of ultracentrifugation as a laboratory tool, and members of its staff, notably Ralph W. G. Wyckoff and Jonathan Biscoe, joined forces with Pickels and Bauer in the developmental research.

Primarily the research was aimed at duplicating with the air drive the kind of performance that Svedberg had already accomplished with the more costly oil-driven machine. Within less than a year the group was able to report the attainment of an ultracentrifuge built in their own shop with materials (including the lenses) which cost less than $1,000. It had an oval-shaped seven-inch rotor made of aluminum alloy. "In designing this ultracentrifuge," Pickels, Wyckoff and Biscoe reported, "we have sought to utilize, to the greatest possible extent, the experience embodied in the publications of Svedberg. We have therefore copied his optical system directly and have employed large rotors capable of giving the 6.5-centimeter distance between cell center and rotation axis which he has adopted for his most accurate measurements." This was the

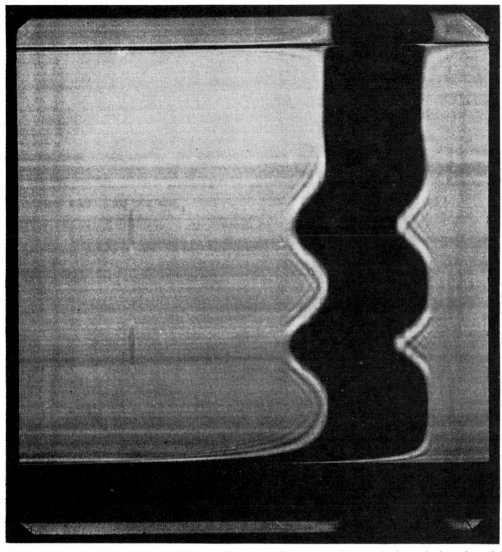

REFRACTIVE-INDEX METHOD of photographing the sedimentation of a protein shows the boundary of the protein as a horizontal peak. At the right is a photograph of two protein constituents by the absorption method. The corresponding photograph made by the refractive-index method, showing two peaks, is at the left.

first time that a full-size rotor had been driven by air, for even in their vacuum-type machine of the previous year Beams and Pickels had tried nothing larger than a 3.5-inch rotor. With their new seven-inch whirler molecular sedimentation was easily demonstrated, first with horse hemoglobin and then with the tobacco-mosaic virus which had recently been crystallized by Wendell M. Stanley.

By certain modifications in the oval shape of the rotor Bauer and Pickels increased its safe operating velocity from 54,000 r.p.m. to 60,000. They also improved the air drive and the air bearing which supported the entire rotating assembly. By adding reverse vanes to the turbine they provided an air brake by which the rotor could be brought to rest from full speed in 20 minutes. By these and other modifications of the basic design, the air-driven ultracentrifuge was made more nearly an instrument of high precision for biological research, especially for virus research.

The Bauer-Pickels collaboration came to an end in 1945. Bauer went abroad on a war-relief mission, and a few months later Pickels resigned to engage in the commercial manufacture of ultracentrifuges. His company, the Specialized Instruments Corporation, which opened for business in California in 1946, has built more than 70 analytical ultracentrifuges for laboratories in this country and abroad. These Spinco machines are of the vacuum type but are driven by electric motor instead of compressed air.

The Next Decimal Point

Ultracentrifugers are perfectionists, continually striving to heighten the vacuum, lower the friction, reduce the fluctuation. Bauer and Pickels, never satisfied with the precision of their instrument, were always reaching for the next decimal point. So too with Beams. A recent visitor to Charlottesville found Beams engrossed in a coasting type of ultracentrifuge. Its seven-inch rotor is supported by magnetism and is driven by air to the top speed of 60,000 r.p.m., after which the air turbine drops off and the rotor whirls in the vacuum under its own momentum. Friction is so slight that at the end of 10 hours the rotor has lost only one revolution per second, and more than a year would be required for it to come to rest. Since it rotates by inertia, with no propulsion from outside, this coaster represents the ultimate in steadiness of motion, Beams believes.

Meanwhile John C. Bugher, who two years ago was placed in charge of the Rockefeller Foundation ultracentrifuge, has been exploring ways to better that instrument's performance. The vacuum in which its rotor whirls has been improved and is now down to one ten-millionth of an atmosphere. During an hour of running in this highly rarefied air at 60,000 r.p.m., the rotor temperature rises by only half a degree C. For most of the virus studies, however, the speeds required are not over 45,000 r.p.m., generating forces of 147,000 g.

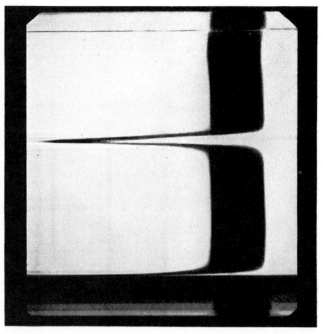

SEDIMENTATION OF A PROTEIN is shown in four photographs made by the refractive-index method. The first photograph shows the hemocyanin of *Limulus polyphemus* after centrifugation for an hour; the second, after an hour and a half; the third, after two hours; the

Steadiness of rotation in this machine has been greatly enhanced by the addition of an electronic control. Under the old system it was necessary for an operator to keep a weather eye on the air drive and give it attention whenever the rotor began to speed up or slow down. Bugher has surrounded the air turbine with an electromagnet controlled through electron tubes which are synchronized with time signals received by radio continuously from the Bureau of Standards in Washington. If the rotor tends to race, the controlled magnetic field restrains it by dragging against the turbine; if it tends to slow down, the field puts in added power and speeds up the turbine. The pulsations necessary to provide these corrections are so slight that their heating effect is insignificant. The entire hookup is automatic and, once set in motion, requires no attention. Under the earlier system the rotational speed was known to an accuracy of 99 per cent; with the electronic control it is known to an accuracy of 99.999 per cent.

This never-ending solicitude for constancy is understandable when we remember that the whole purpose of running the ultracentrifuge is to measure the rate at which the particles sediment, and that the accuracy of this measurement depends on correct knowledge of the rotational speed. The measurement is made by photographing the state of sedimentation at intervals of time. The state of sedimentation is indicated by the position of the boundary between the zone of concentration of heavy particles and the zone of lighter particles left behind. As each picture is taken, it records the position of the moving boundary at that moment. Thus it discloses how

far the boundary has progressed toward the rim of the rotor in each interval.

There are two ways of photographing the boundary. One, known as the absorption method, makes use of the fact that any given substance selectively absorbs light of a certain wavelength or band of wavelengths. To photograph the boundary of yellow-fever virus concentrated in the rotor, for example, one sends a beam of the particular "color" of light that the virus is known to absorb most strongly through the quartz windows which face each other on opposite sides of the rotor. Since the virus absorbs this light, the zone it occupies appears black on the photograph.

Absorption photographs usually show a fuzzy band rather than a sharp boundary. But even more serious is the wide margin of probable error in interpreting the optical density in terms of the actual concentration of material. These uncertainties have led to another stratagem: the refractive-index method. It makes use of the fact that when light passes through a transparent fluid which has zones of different density, the boundary between these zones bends the light. The boundary between heavy-weight and light-weight particles operates like a refraction lens; when the rays that pass through it are photographed, they describe a "peak." In practice the optical system turns the refracted image in such a way that it is projected on the photographic plate with the peak pointing to the left, and as sedimentation progresses the peak moves downward. The rate at which the peak moves is a measure of the rate of sedimentation.

The absorption method was introduced by Svedberg; the refractive-

index method by a Swedish colleague, O. Lamm. Both techniques are widely used, but the refractive index is more often preferred, particularly when complex mixtures are to be investigated and it is desired to measure the concentrations of several components. Bugher is using this refractive-index method in current studies of 18 unknown viruses which the International Health Division picked up in the course of its yellow-fever surveys in Africa and South America. Two of the unknowns—Bwamba fever virus and Semliki Forest virus—have been subjected to preliminary ultracentrifugation. Bwamba shows up as a particle approaching the large virus of influenza in weight and size, whereas Semliki Forest virus is smaller, like that of yellow fever. Bugher reports that in 15 minutes of ultracentrifugation at 12,000 r.p.m., the boundary of Bwamba fever virus moved four millimeters, whereas that of yellow fever took 100 times as long to sediment that distance.

Mathematics

This business of weighing molecules requires more than centrifuging them at constant speed and measuring the rate at which their boundary moves. All that such a measurement gives is the sedimentation velocity, which is but one of six terms in the equation that determines the molecular weight. The other terms are the gas constant, the absolute temperature, the diffusion coefficient, the partial specific volume of the material being studied and the density of the solution.

In the summer of 1950 a three-day conference on ultracentrifugation was

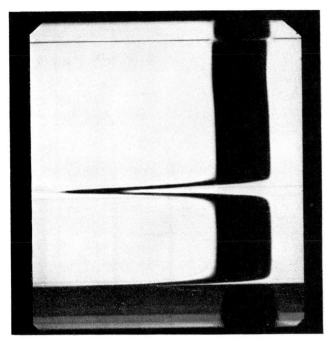 

fourth, after two hours and a half. The line at the top was made by the surface of the solution containing the protein; the long horizontal peak, by the protein itself.

held at Shelter Island, N. Y., under the auspices of the National Academy of Sciences. Leading workers in this field of physical chemistry participated, and most of the time was occupied by discussion of the terms of the equation. The gas constant you get out of a book, but all the other factors have to be determined by actual measurement, as the sedimentation velocity is. For example, to obtain the diffusion coefficient you place a solution of the protein in contact with a buffer solvent with a sharp boundary between them, and then measure by optical means the rate at which the boundary disappears as the protein diffuses into the solvent. The partial specific volume and the density are particularly critical terms in the equation, for a slight error in these measurements can make a very large difference in the result. The partial specific volume of the protein, which is the reciprocal of its density in the solution, is commonly calculated from measurements of the density made with a pycnometer (measuring flask) at different concentrations. Karl Drucker, of Svedberg's group, has increased the precision of this measurement by recent improvement of the technique, but here as in other areas the biochemists are still reaching for the next decimal point.

One of the subjects discussed at the Shelter Island conference was Svedberg's equilibrium method of determining molecular weight. This method, which employs a somewhat different mathematical equation from the sedimentation-velocity method, has the advantage that the diffusion as well as the sedimentation data are provided by the ultracentrifuge itself. The great disadvantage is that it requires long periods of low-speed rotation. However, one of the participants in the conference, W. J. Archibald of Dalhousie University in Nova Scotia, reported that he had worked out a scheme for shortening the process. He does not wait for the attainment of equilibrium between sedimentation and diffusion but takes observations of the concentration gradient at fixed intervals in the early stages of ultracentrifugation. These readings provide the points for defining a series of curves, and from these by extrapolation the equilibrium figure is computed. In a few hours Archibald's method can get a result which would take weeks by the full equilibrium process.

Giants and Supergiants

A complete review of the gains that have accrued to science from the development of the ultracentrifuge would fill many pages. The primary contribution is the higher precision it has brought to our knowledge of molecular biology. In some cases it has corrected, in others confirmed or extended, the testimony of other instruments and techniques. The biochemist, indeed, has many tools for studying molecules: osmotic pressure measurements, filtration through pores of progressive sizes, electrophoresis, electron microscopy, spectroscopy, chromatography, X-ray diffraction. Each is, in a sense, a different window on the world. And the view from one window is not enough; usually it opens in only one direction. The biochemist wants to see in every direction, to examine all properties of molecular structure and molecular reaction, to measure their magnitude and elucidate, if possible, the part they play in the scheme of life. The ultracentrifuge is only one of these instrumentalities, but it is a powerful one.

Before the development of the ultracentrifuge there was widespread doubt among chemists as to the existence of giant molecules. Diffusion data, viscosity measurements, osmotic pressure determinations and other indirect evidence had indicated that protein particles were of tremendous size. But the general opinion held that proteins were not strictly unitary structures, like the molecules of acids and sugars, but were merely clusters of small molecules bunched together to form composite particles of heterogeneous mass. Svedberg's very first determination showed that the particles of hemoglobin were of uniform weight and that their weight was around 69,000 times that of a hydrogen atom. A little later he subjected hemocyanin to ultracentrifugation and found that its weight was about five million units. This measurement seemed utterly unbelievable. Such a molecule would consist of hundreds of thousands of atoms, and how could so huge a structure hold together?

But Svedberg and his associates continued to spin solutions of proteins and weigh their molecular masses, and soon the new technique of electrophoresis began to confirm the testimony of the ultracentrifuge. When the Royal Society of London held its symposium on "The Protein Molecule" in 1938, it invited Svedberg to open the discussion. He was able to point to some 60 proteins which had been run through his molecular weighing machine with results so consistent that he could declare: "The pro-

COMMERCIAL ULTRACENTRIFUGE was designed by Edward G. Pickels, one of the pioneers of ultracentrifuge development. It is driven by a geared-up electric motor instead of an air or oil turbine. The rotor of the machine, which is located at the Sloan-Kettering Institute for Cancer Research, may be seen at left center.

teins are built up of particles possessing the hallmark of individuality and therefore are in reality giant molecules."

In addition to the Uppsala determinations, important ultracentrifugal measurements of proteins have been made by chemists in Stockholm, London, Oxford and Berne. Among the successful protein weighers in the U. S. are J. L. Oncley of the Harvard Medical School, who employs an air-driven instrument of the Beams-Pickels-Bauer type, and J. W. Williams of the University of Wisconsin, who uses an oil-driven ultracentrifuge of the Svedberg type. There is only one other oil-driven ultracentrifuge in the U. S.—at the Du Pont Laboratories in Wilmington, where it has been operated for a number of years by Svedberg's former student, J. Burton Nichols. Nichols specializes in the study of high polymers, those long chainlike molecules of which rubber and cellulose are examples. They present a striking contrast to the predominantly globular or ovoid shape of the proteins, but Nichols and his associates find that "whirl is king" among the filamentous molecules no less than among those of more rounded form.

Apart from a few striking exceptions, such as tobacco-mosaic virus and the vaccinia of smallpox, the viruses have presented peculiar problems in sedimentation studies. Their extreme biological activity may be a reason for this. Bauer and Pickels found the sedimentation of yellow-fever virus exceedingly difficult. Often, after centrifuging the serum of sick monkeys to obtain a workable concentrate, they would end with an amount of active virus that was little more than they had started with. It was only by centrifuging the material again and again that they were able to obtain a sufficient concentration to put in the ultracentrifuge for sedimentation measurements. Their determination gave the molecular weight of the yellow-fever virus as 2,930,000. This truly is a giant when compared with familiar proteins such as hemoglobin, weight 69,000, and egg albumin, 44,000. But it is dwarfed in turn by the influenza virus, a supergiant weighing 650 million units.

Separators

The ultracentrifuge is often confused

with the centrifuge; strictly speaking they are different tools. The ultracentrifuge, as defined by Svedberg, who coined the term, is "an instrument by means of which sedimentation in a centrifugal field is measured quantitatively." A centrifuge, on the other hand, separates substances of different particle weight but cannot weigh or measure them. No matter how high its speed or how elegant its mechanical design and operation, the centrifuge is simply a glorified cream separator. It has no optical system, requires no observation of the rate of sedimentation, is not seriously affected by fluctuations in its rotational velocity. The development of the ultracentrifuge has, however, contributed enormously to the improvement of centrifuges. High-speed ordinary centrifuges capable of 40,000 or more r.p.m. are becoming quite common items of biochemical research equipment. In the current lingo of the laboratories these centrifuges are called "preparative ultracentrifuges," meaning that they serve to concentrate substances out of mixtures in preparation for more refined analysis in the ultracentrifuge. The true ultra-

MAGNETICALLY SUSPENDED ROTOR is the principal feature of the new ultracentrifuge designed by Jesse W. Beams of the University of Virginia. The rotor is driven by an air turbine, disengaged and allowed to coast.

centrifuges are called "analytical ultracentrifuges." Some manufacturers are making combination machines which can be used either to weigh molecules or as preparative separators.

An interesting example of the use of centrifugation in medical research comes from the University of California. In the Donner Laboratory of Medical Physics there, John W. Gofman and a group studied the blood of patients suffering from arteriosclerosis. They found that at high speeds a centrifuge was able to separate from the blood of these patients a substance that was rarely found in the blood of healthy persons. The new-found substance was put through an analytical ultracentrifuge and found to have a molecular weight of about three million. Other medical researchers are studying this phenomenon. If it should turn out that the presence of the Gofman factor provides early evidence of the onset of disease, the results will be far-reaching. Perhaps eventually we may have diagnostic centrifuges.

The marvel is that despite the tremendous forces to which molecules are subjected by centrifugation, they suffer no mutilation. This is especially surprising because proteins, for example, are notoriously vulnerable to many of the chemist's customary reagents, such as acids, alkalies and heat. Under the circumstances, the centrifuge is rapidly gaining favor in laboratory use as a reliable means of extracting protein substances intact from a mixture.

It is also remarkable that living cells can survive these forces. At the University of Iowa T. J. MacDougald, H. W. Beams and R. L. King centrifuged fragments of an embryonic chick heart at velocities which generated 400,000 g. After half an hour of this treatment the heart tissue continued to pulsate; when cultured, it grew at the same rate as other fragments of the same heart which had not been centrifuged.

Centrifugation can separate mixed gases of different weight. The possibility of using this property to separate fissionable uranium 235 from uranium 238 was explored in the wartime atomic bomb project. In *Atomic Energy for Military Purposes* Henry DeWolf Smyth reported: "The first experiments with centrifuges failed. Later development of the high-speed centrifuge by J. W. Beams and others led to success. H. C. Urey suggested the use of tall cylindrical centrifuges with countercurrent flow; such centrifuges have been developed successfully." But in the final competition of techniques for separating uranium, it was the gaseous exchange method and the electromagnetic method that won acceptance for large-scale application. Although a pilot plant for the centrifuge method was constructed at Bayway, N. J., and "operated successfully and gave approximately the degree of separation predicted by theory" this plant was later shut down.

Before the war Beams had experimented with cylindrical rotors and declared them preferable to conventional rotors for processing a large quantity of material. His studies showed that a "tubular centrifuge," as he called it, would do the work of 40 ordinary centrifuges. But since the tubular centrifuge has become a subject of classified military research, Beams has ceased to work with this type of rotor.

His laboratory at Charlottesville is giving much attention today to experiments with an electromagnetic drive with which he began to work in the 1930s. The rotor is a solid steel ball. Suspended in a vacuum and supported by a magnetic field, it floats in space like Mohammed's coffin. It has no need for axles, bearings, lubrication, air valves, mechanical switches or other ponderable parts. It is driven simply by the pulsations of a rotating electric field generated by electron tubes and controlled by vibrating crystals. Thus suspended, supported and driven, with friction almost zero, the ball whirls millions of revolutions per minute.

Beams is using these rotating balls to study the properties of solids. "Metals break at a tensile stress below that predicted by theory," he explained, "and to find out why they do so we are coating steel balls with other metals by electrolysis and then rotating the balls at increasing velocity until the plated film is thrown off. In an experiment with a coating of antimony it took 400,000 g to fling the antimony off the steel."

Beams is still trying to see how far he can push rotation. Recently he made the electric pulses drive one of the diminutive steel spheres to the unbelievable velocity of 48 million r.p.m.! The microscopic ball, only one twelve-thousandth of an inch in diameter, generated 500 million g. At that acceleration a drop of water would press out with a force equal to the weight of a locomotive.

This is the top attainment in rotational speed to date, so far as any published record shows. Beams has set his goal at 60 million r.p.m.—a million rotations a second—and admits that he won't be happy until he achieves it.

18   Density Gradients

by Gerald Oster
August 1965

*Devices incorporating this simple concept have
recently found application in fields as diverse as
solid-state physics, the manufacture of plastics and
the chemistry of the genetic code*

The term "density gradient" has a somewhat technical ring, but there is nothing esoteric about the concept. Examples of density gradients will be immediately familiar to the reader. The earth becomes progressively denser toward its center, with a sharp rise in density at a boundary some 1,800 miles below the surface. The density of the air we breathe decreases exponentially with increasing altitude up to about three miles, where fluctuations in temperature caused by the absorption of the sun's ultraviolet radiation disrupt its smooth decline. The density of the water in the oceans increases roughly as a linear function of increasing depth. Within the human body—indeed, in all biological systems—density gradients in the form of gradients in the concentration of various molecules are essential for the distribution of these molecules to their respective sites of action.

In the laboratory, devices that incorporate density gradients are used to measure tiny variations in the density of samples of matter. The standard device of this type, called a density-gradient column, is capable of detecting variations in density as minute as one part in 10 million, and can do so with a sample weighing only a millionth of a gram! The wide utility of this simple device can be appreciated when one considers that virtually every chemical reaction involving solids or liquids is accompanied by a change in density. In addition to serving as a powerful analytical tool, the density-gradient column is employed extensively in both research and industry to separate substances that could be isolated only with great difficulty, if at all, by other techniques. This article deals with the theoretical and practical aspects of the standard density-gradient column and also describes some of its more interesting recent ap-

plications in fields as diverse as solid-state physics, the manufacture of plastics and the chemistry of the genetic code.

The easiest way to produce a density gradient in the laboratory is to pour a lighter liquid gently over a heavier liquid, as a bartender does in making a many-layered pousse-café [*see illustration at left on page 188*]. At first there is a sharp boundary between the two liquids, but if the liquids are miscible the boundary soon begins to spread out to form a more gradual transition zone. If a sample of matter (either solid or liquid) with a density somewhere between the densities of the two original liquids is now placed in the density-gradient column, the sample will come to rest at a point where the density of the liquid in the column exactly matches its own. By carefully calibrating the column beforehand the density of the sample can be read off with considerable precision.

If a collection of samples with different densities is placed in the density-gradient column, each sample will come to rest at a different height in the column; by successively removing samples at different heights the density-gradient column can be made to serve as a highly reliable separation device. Moreover, such a column is a self-stabilizing device: any object placed in it will return to its proper level after being disturbed.

These three important features of a density-gradient column were recognized more than 300 years ago by Galileo Galilei, who described the following experiment in his *Dialogues concerning Two New Sciences:* "In the bottom of a vessel I placed some salt water and on this some fresh water; then I showed them that the ball [of wax] stopped in the middle and that

when pushed to the bottom or lifted to the top it would not remain in either of these places but would return to the middle." Noting that a single drop of the sample sufficed for the experiment, Galileo went on to suggest that his invention might have some practical applications, perhaps in medicine.

A density-gradient column constructed by Galileo's method has one serious drawback: normally the interdiffusion of the two liquids proceeds so slowly that it takes days or even weeks to achieve a useful linear gradient, that is, a gradient in which the density varies in direct proportion to the height. The diffusion process can be speeded up, however, by several techniques. One is simply to agitate the boundary region with smooth up-and-down stirring movements of increasing amplitude [*see illustration at right on page 188*]. In order for such a column to be stable for at least a month the ends of the column must be provided with reservoirs of the two liquids whose densities represent the extremes of the density range of the samples to be tested. Since the gradient can be disrupted by fluctuations in temperature, it is also desirable that the entire apparatus be placed in a constant-temperature bath. "Instant" density-gradient columns can be made by pouring the two liquids into the column at controlled rates [*see top illustrations on pages 190 and 191*].

Another way to produce a density gradient is to subject any solution to a substantial centrifugal force. Modern ultracentrifuges rotate at speeds in excess of 1,000 revolutions per second and generate forces enormously greater than gravity. As a result the molecules in solution distribute themselves in the centrifuge cell in such a way that the density is higher the greater the distance from the center of the rotor.

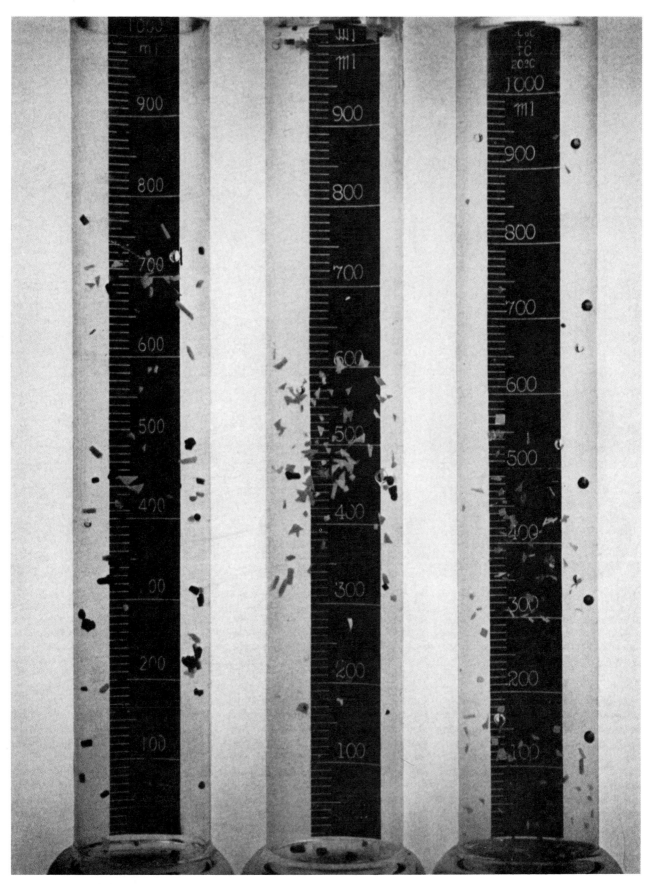

DENSITY-GRADIENT COLUMNS used for measuring the density of samples of polyethylene were photographed at the plastics-manufacturing plant of the W. R. Grace & Co. in Clifton, N.J. The columns contain a mixture of water and isopropyl alcohol, with the water on the bottom. The samples are dropped into the columns from the top one at a time and come to rest at points where their density exactly matches that in the columns. Dark spheres at right and clear spheres elsewhere in the columns are glass floats used for calibrating the density gradients. Many samples from completed tests remain in the columns. Columns are about three feet high.

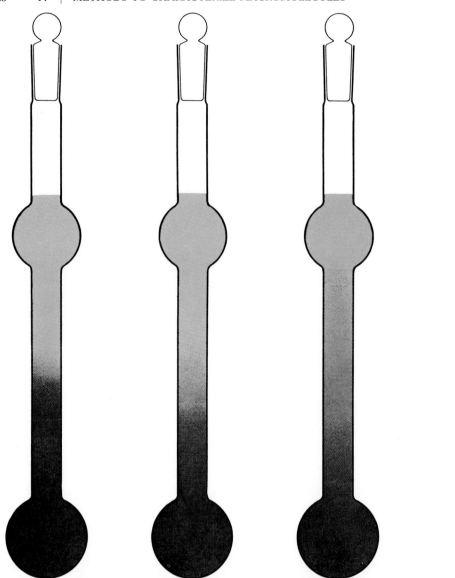

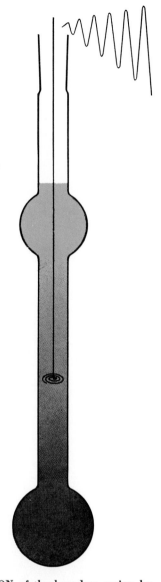

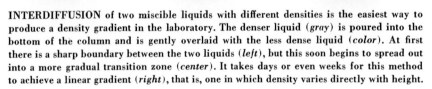

INTERDIFFUSION of two miscible liquids with different densities is the easiest way to produce a density gradient in the laboratory. The denser liquid (*gray*) is poured into the bottom of the column and is gently overlaid with the less dense liquid (*color*). At first there is a sharp boundary between the two liquids (*left*), but this soon begins to spread out into a more gradual transition zone (*center*). It takes days or even weeks for this method to achieve a linear gradient (*right*), that is, one in which density varies directly with height.

AGITATION of the boundary region between the two miscible liquids at left is one way to speed up the interdiffusion process. To achieve a good linear gradient the mixing loop should make smooth up-and-down movements of increasing amplitude.

A density-gradient column can be calibrated in several ways. One method is to introduce hollow glass floats or liquid droplets of known density into the column. (In the case of liquid droplets it is important that they be immiscible with both liquids in the column.) Another method is to determine the variation in the refractive index of the liquid in the column; this method takes advantage of the fact that light is refracted, or bent, at a larger angle when it passes through a denser medium than when it passes through a less dense medium. In an ultracentrifuge this type of calibration is usually accomplished by means of the "schlieren" optical technique [*see bottom illustra-* *tion on page 190*]. In a recent series of experiments performed in our laboratory at the Polytechnic Institute of Brooklyn, Yasunori Nishijima and I found that the variation in refractive index with density could be determined far more simply by the use of moiré-pattern gratings [*see bottom illustration on page 191*].

A sensitive density-gradient column, that is, one in which there is a small but constant variation in density over a workable distance, has many applications in research. For example, the column can be used to record the dislocations that occur in certain crystals as a result of ionizing radiation. When potassium chloride crystals, say, which are normally colorless, are treated with X rays, they turn purple; this is because some of the chlorine atoms are pushed out of their positions in the crystal lattice and are replaced by electrons. The change in color is accompanied by a minute decrease in density, which can be observed by placing the irradiated crystals in a density-gradient column consisting of tetrabromoethylene and benzene. Recently I found that if the purple crystals were subsequently illuminated with yellow light, they sank, demonstrating that their original density had been restored.

The density-gradient technique is also capable of detecting extremely small amounts of impurities in a test sample. One area in which this is particularly important is the analysis of crystalline semiconductors, such as those employed in transistor devices. As little as two parts in 10 million of boron in silicon crystals has been detected in this manner.

Density gradients play a role in almost every separation technique used today in research and industry. The traditional unit operations of chemical engineering, in which the components of a mixture are separated in the course of being transferred from one homogeneous phase to another, all involve the action of a density gradient subject to gravity; these operations are gas absorption, distillation, liquid extraction, leaching, crystallization and drying. The same applies to the newer separation techniques, which include zone refining, electrophoretic convection and thermal diffusion. Incidentally, it is quite easy to separate sodium chloride crystals from potassium chloride crystals by the density-gradient technique, a task that is practically impossible by any other means; the density gradient is provided by the interdiffusion of two liquids in which the crystals are insoluble.

It is now common practice in all polyethylene-manufacturing plants to evaluate samples of the plastic by means of density-gradient columns [see illustration on page 187]. The density of a particular sample of polyethylene is directly proportional to its degree of crystallinity. The more linear the long-chain polyethylene molecules are (that is, the less branched they are), the better the chains can pack and hence the more crystalline the sample is. The crystallinity of the plastic in turn determines its gross mechanical properties, such as stiffness and tensile strength. Thus by merely observing where a sample of polyethylene comes to rest in a calibrated density-gradient column one can immediately determine its suitability for a particular end use.

Crystallization in plastics is a rather slow process; for example, if saran is melted and suddenly chilled to room temperature, it remains amorphous for some time. Apparently the alignment of the long-chain molecules proceeds quite slowly, owing to the high viscosity of the plastic and the length of the molecules. When a piece of amorphous saran is placed in a density-gradient column, it gradually sinks as crystallization proceeds; in this way one can conveniently follow the kinetics of crystallization. Another interesting trick is to place in a density-gradient column a piece of

Mylar that has been freshly cut at one end with a pair of scissors. The cut end will tip down, presumably as a result of the greater crystallinity brought about by the shearing action of the scissors. In time, however, the sample will become level, thereby indicating the particular "relaxation" time associated with the sample.

All chemical reactions in solution involve some alteration in the bonding of the reacting molecules and hence in the distribution of electrons in the sample. This in turn almost invariably means that the density of the sample is changed. A rather extreme case of density change occurs in the conversion of a vinyl monomer into a long-chain polymer: the transformation of the double bond of the monomer into the single bond of the polymer results in an increase in density of about 35 percent. Masahide Yamamoto and I recently succeeded in carrying out such a polymerization reaction in a density-gradient column. Acrylamide, a vinyl monomer, in the presence of riboflavin can be made to link into polymer by light; we illuminated a droplet of the monomer in the column and observed the rate of fall of the droplet and thus the rate of polymerization of the sample.

The late Danish investigator Kai U. Linderstrøm-Lang, who invented many ingenious experimental techniques in biochemistry, was the first to apply the density-gradient technique to the study of the process by which proteins are digested in the intestines. When a protein molecule is attacked by digestive enzymes, its chain (or chains) of amino acid subunits is split at certain points, yielding mostly single amino acids or combinations of two amino acids. These comparatively simple units are then ingested into the bloodstream. At one end of each unit is an amino group (NH_2); at the other is a carboxyl group (COOH). Both groups are electrically charged and are capable of causing a considerable increase in the density of water by breaking down its rather open molecular structure—a process known as electrostriction. Linderstrøm-Lang was able to follow the digestion of proteins by periodically placing droplets of protein mixed with digestive juice in a density-gradient column consisting of hexane and nitrobenzene. He found in the case of the milk protein lactoglobulin that the electrostriction was greater than it should have been if only the bonds linking the amino acid subunits in the protein chain had been split. Evidently before the splitting of the bonds some kind of unfolding of the protein chain occurs that also contributes to electrostriction.

Biochemists who work with the cells of living tissues generally use density gradients in conjunction with the centrifuge. When a cell membrane is broken, many cell components—nuclei,

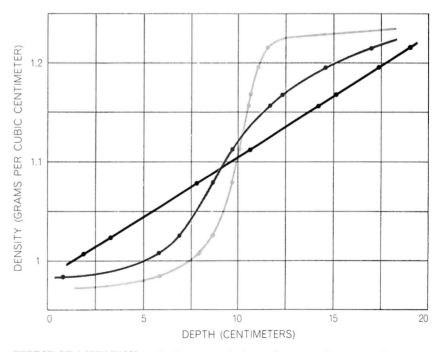

EFFECT OF AGITATION on the formation of a linear density gradient in a column consisting of benzene and bromobenzene is depicted here. The three curves represent the original density distribution (*light gray*), the distribution after 25 strokes with the mixing loop (*dark gray*) and the distribution after 50 strokes and standing for 10 hours (*black*).

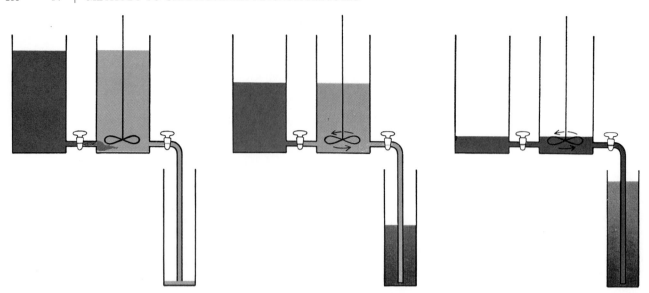

"INSTANT" DENSITY GRADIENT can be set up by the following method: The denser liquid (*gray*) is introduced into the bottom of a vessel containing the less dense liquid (*color*), where the two liquids are mixed by a propeller-like agitator. The resulting mixture flows out of the second vessel through a tube that leads to the bottom of the density-gradient column. If the rate at which the denser liquid flows from the first to the second vessel is exactly half the rate at which the mixture flows out of the second vessel, a linear gradient will be formed in the column, with the denser liquid at the bottom and the less dense liquid at the top.

mitochondria and so on—are released more or less intact. Such particles are too small to settle in a reasonable time in an ordinary gravitational density gradient. Their separation can be accelerated, however, by installing the gradient in a centrifuge. One instrument developed specifically for this purpose by Norman G. Anderson of the Oak Ridge National Laboratory allows the continuous separation of subcellular particles while the machine is operating; the method is particularly suited for collecting viruses and other particles needed in large quantities for research [*see top illustration on page 192*].

The idea of employing a density gradient inside the cell of an ultracentrifuge (a very-high-speed centrifuge) was conceived about 10 years ago by Matthew M. Meselson, Franklin W. Stahl and Jerome R. Vinograd, who were then working at the California Institute of Technology. They dissolved deoxyribonucleic acid (DNA) in a water solution of the salt cesium chloride and proceeded to subject the mixture to centrifugation. At high speeds of rotation the cesium chloride solution forms a density gradient, with the concentra-

tion of the salt (and therefore the density of the solution) highest at the outer end of the centrifuge cell. As the centrifugation proceeds the DNA molecules move outward in the cell until they come to rest at a point where their density is exactly matched by the density of the cesium chloride solution. To achieve this condition of equilibrium a rather long period of centrifugation is required, but the process can be considerably accelerated by using a very short centrifuge cell.

This technique was utilized by Meselson and Stahl to perform an ex-

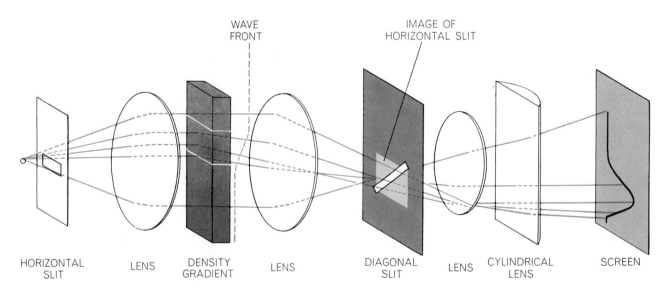

| HORIZONTAL SLIT | LENS | DENSITY GRADIENT | LENS | DIAGONAL SLIT | LENS | CYLINDRICAL LENS | SCREEN |

"SCHLIEREN" TECHNIQUE can be used to calibrate a density-gradient column by determining the variation in the refractive index of the liquid in the column. This method takes advantage of the fact that light is refracted, or bent, at a larger angle when it passes through a denser medium than when it passes through a less dense medium. The essential optical parts of the system are shown.

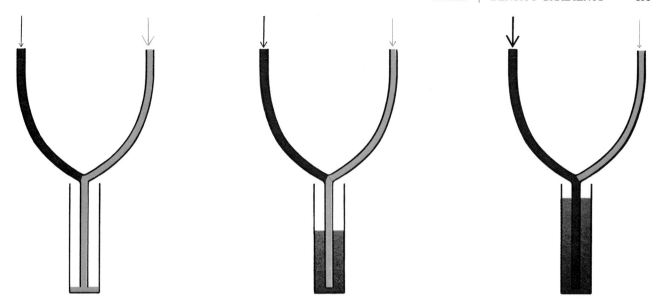

ANOTHER METHOD by which a linear density gradient can be set up almost instantaneously is to introduce the two liquids simultaneously into the column at controlled rates of flow. The flow of the less dense liquid begins at a maximum and decreases to a minimum, whereas the flow of the denser liquid begins at a minimum and increases to a maximum. The total flow at any time remains constant. One way to control the flow of the two liquids is to use syringes whose pistons are operated by helical cams of variable pitch. Vessels containing the liquids can also be suspended on a pulley that raises one at the same rate at which it lowers the other.

periment of great importance to biology. They set out to test the hypothesis, put forward by James D. Watson and F. H. C. Crick in 1953, that the molecule of DNA is a double-strand helix that replicates itself by becoming untwisted and providing two separate templates for the formation of two new complementary strands. Meselson and Stahl began by growing bacteria in a medium containing the heavy isotope of nitrogen (N-15); the heavy nitrogen was incorporated into the bacteria's DNA. The heavy-nitrogen bacteria were then placed in a medium containing the lighter common isotope of nitrogen (N-14). Shortly thereafter samples of DNA were extracted from the growing bacterial population and placed in the density-gradient ultracentrifuge. Three types of DNA were separated in the centrifuge cell: (1) molecules containing only N-14, (2) molecules containing only N-15 and (3) molecules containing both N-14 and N-15. The isotopes in the last category were equally divided, suggesting that the replication of DNA in the bacteria involves two continuous molecular units. Meselson and Stahl also found that when the hybrid DNA was heated, it split into two pure samples respectively containing only N-14 and only N-15, thus demonstrating the presence of two separable units of the original genetic material. These results provided strong support for the Watson-Crick hypothesis.

Much further work has been done on DNA with the density-gradient ultracentrifuge. For example, the technique has demonstrated that the density of a DNA molecule varies with the proportion of its subunit "bases" (specifically with the ratio of, on the one hand, the paired bases adenine and thymine and,

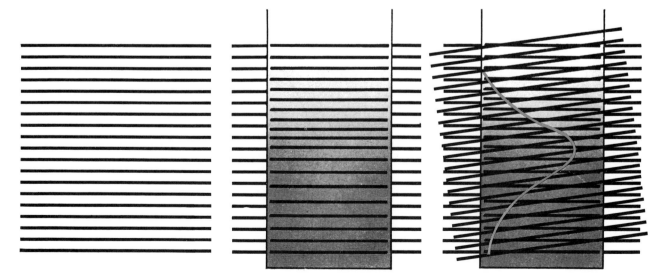

MOIRÉ TECHNIQUE also measures the density of the liquid in the column by determining the variation in refractive index. The distortion in the image of a regular grating seen through the liquid is magnified by viewing it through a slightly rotated, identical grating. The resulting moiré pattern (right) contains a curve (color) that indicates directly the variation of refractive index with density.

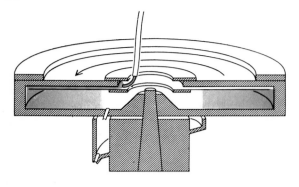

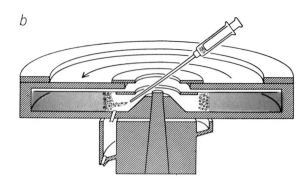

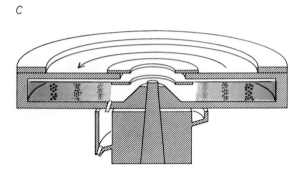

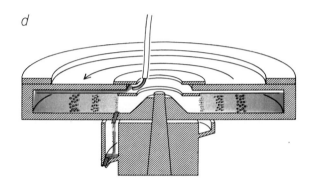

DENSITY-GRADIENT ULTRACENTRIFUGE developed by Norman G. Anderson of the Oak Ridge National Laboratory allows continuous separation of subcellular particles while the machine is operating. The gradient liquid flows into the spinning rotor from the center through small tubes to the rotor's outer edge, where it builds up a density gradient with the densest liquid outermost (a). The cell fragments are then injected into the inner, least dense edge of the gradient (b). After more spinning the particles are separated (c) and then removed by introducing a dense solution to outer edge of rotor, pushing gradient out of an inner drain tube (d).

on the other, the paired bases guanine and cytosine). This observation has been particularly interesting to workers engaged in deciphering the genetic code. Chemists who work with synthetic long-chain polymers have also found a use for the density-gradient ultracentrifuge: it is now possible to separate polymers that differ in chemical composition, degree of branching or general three-dimensional geometry.

The stabilizing action of a density-gradient column can be used to great advantage in the separation of substances by the technique of electrophoresis. In the conventional electrophoresis apparatus a mixture of substances in solution—proteins, say—is placed in a tube and subjected to the influence of an electric field. If the molecules of one of the substances have a characteristic charge, they will migrate through the tube at a characteristic rate and form a band that is separate from the rest of the mixture. The leading edge of the band is denser than the liquid behind it, which gives rise to some ambiguity in observation. If the entire process is carried out in a preformed density gradient of sugar and water, however, the band is much sharper. The Swedish investigator Harry Svensson has shown that density-gradient electrophoresis yields much finer resolution of the proteins in blood serum than had ever been thought possible. Work along these lines could lead to the discovery of new blood proteins and, it is hoped, to a more precise diagnosis and analysis of certain diseases.

In an age of increasingly complicated instrumentation it is refreshing to know that such a simple concept as the density gradient continues to find new uses in so many fields of investigation.

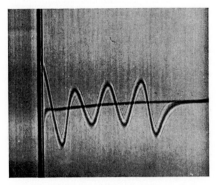

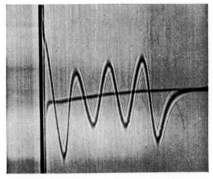

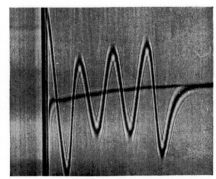

SCHLIEREN PHOTOGRAPHS of a density gradient consisting of four copolymers of styrene and iodostyrene were made by J. J. Hermans and H. A. Ende of the Chemstrand Research Center in Durham, N.C. The gradient was produced in an ultracentrifuge; the dark vertical line at the left in each photograph is the meniscus, or surface, of the liquid. The four cycles in the curve indicate the variation in refractive index for each of the copolymers, which differ in their percentage of iodine and hence in their density.

Chromatography

by William H. Stein and Stanford Moore

March 1951

*Some 50 years ago a Russian botanist devised a way
of separating the pigments of a green leaf. Now the
method has been developed to fractionate a host of
mixtures of subtly different substances*

If a petroleum ether solution of chlorophyll is filtered through a column of an adsorbent (I use mainly calcium carbonate which is stamped firmly into a narrow glass tube), then the pigments, according to the adsorption sequence, are resolved from top to bottom into various colored zones. . . . Like light rays in the spectrum, so the different components of a pigment mixture are resolved on the calcium carbonate column according to a law and can be estimated on it qualitatively and quantitatively. Such a preparation I term a chromatogram, and the corresponding method, the chromatographic method. It is self-evident that the adsorption phenomena described are not restricted to the chlorophyll pigments, and one must assume that all kinds of colored and colorless chemical compounds are subject to the same laws.

IN THESE FEW words a 34-year-old Russian botanist named Michael Tswett in 1906 described a new technique which, to quote a later comment, "was destined to influence the life of man and beast the world over." As the author of this comment, H. H. Strain of the Carnegie Institution of Washington at Stanford University, went on to observe, Tswett's chromatographic technique provided scientists with a "particularly efficient procedure for the preparation of chemical compounds in a high state of purity. Isolation and identification of chemical substances, prerequisites to investigations of composition and molecular structure, were brought to a new state of perfection." Chromatography eventually made feasible the isolation of "numerous ephemeral substances" such as vitamins, drugs and pigments, and revealed "undreamed-of reactions" in living cells.

It would be pleasant to record that Michael Tswett received in his lifetime the acclaim warranted by his discovery. Such, however, is not the case. A quarter of a century elapsed before science began to make wide use of his findings. Since 1930, however, the applications

of chromatography have been legion. By now literally thousands of papers have appeared describing investigations in which the method has been used in one way or another. This article will attempt to trace in broad outline the development and application of the technique.

Chromatography utilizes the phenomenon of adsorption. This may be defined as the adhesion of a thin layer of molecules of a gas, a dissolved substance or a liquid to the surface of a solid body. Adsorption is well known as a method for removing impurities. For example, the common gas mask contains charcoal to adsorb the molecules of a noxious gas contaminating the air breathed in by the wearer. In the sugar industry charcoal is employed to remove colored impurities from concentrated cane-sugar solutions so that a clear white crystalline product may be obtained.

Adsorption is also frequently used for the concentration of a desired substance as, for example, in the isolation of the antibiotic drug streptomycin from its mold culture. The drug was adsorbed on charcoal, along with some other substances, and was separated from many of the unwanted products in the culture by simply filtering off the charcoal-streptomycin combination. The conditions were adjusted so that streptomycin would be almost completely adsorbed in a single "batch adsorption" step. A concentrated but still rather impure preparation was obtained when the drug was dissolved from the charcoal with a solvent. If only this batch method were employed, the operations of adsorption, filtration and elution (dissolving off) would have to be repeated many times, employing fresh adsorbent and fresh solvent each time, to obtain material of a high degree of purity.

Tswett's Column

The great value of Tswett's chromatogram is that it provides a very simple device for carrying out repeated adsorptions and elutions continuously. The chromatogram consists essentially of a

glass tube filled with a specially prepared adsorbent. A solution of the material to be fractionated is added to the top of the column of adsorbent. After it has drained into the adsorbent, fresh solvent is percolated through the column. The granules of each layer of adsorbent, and the small volume of liquid that surrounds them, act as a single batch adsorption stage. As the solvent flows away from the first layer of granules and surrounds the granules of the next layer, a second batch adsorption occurs, and so on down the column to give the equivalent of hundreds or thousands of individual stages. In this manner a liquid flowing slowly through a packed column performs continuously a number of operations which would require an impossible amount of time and labor to perform individually.

One important difference should be noted between the single batch adsorption in the process we have described for the purification of streptomycin and a single adsorption stage in the chromatogram. Under these particular conditions the adsorption of streptomycin from the mold culture is essentially a one-way process. In order to dissolve the drug from the adsorbent, a solvent (acidic alcohol) that is different from the original medium is required. In the chromatogram, on the other hand, adsorption and elution frequently take place with the same solvent. Thus the molecules of a given compound in the solution are continuously shuttling back and forth in a reversible equilibrium between the flowing solvent and the stationary adsorbent. When the advancing front of solvent carries these molecules down to an unoccupied layer of adsorbent, the molecules leave the solution and are bound at the surface of the adsorbent. As soon as the influx of new solvent from above dilutes the dissolved material to less than a certain concentration, the adsorbed molecules are again released into solution. Thus a given compound travels down the chromatograph column as a zone, the leading edge of which is continuously inching

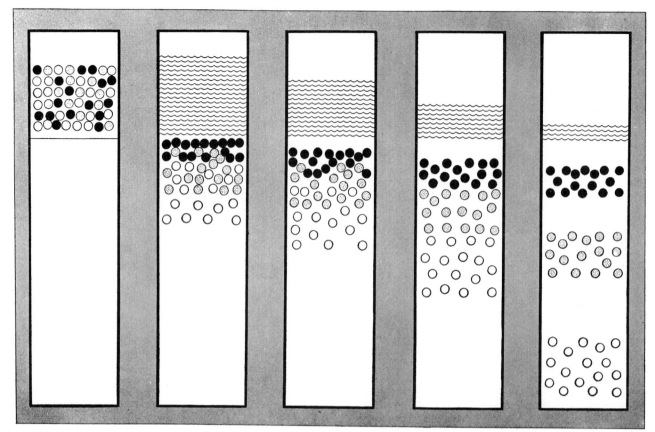

CHROMATOGRAPHIC SEPARATION is illustrated by schematic diagrams. In the first diagram a solution containing a mixture of three substances has been placed on top of an adsorbent packed into a glass tube. In the second the mixture has drained into the adsorbent and fresh solvent has been added. In the third, fourth and fifth the three substances have separated into zones as the solvent has percolated through the column.

ahead while the trailing edge is being gathered up.

The rate of travel of this zone is governed by the relative affinity of a given compound for the solvent and for the adsorbent. If the molecules in question have a great tendency to stick to the adsorbent, the zone will move slowly. If they greatly favor the flowing solvent, on the other hand, a rapidly moving zone will result. The degree of preference of a particular substance for either the solvent or the adsorbent can be measured quantitatively and given a numerical value known as the distribution coefficient. If the molecules of a substance tend to be distributed equally between the solvent and the adsorbent, the distribution coefficient is 1, and a moderately fast-moving zone results. If the coefficient is smaller than 1, the zone moves more slowly; if larger, it moves more rapidly. The actual value of the distribution coefficient depends upon the nature of the adsorbent and of the solvent as well as upon the chemical structure of the compound itself. For a given combination of adsorbent and solvent, however, the coefficient will vary only with the structure of the substance.

Sensitive Separations

It has been found that the magnitude of the coefficient is frequently very sensitive to slight differences in chemical structure, and it is for this reason that adsorption has been found so useful as a means for separating compounds. In the operation of the Tswett column very small differences in distribution coefficient are translated into appreciable differences in the rate of travel down the column. Thus various substances in a solution, even when very similar to one another chemically, may move down the column at different rates and form separate zones. Within recent years it has been shown that the behavior of zones on a chromatogram can be expressed in mathematical terms, and doubtless this is the "law" whose existence Tswett guessed and mentioned in his first paper.

After the pigments or other substances in a mixture have been separated into zones on the chromatogram, it is necessary to recover the material in each zone for chemical identification. Often this is done simply by pushing the column of adsorbent out of the chromatograph tube with a plunger. As each colored zone of adsorbent emerges from the glass tube, it is cut off with a knife. The segment cut off is then treated with a solvent that removes the desired pigment from the adsorbent, and this solution in turn is evaporated to yield the purified material. There are few more dramatic events

in a laboratory than the crystallization of a much sought compound after this simple procedure.

In recent years the Swiss chemist Tadeus Reichstein, and others, have preferred to pass solvent through the column until the various zones emerge successively from the bottom in solution, instead of pushing the entire column out of the chromatographic tube. Each zone is collected as a separate fraction.

When pigments are being separated, forming zones of different colors, the process can easily be followed visually, as Tswett did in his pioneer experiments. Fortunately, as Tswett predicted, chromatography is not restricted to pigments. The technique works equally well for colorless substances, although some physical or chemical method must be employed to make them distinguishable. For example, some types of compounds show characteristic fluorescent colors when viewed under ultraviolet light. In other cases the fractions from the column, or the extruded column itself, may be treated with various chemical reagents that react with the material in the different zones to produce colors. A radioactive material can be located, after separation, with a Geiger counter; a biologically active fraction can be identified by bioassay, and so on. Actually there is a vast number of ways in which the progress of chromatographic fractionations can be followed, and new ones are continually being devised.

One of the advantages of chromatography is that it requires no elaborate or expensive equipment. All the essential items are readily available in any chemical laboratory. Much of the early and extremely useful work in chromatography was done simply with a long glass tube narrowed at one end and with a few flasks and beakers. Recently additional equipment and procedures have been devised that permit greater efficiency and ease of operation, but few of the modern innovations have modified in any fundamental manner the elegant simplicity of Tswett's original column.

Permutations and Computations

Chromatography is a versatile method. It can be used to fractionate mixtures of gases, liquids or dissolved solids. The method is also a mild one. Although occasionally compounds are altered chemically during the procedure, chromatography has been employed successfully for the separation of some of the most fragile and elusive of substances. It appears increasingly likely that it will eventually be possible to handle chromatographically any group of compounds that can be brought into solution. All that is required is the right combination of solvent and adsorbent.

But that is the difficulty: one of the prime shortcomings of chromatography is that it is not yet possible to predict with assurance what combination of solvent and adsorbent will prove suitable in any given case. The number of possible combinations of solvents and adsorbents is staggering. A partial list of the substances used as adsorbents includes the carbonates of calcium, magnesium and sodium; various forms of charcoal activated in special ways; fuller's earth, bentonite, Lloyd's reagent, talc and innumerable other clays and diatomaceous earths; alumina; silica gel, and a whole host of organic substances such as cellulose, starch, sucrose, inulin, benzoic acid and, more recently, ion-exchange resins. The solvents, too, form an imposing list: water; aqueous solutions of various acids, alkalies and salts; methyl, ethyl, benzyl, propyl, butyl and amyl alcohols; ketonic solvents such as acetone, methyl ethyl ketone and cyclohexanone; hydrocarbons such as benzene, toluene and hexane; the chlorinated hydrocarbons chloroform and carbon tetrachloride; various ethers; the newer derivatives of ethylene glycol, and countless more, alone or in combination.

It is not surprising that for a long time the choice of a chromatographic system was based largely on trial and error. Experience and a few general rules, such

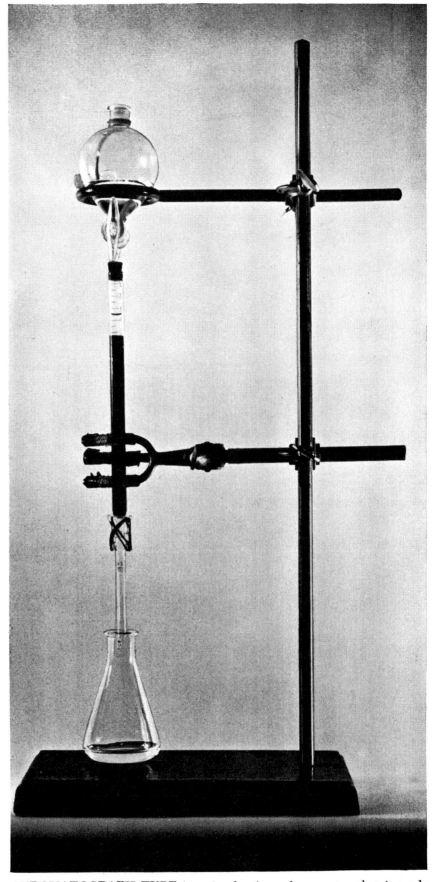

CHROMATOGRAPH TUBE is a simple piece of apparatus that is made in a variety of sizes for different purposes. This small tube is filled with an ion-exchange resin as an adsorbent. At the top of the column is a reservoir of solvent. A small amount of the solvent may be seen on top of the adsorbent.

AUTOMATIC MACHINE is used to collect fractions containing the constituents of a complex mixture. The chromatograph column is in the center; solvent drips from it into collecting tubes on the turntable at the bottom. In the arm beneath the column is a photoelectric cell; when a given number of drops pass it, the turntable rotates and brings an empty tube into position.

as that like tends to adsorb like, have gradually provided some guidance in choosing adsorbent-solvent combinations, and progress is continually being made toward a less empirical procedure. The necessity for much tedious experimentation still is a basic difficulty, and is probably responsible for the fact that the development of the chromatographic method did not proceed more rapidly after Tswett's discovery in 1906.

Milestones

From 1906 until 1931 the Tswett column was used in only a few sporadic investigations. Then R. Kuhn and E. Lederer, two chemists working in Germany, found out with the help of chromatography that crystalline carotene, the yellow pigment of carrots, which for a century had been thought to be a pure substance, was in reality a mixture of two compounds. This important discovery focused attention upon chromatography as a laboratory tool.

Carotene is a precursor of vitamin A, and at that time chemists the world over were busily engaged in attempts to isolate and determine the structure of this and many other vitamins. Soon many investigators in the field of the carotenoids began to use the Tswett column. With its aid Paul Karrer and his associates in Switzerland made important progress in studying the chemistry of vitamin A, and Laszlo Zechmeister of Hungary (now at California Institute of Technology), working with L. Cholnoky, made significant contributions both to carotenoid chemistry and to chromatography. Strain at Stanford, whose comments were quoted at the beginning of this article, did pioneer work in the U. S. in the same fields.

Meanwhile the new tool was also producing major advances in the field of sterol chemistry. The sterols are a large group of substances that play key roles in the life processes of animals and plants. Within this group are found cholesterol, the bile acids, the male and female sex hormones, the plant sterols, vitamin D and the hormones of the adrenal cortex, including desoxycorticosterone and the now much sought cortisone. The sterols have a complicated structure that is capable of innumerable variations. To isolate them in pure form from the intricate mixtures in which they occur in the body, and to elucidate their chemical structure, presented a tremendous challenge to chemists. The eventual successful solution of the problem constitutes one of the truly brilliant chapters in modern chemistry. And in much of this work, destined to be of crucial importance to biochemistry, biology and medicine, the Tswett column was an indispensable aid.

These applications of chromatography all involved the fractionation of mix-

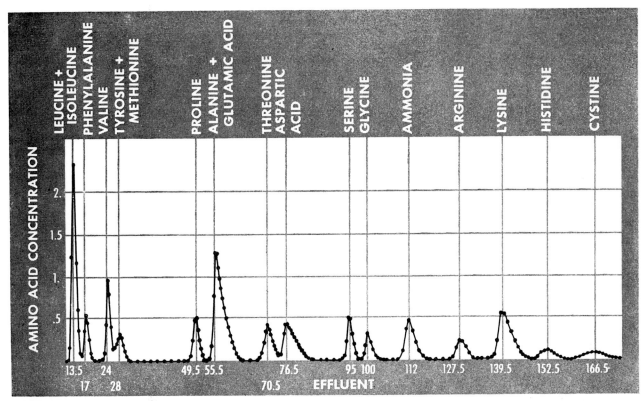

DIAGRAM shows the results of a fractionation by the machine on the opposite page. The mixture was composed of amino acids resulting from the breakdown of the protein bovine serum albumin. Each dot on the curve represents the concentration of one or more amino acids found in one collecting tube. The identity of the amino acids separated is indicated at the top of the page; their relative quantity is shown by the area under the peaks. The concentration of the amino acids is given in millimoles; the volume of effluent, in cubic centimeters.

tures of substances soluble in fat solvents. The widespread use of chromatography for the separation of compounds soluble in water constitutes another chapter in the history of the method. This chapter was opened about 10 years ago by two young British chemists, A. J. P. Martin and R. L. M. Synge, who devised a technique they have called "partition chromatography."

In partition chromatography the column is packed with a porous solid, such as silica gel, cellulose or starch, which is capable of holding water; the water is usually added to the solid before the chromatograph tube is packed. The substances to be separated are put through the column dissolved in a solvent, such as butyl alcohol, that does not mix with water. In such columns compounds separate into zones just as they do in the Tswett column. As Martin and Synge explained their principle, the compounds were separated by virtue of their distribution between the water, held within the pores of the packing as in a sponge, and the organic solvent flowing through the column. In other words, the water took the place of the adsorbent in the classical Tswett column. Hence the new process was called "liquid-liquid partition" chromatography, to distinguish it from the solid-liquid distribution operating in the Tswett column. Actually it now seems likely that

the two processes do not differ as much as was first thought.

The application of this technique made it possible to separate a number of water-soluble compounds of particular interest to the biochemist, notably the carbohydrates and the amino acids. The new approach also led to improvements in methods for the separation of fat-soluble substances.

Paper Chromatography

Perhaps the most important development that has arisen from the concept of partition chromatography is the novel and simple technique of paper chromatography. In this method, developed by Martin and his colleagues R. Consden and A. H. Gordon, the separations are carried out on strips or sheets of paper instead of on an adsorbent in a glass tube. Chromatography on paper had been tried before, but the British investigators made it work so effectively that within a few years it has become by far the most popular of all chromatographic methods, and is now being used in chemical laboratories the world over.

An entire paper chromatogram ca.. be carried out on a sheet of paper somewhat larger than a page of this magazine. This corresponds to the chromatographic column. A single drop of the solution to

be fractionated is deposited on a corner of the sheet, and the sheet is then hung, with the drop at the top, in a closed box containing an atmosphere saturated with water vapor. The top edge of the paper is bent over to dip into a trough containing an organic solvent mixture (*see photograph on the next page*). The solvent creeps down the paper by capillarity. After the solvent has traveled the length of the paper, the compounds that were mixed together in the original drop of solution are separated and distributed as a series of spots in a vertical line along the edge of the paper. If, as is frequently the case, the spots still contain a mixture of compounds, making additional purification desirable, the paper may be turned around so that this edge is at the top and a new solvent is then allowed to flow down the sheet. Thus each spot serves as the starting point for a new chromatogram and may yield a series of spots across the sheet. The end result is a so-called "two dimensional" chromatogram, in which compounds are distributed more or less over the whole sheet. This procedure substantially increases the resolving power of the method.

A complex mixture of amino acids, for example, can be separated into spots which give the pattern shown in the background of the illustration on the cover of this issue. Here the spots have been treated with a reagent that reacts

PAPER CHROMATOGRAPHY uses paper in place of the chromatograph column. Here the paper is hung from a trough containing a solvent. As the solvent travels down the paper (*translucent area at top*), the constitu-ents of a mixture placed near the top will be carried various distances. The paper is suspended in a glass-enclosed box. Within the box a saturated atmosphere is maintained by water and solvent in the dishes at the bottom.

with amino acids to produce a reddish blue color. This is the usual procedure when colorless substances are chromatographed; the paper is sprayed at the end of the experiment with a specific reagent that reacts with the particular class of compounds to yield colored products. Some types of compounds can be seen by examination of the paper under ultraviolet light. Sometimes the spots may be cut out and the separated compounds washed off the bits of paper for further tests.

How Chemists Use It

The great advantages of paper chromatography are its simplicity, rapidity and high resolving power. It is being applied to the study of the constituents of human blood and urine in health and disease, to the separation of antibiotics, to the discovery of new amino acids in bacteria and to a great variety of other chemical problems.

Recently paper chromatography has been employed by Melvin Calvin and

Andrew Benson at the University of California in very beautiful investigations of the mechanism of the fixation of carbon dioxide during photosynthesis. They exposed green algae to radioactive carbon dioxide and later chromatographed on paper the mixture of photosynthetic products extracted from the plant cells. The paper chromatogram was then covered with a sheet of X-ray film and left in a dark room for several days. Any products of photosynthesis that had taken up radioactive carbon dioxide would of course show their radioactivity by making spots on the film, which could be seen when the film was developed. From the positions of the spots on the film laid over the paper, the nature of the products that had taken up carbon dioxide could be inferred.

The ultramicro scale upon which paper chromatography can be performed is one of the factors contributing to its usefulness. With the expenditure of only a fraction of a milligram of material, an investigator can frequently obtain information of incalculable value. The small amounts involved mean that it is not possible to isolate a substance in sufficient quantity for identification by conventional chemical procedures. In paper chromatography, identification rests primarily upon the position of the compound upon the paper sheet. Control experiments are required to show that the position of an unknown substance matches exactly that of a known. For this reason paper chromatography cannot provide identification of a completely new compound never before handled by the investigator. Such a substance may appear as a spot in a position not normally occupied by any known compound. When that occurs, the chromatographer must resort to other methods, frequently column chromatography, to isolate and identify the substance. Paper chromatograms do not always provide quantitative results. Frequently they reveal the nature, but not the exact amounts, of the substances present in the mixture chromatographed.

Separation of Amino Acids

Within the past five years certain refinements in the techniques for column chromatography have been developed. These were stimulated by the long-standing need for quantitative work in the field of protein chemistry. They were touched upon in Joseph S. Fruton's article, "Proteins," (SCIENTIFIC AMERICAN Offprint 10).

As Fruton pointed out, it is of great importance in protein chemistry to find out precisely the quantities of each of the 20-odd amino acids yielded by the

breakdown of a protein. In the laboratory of the late Max Bergmann at the Rockefeller Institute for Medical Research the authors of the present article had been concerned with this problem for some time. In 1945 they undertook quantitative amino acid analysis with the aid of chromatography. The procedure finally evolved employed columns packed with potato starch and solvents composed of mixtures of certain alcohols with water or aqueous hydrochloric acid. After the amino acid mixture to be analyzed had been added to the prepared chromatograph column, the solvent was run through very slowly until the amino acids had progressed the length of the column and emerged in the effluent issuing from the bottom of the column. The effluent was collected in a large number of successive small fractions of accurately known volume (.5 cubic centimeters per fraction). A quantitative analysis of each fraction was made to determine its amino acid concentration. From these analyses a curve was drawn, showing the concentrations of the various amino acids in the successive fractions (see diagram on page 197).

The amino acids, which emerge serially from the bottom of the column, appear on the curve as peaks more or less separated from one another by valleys. Complete separation of all the amino acids in the original mixture is not obtained in a single chromatogram; several emerge together in one peak. Those that overlap can be separated, however, by running additional chromatograms with different solvents. Moreover, the quantity of amino acid present is given accurately by the area under each of the peaks. It has been possible, therefore, to separate and to determine quantitatively each of the 20 or more amino acids found among the cleavage products of a protein. Starch chromatography has proved useful in several laboratories for determining accurately the composition of blood proteins such as hemoglobin and of important hormones of the pituitary gland. It is being employed also in investigations of the metabolism of amino acids using isotopic tracers.

An efficient chromatogram often requires a very slow rate of solvent flow through the column. It takes seven days, for example, to complete an experiment of the type we have been discussing. Obviously it would be too tedious to stand by all this time and collect each fraction by hand. A simple automatic collecting machine has therefore been built. It consists essentially of a circular rack holding the receiving tubes and a photoelectric counter to count the drops emerging from the chromatogram (see photograph on page 196). After a set number of drops have emerged, the

turntable moves one notch to bring a new tube under the column. The machine is capable of operating 24 hours a day, thus making practical the performance of experiments of long duration.

Recent Refinements

This article would not be complete without calling attention to other recent developments which appear likely to broaden even further the scope and utility of the chromatographic method. Arne Tiselius and S. Claesson in Sweden have described two new chromatographic techniques known as "frontal analysis" and "displacement development." These procedures, based on new principles, give curves consisting of a series of steps, instead of peaks and valleys. The Swedish workers have devised recording instruments for measuring the stepwise changes in concentration of the effluent. The development of these methods has provided the chemist with additional valuable tools to aid him in his never-ending quest for better ways of resolving complex mixtures. They have already been used to fractionate such divers mixtures as gaseous hydrocarbons, amino acids, peptides, fatty acids and sugars.

The recent development of organic ion-exchange resins also promises to have profound effects upon the chromatographic method. These ion exchangers were originally prepared to remove dissolved salts from water ("Ion Exchange," by Harold F. Walton; SCIENTIFIC AMERICAN, November, 1950). It soon became apparent, however, that they are capable of much wider application. During the war, workers at Oak Ridge and associated university laboratories employed columns packed with ion-exchange resins to separate many of the rare-earth elements formed as fission products in nuclear reactors. Columns of ion-exchange resins are now being used in many laboratories to separate organic compounds. Indeed, current investigations by S. M. Partridge in England and by the authors at the Rockefeller Institute suggest that ion-exchange resins may prove much superior to starch or cellulose as adsorbents for the separation of amino acids and peptides.

It has been said by many, notably the eminent 19th-century physiologist Claude Bernard, that progress in science frequently depends upon the development of good methods. Chromatography furnishes a vivid example of the truth of this statement. It is doubtful that even the fertile imagination of Michael Tswett could have foreseen the great advances in scientific knowledge that have already come, and the greater ones that will surely come in the future, from the application of his chromatograms.

Electrophoresis

by George W. Gray
December 1951

*It is the migration of charged particles in a fluid
between two electrical poles. In recent years the
phenomenon has been used with signal success to
study and to separate protein molecules*

NATURE is restless—motion is one of its most fundamental attributes. "All things flow," declared Heraclitus 2,500 years ago, and modern science is continually turning up new evidence in support of his generalization. On the grand scale we have the testimony of the stars and nebulae; at the microscopic level we can see the eternal Brownian movement of tiny particles. Robert Brown, the 19th-century botanist who discovered the cell nucleus, first observed this perpetual motion in the dancing of pollen grains in a drop of water. In our century the Brownian movement has been studied by many eminent experimenters and theorists, including Albert Einstein, and we know that the random motion is characteristic of all particles in gaseous and liquid suspensions, from the largest colloids to the smallest molecules.

Brownian movement is responsible for the diffusion of molecules and for the chance contacts among them that result in chemical reactions. Important and helpful as this random motion is to the formation of new compounds, it is obstructive when the chemist desires to isolate one of the compounds from a mixture. Faced with this problem of separating out molecules of a particular composition or species, he has resorted to various schemes. Crystallization is one stratagem for getting a fraction pure; sedimentation, assisted by the centrifuge, is another. A third technique is based on the electrical properties of particles. This remarkable method of analysis is known as electrophoresis. The term, derived from the Greek, means "borne by electricity."

Electrical Migration

The principle of electrophoresis has been known for nearly a century and a half. In 1807 a Russian physicist named Alexander Reuss observed that when electricity was passed through glass tubing containing water and clay, colloidal clay particles moved toward the positive electrode. Michael Faraday in England and E. H. Du Bois-Reymond in Germany

confirmed his discovery and extended it, showing that any negatively-charged particles in solution or suspension moved toward the positive electrode and positively-charged particles in the opposite direction. What was more, particles moved at differing speeds depending on the number of excess charges they carried; the greater the number of charges, the faster the migration. The way was thus opened to use electrophoresis as a means of separating particles out of a mixture according to their electrical properties. Physicists and chemists in half a dozen countries contributed to the further exploration of the phenomenon. Bit by bit over the decades the theory of electrophoresis was built into a logical structure and methods were developed for applying it to laboratory problems.

Today there are three different techniques. We shall take them in the order in which they were developed. The oldest is known as microscopic electrophoresis.

Under the Microscope

The first essential for an electrophoretic experiment is an electrically active particle. This means a particle with more positive than negative charges on its surface—in other words, an ionized particle. Many particles, *e.g.*, those of certain gums, sugars, and the huge composite molecules known as polysaccharides, carry no excess charges on their surfaces. They therefore remain essentially stationary in the electric field, and it is useless to attempt to study them by electrophoresis. But practically all proteins carry unbalanced distributions of surface charges, and electrophoresis is ideally adapted to the analysis, separation and identification of these biologically important substances.

The second requisite is a solution that conducts electricity. Distilled water is a poor conductor, but the addition of a pinch of salt, acid, alkali or other ionizing compound (known as an electrolyte) will quickly render it conducting.

In the microscopic method of electrophoresis the experimenter watches and

measures the migration of particles in a solution or suspension contained in a glass tube placed horizontally on the stage of a microscope. Only relatively large objects that can be seen under the microscope, such as blood cells, protozoa, bacteria and colloidal particles, lend themselves to investigation by this method. This would seem to rule out molecules, inasmuch as even the largest are invisible, but chemists have found it possible to study protein molecules indirectly by introducing tiny quartz spheres into the solution. The molecules attach themselves to the spheres in an adhering layer one molecule thick, and thereafter the protein-surfaced quartz responds to the electric current in terms of the charges on the protein.

The microscopic method has been used to study surface phenomena and immune bodies in liquids. John H. Northrop and Moses Kunitz found it useful in their investigation of enzymes at the Rockefeller Institute for Medical Research. Harold A. Abramson employed it at Columbia University in studies of red blood cells and soluble proteins. Recently G. L. Ada and Joyce D. Stone, at the Walter and Eliza Hall Institute in Australia, turned to the microscopic method to study the effect of certain agglutinizing viruses on human red cells.

But the microscopic technique is not often used today. It has been largely supplanted by a newer method which is better suited to the separation of particles, is applicable to large quantities of material and does not require visibility, so that it can be applied to the smallest molecules and even the molecular fragments known as ions. This method is called moving-boundary electrophoresis.

The Moving Boundary

Instead of observing individual particles, the experimenter measures the movement of the boundary of a mass of particles. The material to be studied is poured into the bottom of a U-tube. On top of this, in the two arms of the U, is placed an electrolytic ("buffer") solu-

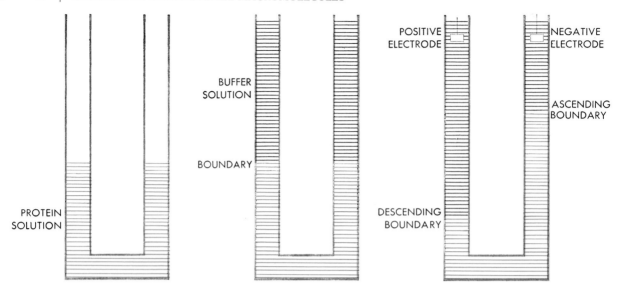

MIGRATION OF A PROTEIN by electrophoresis is shown in these three diagrams. At the left a protein solution (*red*) is placed in a U-shaped tube. In the center a buffer solution is carefully laid atop the protein solution. At the right a current is passed through the solutions and the protein migrates toward one of the electrodes.

tion. It is laid on carefully so that there is a sharp boundary between the two solutions (*see drawings at the top of this page*). An electric current is then passed through the solutions by means of a positive electrode inserted in the top of one arm of the tube and a negative electrode in the other. The object is to see how fast the boundary moves under the urge of electricity—and in what direction.

If the material under study is a protein, carrying an excess of positive charges on its molecular surface, the boundary moves toward the negative electrode. Each charge serves as a little tugboat to tow the huge molecule in the direction of motion of the positively-charged stream, and the more tugs there are, the faster, of course, the molecule moves.

It is possible by changing the acidity of the solution to slow down this movement. As the *p*H (concentration of hydrogen ions) is progressively raised, the boundary moves slower and slower, until finally a point is reached at which it becomes stationary. If the *p*H is further increased to make the solution alkaline, the negative charges on the protein will begin to assert themselves, operating as tugs pulling in the opposite direction, and then the boundary will move toward the positive electrode. The composition of the electrolytic solution is therefore a matter of the greatest importance in electrophoresis. Its degree of acidity or alkalinity must be regulated with high precision. The balance between acid and base is so delicate that the slightest interference—such as absorption of carbon dioxide from the air or of a trace of alkali from the glass tube—can render the electrophoretic calculations inaccurate. To protect the precision of the con-

centration the chemist adds certain salts which act to maintain the *p*H at a constant level, the salts themselves absorbing any tendency to variation. These substances, cushioning the chemical blows to which a solution is subject, are called buffers. Some of the recent advances in moving-boundary technique must be credited to the development of more satisfactory buffer systems.

The stage of buffered concentration at which a substance reaches electrical equilibrium and ceases to move toward either electrode is a measure of its isoelectric point. Every electrically responsive compound has its own isoelectric point, and this characteristic is important because it tells something about the substance's structure and chemical properties. The moving-boundary method has established the isoelectric points of hundreds of proteins and other biologically active molecules. The method has scored its greatest triumphs in identifying hidden substances, whose presence in some instances was not even suspected, in complex materials such as blood or milk.

The moving-boundary method has brought in a new era of biochemical research in the past 15 years. It owes its coming-of-age principally to a group of experimenters in Sweden, and most of all to the famous chemist Arne Wilhelm Kaurin Tiselius. Indeed, Tiselius is so closely associated with the development that the modern type of moving-boundary equipment is commonly called "the Tiselius apparatus."

Exciting Invention

Tiselius was born in Stockholm in 1902, fell in love with chemistry in high school and selected Uppsala as his uni-

versity because there he could study under The Svedberg. Svedberg, then engaged in developing the ultracentrifuge, called the attention of his student to electrophoresis as another means for studying proteins. After graduation Tiselius joined Svedberg in a research project which introduced him to the moving-boundary method. Certain anomalies in the experiments led Tiselius to investigate possible sources of error in the existing apparatus and the prevailing technique, and he spent the next five years testing the possibilities and limitations. When he presented himself for the doctor of science degree in 1930, he submitted a comprehensive review of electrophoresis as his thesis. And with this account Tiselius apparently bade electrophoresis adieu.

He turned to the investigation of adsorption and diffusion. The new interest took him on an exploratory trip to the Faroe Islands in search of zeolite crystals for diffusion experiments and in 1934 to Princeton University on a year's Rockefeller Foundation fellowship. In Princeton he met John H. Northrop and M. L. Anson of the Rockefeller Institute staff. Close association with these biochemists rekindled his interest in electrophoresis. Nobody needed to point out its possibilities to Tiselius; he was keenly aware from his previous studies that it could be enormously helpful, provided only that the technique were made exact and discriminating. When he got back to Uppsala, he resumed experiments with the moving-boundary method.

The first fruit of this study came in the summer of 1937. Tiselius announced in the *Transactions* of the Faraday Society of London the development of a new apparatus for electrophoretic analysis. His paper not only described the design and operation of this equipment but reported a discovery which gave evidence of its value as an analytical tool. Employing it to analyze the serum of blood, he had found that the blood fraction known as globulin was in reality a mixture of three substances. He named them alpha globulin, beta globulin and gamma globulin.

The publication of this paper fired the imagination of biochemists all over the world. Its arrival in a laboratory was like the touching of a spark to tinder, for immediately researchers began to inquire how they might buy, borrow or build one of these powerful machines.

Chain Reaction

When Tiselius' paper arrived at the Rockefeller Foundation's International Health Division, which was then studying the viruses of yellow fever and influenza, its staff decided that the new tool might be useful in the virus program. They sent Frank L. Horsfall to Uppsala in January, 1938, and he brought back a Tiselius apparatus—one of the first to reach the U. S. Meanwhile, the Tiselius report had also attracted the attention of Duncan A. MacInnes and Lewis G. Longsworth at the Rockefeller Institute. Longsworth began to build an apparatus according to the Tiselius design. The homemade instrument he constructed launched him on a career which was to make his laboratory one of the world's leading centers of electrophoretic research.

Longsworth showed his new apparatus to Karl Landsteiner, who was working at the Institute with several substances so similar to one another chemically that they could be distinguished only by their differing effects on experimental animals. Landsteiner offered Longsworth a mixture of two substances—the egg albumins of the guinea hen and the duck—and asked: "Can your electrophoresis demonstrate any physical difference between these materials?" Longsworth put a solution of the mixture through the apparatus and found at once that the two albumins separated, forming boundaries that moved at different speeds. This made a profound impression on Landsteiner. From being skeptical and lukewarm he now became an ardent proponent of electrophoresis, and wanted everything that came his way submitted to its analysis. He was constantly on the lookout for new test material. A man driving a donkey along a New York street was stopped and asked for permission to sample a little of the animal's blood for testing. "No," retorted the driver, "this donkey is going on the stage of the Metropolitan Opera and can't afford to lose any blood!"

The anatomical laboratory of Columbia's College of Physicians and Surgeons and the department of physiological chemistry at the Yale Medical School also adopted the Tiselius apparatus. The Yale group, consisting of Reginald A. Shipley, Kurt G. Stern and Abraham White, decided in 1937 to use electrophoresis to study the hormones of the anterior lobe of the pituitary gland. They started to build an apparatus based on Tiselius' 1930 doctoral thesis, not knowing about the new instrument he had just developed. At this opportune moment Svedberg arrived in the U. S. for a visit. Newspaper reporters who met his ship in New York asked Svedberg what was new in science. "The most exciting thing in Sweden is an apparatus developed by my colleague Tiselius for separating proteins electrically," he answered. The reporters clamored for details, but the professor protested that they were too technical for a newspaper story. "Full specifications are being published in the *Transactions* of the Faraday Society," he added. When the Yale group read this interview in *The New York Times*, they made a beeline for the *Transactions*. From the description and references given there Stern built the Yale Medical School's first Tiselius apparatus.

In like manner the Faraday Society paper started a chain reaction of electrophoretic interest in laboratories in England and on the Continent.

Tiselius' Stratagems

Before Tiselius, all moving-boundary systems used in electrophoresis had been deficient in "resolving power." No matter how well defined the boundary was at the beginning of an experiment, as soon as the substances in a mixture began to migrate, the boundary became fuzzy. It was difficult to determine where one component left off and another began. Everyone familiar with electrophoresis knew the cause of the difficulty. The passage of electricity through any medium generates heat, and in the case of a solution this heating sets up convection currents within the liquid. It was these random currents that disturbed

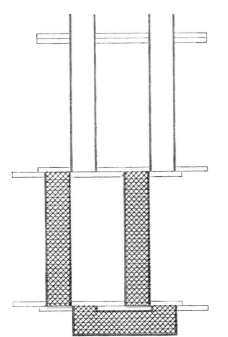 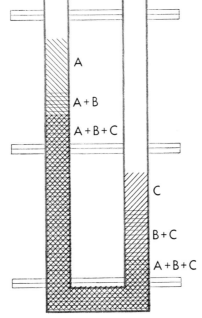 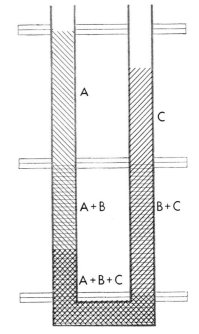

SEPARATION OF A PROTEIN is shown in a U-tube made up of sections. At the left a mixture of three proteins has been placed in the bottom of the tube and a buffer solution in the top. In the center and at the left, when the protein mixture and the buffer solution are in contact, each protein moves in a characteristic way.

ELECTROPHORESIS CELL has a U-tube at bottom center. The electrodes are in the glass columns at left and right. The U-tube is made narrow so that its contents can be cooled to prevent convection currents.

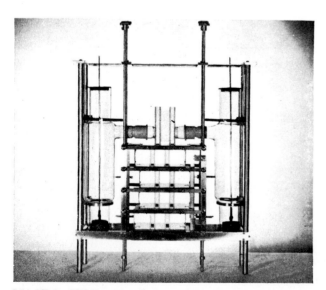

MACRO CELL is used for relatively large-scale separations of proteins. It has four movable sections.

MICRO CELL is used when only small amounts of material are available for the electrophoretic analysis.

and distorted the boundaries. The basic problem was to find some means of dissipating the heat.

Tiselius met this problem with two inventions. First, he used straight-sided tubing, instead of the conventional round kind, for his U-tube. The tubing has a narrow rectangular cross section, like the slot of a mail chute. Such a shape exposes more wall surface for a given volume, and so provides better cooling to the solution within. His second stratagem was to immerse the U-tube in a cooling bath of water held at a certain temperature just above the freezing point. This temperature, about 39 degrees Fahrenheit, is the temperature of maximal density of the solution, meaning the point at which a change in temperature produces the least change in density and therefore a minimum of convection. Hence it is the most favorable temperature for electrophoresis.

The cold-water bath increased the resolving power of the apparatus. But this sharpening of the boundaries would be of little value unless they could be observed and measured precisely. To care for that need, Tiselius equipped his apparatus with an optical system which rendered the boundaries visible. They showed up as shadows or schlieren, from which the system is known as the "schlieren method." Originally developed in France by J. B. L. Foucault about 100 years ago for testing lenses and mirrors, the method is based on the fact that at a boundary between two transparent materials of different density the light rays are bent or refracted, thus casting shadows which mark the place of refraction. The German physicist A. Toepler later adapted Foucault's method to the study of optical inhomogeneities in front of the lens or mirror, and Tiselius' application stems from both Foucault and Toepler.

Tiselius introduced another important innovation in the U-tube which greatly facilitated the trapping of specific fractions after separation of a mixture into its components. Hugo Theorell of the Nobel Institute in Stockholm had had the idea of doing this by means of rubber disks which could be slipped between the different solutions in the tube at the boundaries in such a way as to isolate them in separate compartments. With this device Theorell had obtained relatively pure separations of several proteins. Tiselius improved on it by making the U-tube itself in a series of sections, each of which can be shifted to one side and thus separated from the rest of the tube. This makes it possible to separate a fraction for examination with the minimum disturbance of the contents. The sections slide on glass flanges that are sealed with a chemically inert lubricant. This is the way the apparatus operates in a simple form of the experiment: At the beginning of the experiment the solution of mixed proteins fills the lower sections of the U-tube in both arms, and a buffer solution fills the upper sections. When the electric current is turned on, the proteins travel up the cathode arm. The current is turned off when the upper boundary in this arm reaches the top of a section. The section is then slid on its flanges to the right. Trapped in the cathode arm of the section, especially in its upper part, is a pure fraction of the fastest-moving protein, for this protein is present in purest form at the leading boundary. Similarly a pure sample of the slowest-moving fraction can just as easily be trapped at the trailing boundary in the other limb. An ideal form of the experiment, in which the fastest protein is trapped in the cathode arm and the slowest in the anode, is illustrated on page 203.

The longer the path through which the molecules migrate, the more complete will be the separation of components out of a mixture. This means that the arms of the U-tube have to be extended as high as possible. But Tiselius hit upon a more convenient way to accomplish the same result. He introduced a pistonlike plunger which is slowly lowered by clockwork into the arm up which the migration is ascending. This counter pressure against the moving buffer solution displaces the whole sequence of separated components in the direction opposite to their migration—the effect is similar to the motion of a line of people climbing a descending escalator. The same effect may also be accomplished by flowing fresh buffer solution into the area above the ascending column. The rate of flow is adjusted so that the fastest component will just succeed in climbing up the tube while the slower components will gradually be washed back. In this manner the resolving power of the apparatus is greatly increased.

Peaks and Valleys

Electrophoresis is the most effective technique known for locating and recovering physiologically active substances in a relatively pure state. The Tiselius machine is a more discriminating separator, for example, than the ultracentrifuge. And yet, even with the most refined apparatus, some mixture remains in the fractions. No matter how strong the electric current flowing through the solution, the Brownian movement is still at work, and the path of an individual molecule moving with the current is at best a series of forward-going zigzags, never a straight line. Thus there is always some diffusion from one fraction to the next. Moreover, even in a single compound the molecules are not all exactly alike in the number and distribution of surface charges. This means that some move more slowly than others. Therefore the moving boundary of any given fraction is actually a blurred line rather than a sharp boundary.

J. W. Williams, whose laboratory at the University of Wisconsin has three powerful electrophoresis installations in almost continuous operation, recently said: "I am almost sure that no one has ever obtained in a laboratory, either by electrophoresis, ultracentrifugation or any other means, a pure fraction of a protein—that is, one in which all the molecules are alike."

Despite the lingering impurities, however, electrophoresis remains one of the most discriminating analytical tools of the biochemist. Since its introduction in 1937 numerous improvements have been made in the equipment. The most striking has been in the method of recording the boundaries. The original apparatus, using the schlieren optical system, recorded images which showed the position of each boundary as a dark band. Longsworth introduced a mechanical "schlieren-scanning" device in 1939 which gave a more complete picture in the form of a succession of peaks and valleys. Each peak represents the position of a boundary in the moving column, and the area under the peak indicates the concentration of the chemical fraction responsible for the boundary.

Also in 1939 there came into use another scheme for recording the boundaries—a system known as the "cylindrical-lens" method. It was applied to the Tiselius apparatus by Harry Svensson of Uppsala, utilizing ideas developed earlier by J. Thovert of Paris and J. St. L. Philpot of Oxford. The cylindrical-lens method projects the diagram on a ground glass screen for visual observation as well as on the sensitized film for photography, and the image (*see page 209*) shows as a curve outlined by a jagged line of peaks and depressions—in contrast with the blacked-out solid image projected by the schlieren-scanning method. The schlieren image is illustrated by an electrophoretic analysis of human blood plasma on page 206. The towering peak is the shadow of the albumin boundary. Albumin (molecular weight 70,000) is the plasma protein with the greatest number of surface charges; hence it travels fastest in the electric stream. It is also the most abundant of the plasma constituents. In the normal blood of all animals the albumin peak looms tallest and covers more territory than that of any other fraction. The peaks that follow are those of the globulins.

The globulins are all of approximately the same molecular weight—around 180,000. The ultracentrifuge therefore is powerless to separate them. But fortunately they differ in the magnitude of electric charges on their molecular surfaces, and these differences enabled Tiselius to identify them. He named the fastest-moving globulin alpha, the next in speed beta and the slowest gamma.

Soon after Longsworth devised his

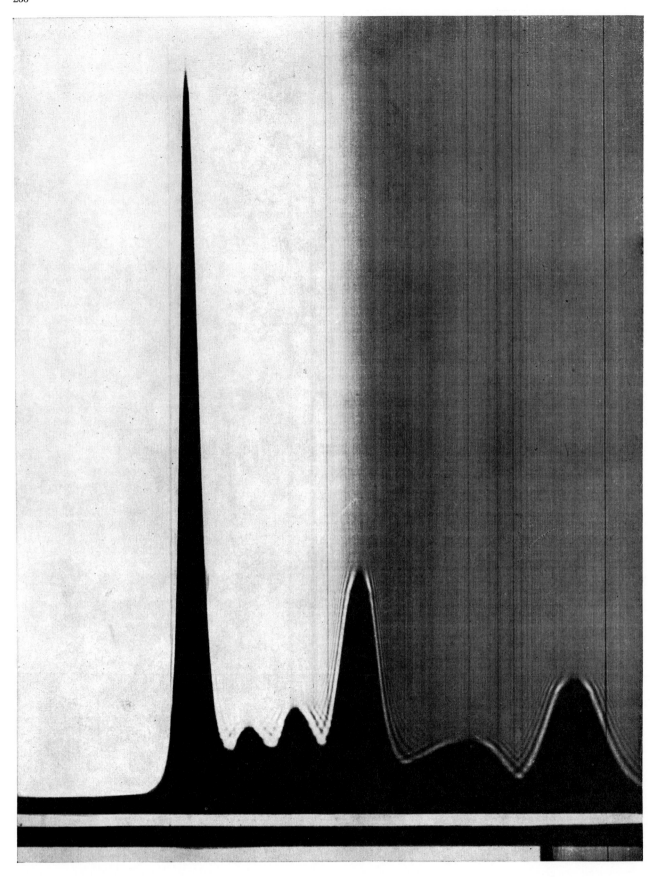

ELECTROPHORETIC DIAGRAM of human blood serum was made with the schlieren-scanning method at the Rockefeller Institute for Medical Research. Each peak on the diagram represents the boundary of a blood protein. The first peak shows the concentration of albu-min; the second, of alpha-one globulin; the third, of alpha-two globulin; the fourth, of beta globulin; and the fifth, of gamma globulin. The pattern was made with serum taken from a pregnant woman, which has a relatively high concentration of globulins.

schlieren-scanning method and installed it in the apparatus at the Rockefeller Institute, he began to experiment with buffer solutions of varying electrolytic concentrations. He found that with a certain combination, using Veronal (the well-known barbiturate) as the buffer, a hitherto unknown globulin appeared in the plasma pattern. Its peak lifted out of the valley between albumin and alpha globulin. Fractions were tested and found to be unquestionably globulin. The new-found material was labeled alpha-one globulin, while the fraction Tiselius had named alpha was now designated as alpha-two. Alpha-one moves electrophoretically at a speed close to that of albumin, and in previous experiments it had been carried along with the albumin and had masqueraded as part of that large fraction.

Longsworth's discovery strikingly demonstrated the strategic importance of the buffer. It was only by employing a buffer composed of univalent ions that he forced the masked globulin to reveal itself. This barbiturate, which has the highest resolving power of any buffer so far known, is in practically universal use today.

In addition to the schlieren-scanning method and the cylindrical-lens method, an optical technique using the phenomenon of interference fringes as the indicator was recently introduced in Switzerland and Germany. But the fringe method is still in an early stage of development.

A further advance in observational methods has just been reported from Uppsala. Harry Svensson, one of Tiselius' chief collaborators, has devised a system which gives a sort of combined interference and schlieren diagram. This has the advantage of showing on the same photograph both absolute concentrations and concentration gradients. Because of its higher sensitivity, the combination optical system permits the study of solutions more dilute than those that could be analyzed by the older methods.

Zone Electrophoresis

Within the last three years a third technique of electrophoresis has developed. In the two described so far—microscopic and moving-boundary electrophoresis—the particles flow in a liquid medium. The newer method flows them through a solid medium. The separated particles do not manifest themselves as moving boundaries but bunch together into zones in the solid; hence the method is known as zone electrophoresis. It was developed independently in Germany, Sweden and the U. S. Among the pioneering investigators are Emmett L. Durrum at the Fort Knox Field Research Laboratory of the U. S. Army; Tiselius, Svensson and Ingra Brattsten at Uppsala, and Theodore Wieland and Fritz

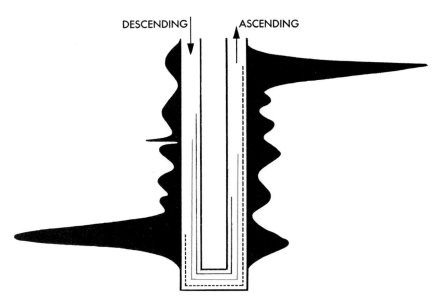

DESCENDING   ASCENDING

GEOMETRIC RELATIONSHIP of the schlieren-scanning diagram to boundaries in a U-tube is shown by this drawing. The pattern is that of human blood serum. For simplicity the boundaries of only three proteins are shown in the U-tube: albumin, beta globulin and gamma globulin. The pattern of the descending proteins differs somewhat from that of the ascending.

Turba in Germany.

The solid most frequently employed in zone electrophoresis is filter paper. A rectangular strip of paper lies horizontally with its two ends dipping into two vessels. These contain a buffer solution and are equipped with opposite electrodes. When a specimen of the mixture to be analyzed is dropped on the surface of the filter paper, it forms a spreading spot. If now a current is passed through the paper between the two electrodes, the components of the mixture will move with the current at speeds determined by their individual surface charges. The effect will be to separate the components and transport them through the paper over varying distances. The final result is a series of spots or zones, each of which is an isolated fraction of one of the components. The distance of each separated spot from the area where the mixture was originally placed is a measure of the mobility of the particles making up that particular spot, and the density and area of the spot provide an index to the quantity of that material in the mixture.

Many of the substances separated in this way have characteristic colors that identify them. Even when they are colorless, it is easy to bring out the spots by dipping the strip of paper in certain dyeing reagents. With the zones thus enhanced, it is possible to plot a line representing the zoning of a mixture into its fractions and obtain a curve which closely approximates the pattern of peaks and valleys obtained by moving-boundary electrophoresis. In a sense zone electrophoresis is like the separation method known as paper chromatog-

raphy except that the separation factor is electricity instead of solubility.

The most obvious advantage of zone electrophoresis is the simplicity and inexpensiveness of the equipment. Besides filter paper, glass powder or silica gel can be used as the solid medium. Tiselius and others claim that zone electrophoresis has several superiorities over the moving-boundary method, to wit: it separates and isolates the individual components of a mixture, instead of merely forming concentration gradients; it makes it possible to study smaller amounts of material at lower concentrations, and it permits development of "techniques of two-dimensional electrophoresis and devices for continuous flow preparative work."

All these advantages have to be bought, however, at the cost of a less well-defined medium, since the filter paper, glass powder or other solid has adsorption and osmotic effects which tend to interfere with the electrophoretic flow. In a recent paper listing the advantages Tiselius and Henry G. Kunkel admitted to a doubt whether it will be possible for zone electrophoresis to achieve "the accuracy obtained with the sensitive optical methods employed in free electrophoresis for locating and quantitating components." In other words, for exactitude in appraising the quantities of protein fractions contained in blood plasma and other body fluids, the moving-boundary method is still the most dependable tool of the biochemist. But zone electrophoresis is in its very young infancy. With dozens of experimenters now taking it up, we may confidently expect radical advances and

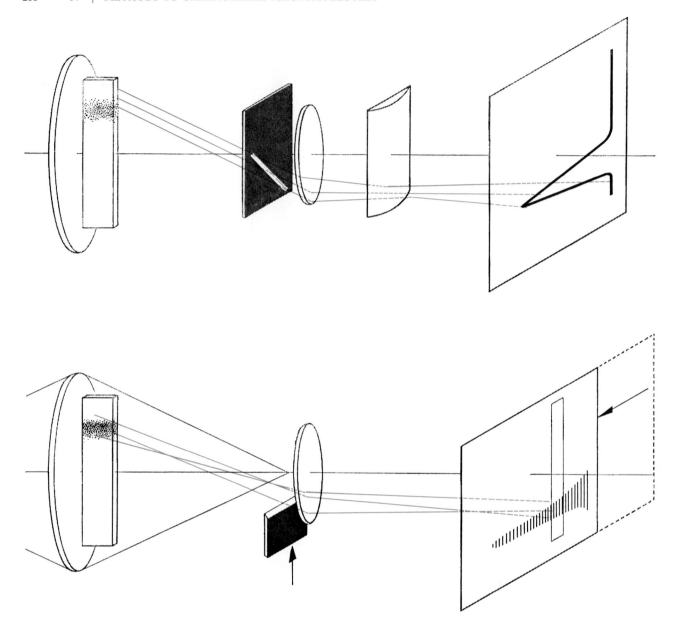

OPTICS of the cylindrical-lens (*top*) and schlieren-scanning (*bottom*) diagrams are compared. One side of a U-tube with a schematic boundary is shown at the left. The nature of the diagrams is illustrated at the right.

productive extensions of the technique in the future.

Electrophoresis is enormously interesting as a demonstration of human imagination and inventiveness, but its prime importance lies in its value as a research tool. Tiselius was spurred to design his ingenious apparatus because of a curiosity about the nature of the invisible particles in blood, just as his teacher Svedberg had invented the ultracentrifuge because he was troubled by ignorance of the molecular weights of proteins. Svedberg's ultracentrifuge and Tiselius' electrophoresis made it possible to isolate the constituents of blood without chemical reactions, and thus to prove that these fractions were individual proteins—not merely fragments split off from a master molecule. The brilliant analysis of the serum proteins that Tiselius accomplished by means of his invention won him a professorship of biochemistry in 1937 and the Nobel prize in 1948. Uppsala made his department into an Institute of Biochemistry and set up funds to construct a suitable building, and the Rockefeller Foundation provided equipment for it.

In Medicine

Electrophoresis has been taken up by the medical chemist and clinician and is now an important part of the equipment of many hospitals. It provides the most convenient and dependable means of analyzing the protein content of the body's fluids and tissues. In the case of a few specific conditions, such as nephrosis, certain liver diseases and the malignant process of the bone marrow known as multiple myeloma, electrophoresis is of direct value as a diagnostic tool.

In multiple myeloma, for example, the electrophoretic pattern of the blood often shows a sharp peak in the region of the gamma globulin. The substance responsible for this peak is a special kind of protein which is found only in multiple-myeloma patients. Electrophoretic analysis of the blood plasma of more than a hundred cases of this disease treated at the Mount Sinai and Montefiore Hospitals in New York showed that half the patients carried the substance in their circulation. The other half did not, but their electrophoretic pattern revealed other abnormalities.

The electrophoretic diagram of a patient's blood plasma or serum is not to be taken as specific for a particular disease but rather as an index to the physiological condition of the patient. For example, if the diagram shows an excess of gamma globulin, the inference is that the body is suffering from an infection, for most of the antibodies evoked by the presence of infectious microbes are gamma-globulinlike proteins. An increase in the alpha globulin, a result of the breakdown of tissue proteins, is likely to herald a fever-producing disease, such as pneumonia, tuberculosis or Hodgkin's. When the blood shows a decrease in albumin and an increase in gamma globulin, the clinician looks to the liver as a possible seat of the disease, because it is the main factory for albumin production. When the liver fails, other tissues try to make up for the lower albumin level, it is believed, by pouring out an excess of globulins.

Since the blood bathes all cells and picks up the products of every kind of tissue, its contents naturally reflect the over-all condition of the body. Electrophoresis thus gives the doctor a picture of what is circulating in the blood, and how much of each component. Its report of what is in excess and what in deficient amount, as well as what is new and unknown, enables him to narrow his search for the trouble to specific organs or tissues. Even though electrophoresis cannot be regarded as a primary prognostic or diagnostic device at the present time, its service as the biological chemist's most reliable analytical stratagem gives it a practical value to clinical medicine.

Its application to clinical medicine has been aided by the development in recent years of compact, self-contained installations of the moving-boundary type of apparatus. Kurt G. Stern, whose part in building the first Tiselius apparatus at Yale was mentioned earlier, and who is now at Brooklyn Polytechnic Institute, has recently been active in the movement to provide moving-boundary equipment in lightweight units that occupy small space and require a minimum of expert operational supervision. Other designers also have contributed to commercial advances in this field. As a result, several types of "packaged" instruments are to be found among the more than 400 electrophoresis machines now in use in the U. S. in hospitals, clinics, medical schools and research institutions. Paper electrophoresis and other forms of zone electrophoresis are also being tested for hospital service, and favorable results are reported. It seems safe to predict that in the course of a few years electrophoresis will be as indispensable to medical practice as X-rays are today.

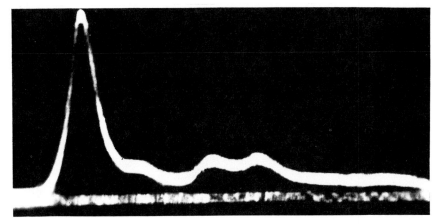

NORMAL subject has this serum pattern. The diagram was made with the cylindrical-lens method for the Lenox Hill Hospital in New York.

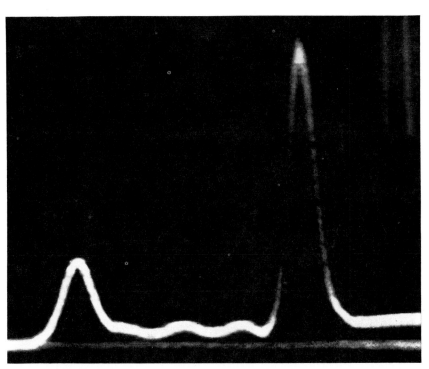

MULTIPLE MYELOMA patient has pattern with a big peak in gamma globulin region. The diagram was made for Montefiore Hospital in New York.

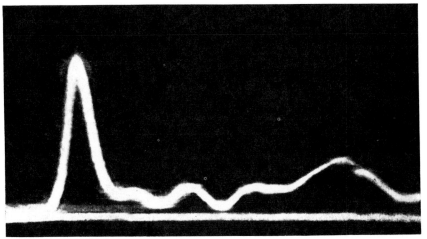

PERIARTERITIS patient has pattern with peak characteristic of infectious diseases. The diagram was made at Mount Sinai Hospital in New York.

21

Relaxation Methods in Chemistry

by Larry Faller
May 1969

By rapidly upsetting the equilibrium of a chemical reaction one can study the important mechanisms that operate in the interval between a thousandth and a billionth of a second

When a chemist has discovered that two substances react to form a third substance, his job has only begun. His next task is to learn something about the mechanism of the reaction. Did substance *A* dissociate into two subspecies before one of them reacted with *B*? Did *A* and *B* first form a temporary complex that rearranged itself to form *C*? Did the reaction require the help of a catalyst such as an enzyme, and if so, what specific regions of the enzyme were involved in the catalytic mechanism? How fast did the reaction go? Because most chemical reactions take place rapidly, it is not easy to obtain answers to such questions. If the reaction takes place in solution, as a great many reactions do, the reaction time is often very short indeed. Hence much ingenuity has been applied to following the details of reactions on an increasingly short time scale.

Fifteen years ago chemists thought they were doing well to study the rate of reactions in solution that were half-completed in a millisecond. Today half-times as short as a few nanoseconds (billionths of a second) can be directly measured. This dramatic progress has resulted from a new approach to the study of reaction rates to which the cryptic term "relaxation" has been applied. In this approach the equilibrium of a chemical reaction is rapidly upset by changing some important condition, such as temperature or pressure; the change in the concentration of reactants or product is monitored while the reaction reequilibrates (relaxes). This basically simple idea, which could have been implemented technically—had anyone thought of it—at least 30 years ago, was introduced in 1954 by Manfred Eigen of the Max Planck Institute for Physical Chemistry in Göttingen, who received a Nobel prize for his

work in 1967. It has provided a new vision into the elementary steps of chemical reactions and is proving a powerful tool for clarifying biochemical mechanisms.

Traditionally chemical reactions have been studied by mixing separate solutions containing known concentrations of the reactants and timing either the disappearance of one of the reactants or the appearance of the product. As long as the progress of the reaction was followed by extracting samples for quantitative chemical analysis, manual mixing was perfectly adequate. The time required to mix two solutions manually depends ultimately on the dexterity of the experimenter, but it is not less than several seconds.

In 1923 Hamilton Hartridge and F. J. W. Roughton of the University of Cambridge devised a way to study much faster reactions. Two solutions containing the reactants were mechanically forced together in a mixing chamber from which the mixture flowed into a long observation tube. There one could measure changes in color that were related to a change in concentration of either the reactants or the product. The time from the start of the reaction depended on the distance the solution had traveled from the mixing chamber and on the velocity of flow. Hence reaction rates could be determined either by observing the extent of the reaction at different points along the observation tube or by observing the reaction at a single point and varying the flow velocity. In this way it was possible to obtain information on reactions within a few milliseconds of their starting time.

In early applications of this method a galvanometer was used to measure the response of a photocell to changes in the

color of the flowing solution. Because several seconds were needed to make a measurement, a constant flow rate had to be maintained for this period, which meant that fairly large volumes of reactants were required. Therefore the method could not be employed with scarce substances such as biochemicals that were difficult to isolate. The development of cathode ray oscilloscopes that can simultaneously detect, amplify, record and time electronic signals led to "stopped flow" instruments that required much smaller quantities of reactants. In this variation of the continuous-flow method the flow was abruptly stopped after mixing and changes in the color of the stationary reaction solution were recorded spectrophotometrically. By 1940 Britton Chance of the University of Pennsylvania had developed an instrument that required only a tenth of a milliliter of reactants to produce enough readings for plotting a useful curve. Since that time there have been refinements in flow methods but none has reduced the minimum time required to mix two solutions to less than a tenth of a millisecond. Further shortening of the time in studying solution reactions required a radically different approach.

The new approach was foreshadowed in the early 1950's by measurements of sound absorption in solutions containing dissolved salts such as magnesium sulfate. Sound waves of certain frequencies were more strongly absorbed than had been expected. It is now known that the anomalous absorption results from the inability of a reaction involving the magnesium ion (Mg^{++}) to keep pace with the pressure variations in the sound wave. It is of historical interest that Albert Einstein, following a suggestion of the physical chemist Walther Nernst, had predicted in 1920 that a dissociating

TEMPERATURE-JUMP INSTRUMENT built by the author at Wesleyan University provides information about fast chemical reactions by means of the relaxation technique. "Relaxation" refers to the natural tendency of a chemical reaction to reach a new equilibrium after it has been perturbed in some fashion. In the author's instrument the perturbation is a sharp rise in temperature.

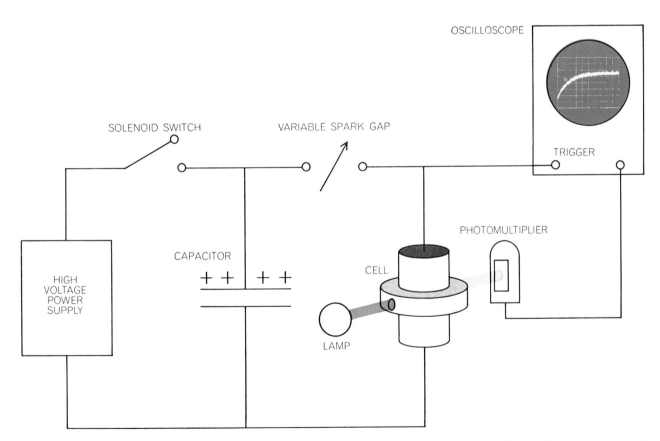

SOURCE OF TEMPERATURE JUMP in the author's instrument is a high-voltage direct-current power supply connected to a capacitor through a solenoid switch. After the capacitor is charged the switch is disconnected. By closing a variable spark gap the energy stored in the capacitor can be discharged through a cell containing the reaction under study. In solution with the reactants are salt ions that carry the discharge current, without entering into the reaction, and rapidly heat the solution. A temperature jump of six to eight degrees Celsius occurs in five microseconds. The temperature jump induces a change in the concentration of reactants and product as the reaction shifts to a new equilibrium. The shift is monitored by a photomultiplier that responds to changes in absorption of one of the reactants. The results are displayed on an oscilloscope, which is triggered by the closing of the spark gap.

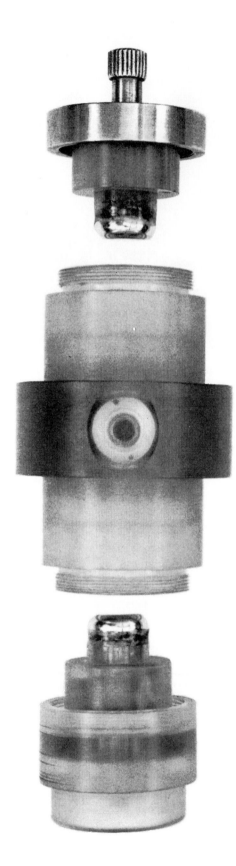

TEMPERATURE-JUMP CELL consists of a cylindrical plastic chamber and two electrodes that fit into the ends of the chamber. When a high current is suddenly passed through the cell, the temperature rises sharply and thus alters the equilibrium of a chemical reaction.

gas should absorb sound at high frequencies because the association and dissociation of the gas molecules would be unable to keep pace with the sharp pressure variations in the sound waves. It remained for Eigen to recognize clearly the importance of these observations for the study of fast reactions. He perceived that fast solution reactions could be investigated by rapidly perturbing a reaction at chemical equilibrium and monitoring the rate at which it shifts to a new equilibrium.

Chemical equilibrium is a dynamic state. Let us consider as a prototype the bimolecular reaction between molecules of A and B to form a new chemical species C [see top illustration on opposite page]. At each instant some molecules of A and B are combining to form C and some molecules of C are dissociating into A and B. The reaction is said to be in equilibrium when the forward and backward rates are equal. The rate at which C forms is proportional to the product of the concentration of A and B. The constant of proportionality (that is, the number that balances the equation) is called the forward rate constant. It is a measure of the probability that the reaction will occur. The rate at which C dissociates to re-form A and B depends on the amount of C present and on the backward rate constant. The relative concentrations of A, B and C are fixed by the equilibrium constant K, which is the ratio of the forward to the backward rate constant.

Two important properties of chemical equilibria should be noted. First, they depend on external conditions, the two most familiar being temperature and pressure. For each temperature and pressure there is a particular set of forward and backward rate constants, and therefore a different value of K. Second, if an equilibrium is disturbed by changing the temperature or pressure, the reaction cannot regain equilibrium infinitely fast.

Suppose our prototype reaction shifts to the right when the temperature is raised. Such a shift is reasonable, because as the temperature increases the molecules of A and B move faster. They collide more frequently, increasing the probability of reaction to form a molecule of C. The dissociation of C does not require collision with another molecule, so that the influence of temperature on the forward and backward rate constants is generally different. If the forward rate constant increases faster with temperature than the backward rate constant,

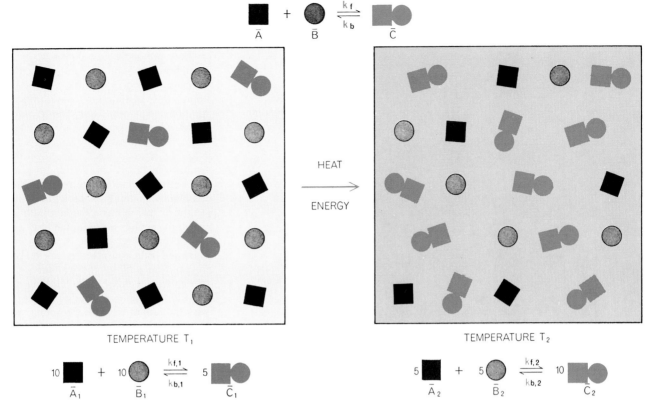

EFFECT OF TEMPERATURE CHANGE on a chemical reaction is usually to shift the concentration of reactants and product that are in equilibrium. Here reactants A and B combine to form C at a certain rate specified by the forward rate constant k_f. The dissociation of C into A and B is governed by the backward rate constant k_b. Letters with bars over them indicate equilibrium concen-

trations of reactants and product. At temperature T_1 10 molecules of A and 10 molecules of B are in equilibrium with five molecules of C. At temperature T_2 the reaction is driven to the right so that five molecules of A and five of B are now in equilibrium with 10 molecules of C. The values of the rate constants at each temperature can be determined by means of chemical relaxation methods.

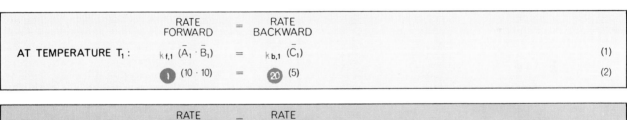

RATE AND EQUILIBRIUM CONSTANTS are computed for the reaction depicted at the top of the page. At each temperature, when equilibrium is reached, the rate forward equals the rate backward. The values of $\bar{A}_1$, $\bar{B}_1$ and $\bar{C}_1$ at temperature T_1 are substituted in equation 1. If the forward rate constant $k_{f,1}$ is set equal to 1, it is clear that the backward rate constant must be 20 to make the equation balance (2). A similar substitution is carried out in equation 3 for temperature T_2, giving equation 4. In equa-

tion 4 the backward rate constant is kept the same as in equation 2 to indicate that the tendency of molecule C to come apart is not greatly affected by a small change in temperature. The rate at which A and B combine, however, is apt to be significantly accelerated because it depends on the frequency with which they collide. Thus in this hypothetical example $k_{f,2}$ is eight times larger than $k_{f,1}$. With this information for each temperature the equilibrium constants K_1 and K_2 can be computed (equations 5, 6).

RELAXATION METHODS fill the gap between more traditional methods for studying the rates of chemical reactions. Reactions with half-times longer than a second can be studied by manually mixing reactants. Mechanical-flow methods extend the range of half-times accessible to study to about a millisecond. At the other extreme, spectroscopy gives information about simple radiation-absorption reactions, which can take place in less than 10^{-10} second (a tenth of a nanosecond). Relaxation techniques fill the gap between 10^{-10} and 10^{-3} second, where many of the elementary steps in chemical reactions take place.

then the relative amount of C in an equilibrium mixture of A, B and C will be greater at higher temperatures.

If our prototype reaction is initially at equilibrium at some temperature T_1, the concentrations of A, B and C will be the appropriate equilibrium concentrations for that temperature [*see bottom illustration on preceding page*]. Now, suppose the temperature is suddenly increased to a new value T_2. As soon as the temperature is raised the appropriate equilibrium values for the reactants are those corresponding to the higher temperature. In contrast, the net rate at which the actual concentration of C can change is limited by the forward and backward rate constants at the higher temperature and by the instantaneous concentrations of A, B and C. If the temperature is increased rapidly compared with the rate at which the reacting system can respond, the change in the actual concentration of C will lag behind the change in its equilibrium concentration.

For small, stepwise perturbations of the temperature, the reactants are observed to approach their equilibrium values at the higher temperature exponentially, that is, the rate at which C changes is approximately proportional to the difference between its equilibrium concentration and its actual concentration. The reciprocal of the constant of proportionality has units of time and is called the relaxation time (designated by the Greek letter tau). In numerical terms it is the time required for C to approach within approximately a third of its new equilibrium value.

The exact relation between the relaxation time, the forward and backward rate constants and the equilibrium concentrations of the reactants depends on the form of the perturbation and on the reaction mechanism. For a simple bimolecular reaction the reciprocal of the relaxation time is equal to the forward rate constant times the sum of the equilibrium concentrations of A and B plus the backward rate constant. All rate constants and concentrations refer to the higher temperature. The forward and backward rate constants can be evaluated by measuring the relaxation time at different concentrations of A and B. A plot of the reciprocal of the relaxation time against the sum of the equilibrium concentrations of A and B yields a straight line whose slope is the forward rate constant and whose intercept is the backward rate constant [*see top illustration on page 217*].

If we had chosen as our prototype re-

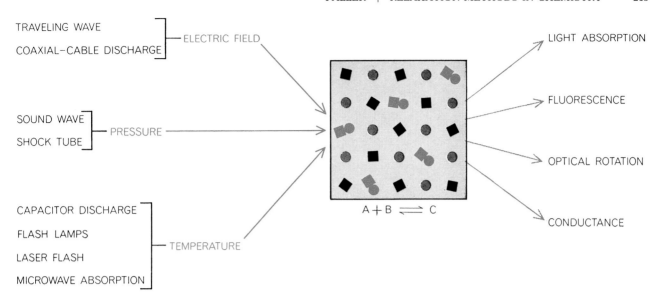

TRAVELING WAVE
COAXIAL-CABLE DISCHARGE — ELECTRIC FIELD

SOUND WAVE
SHOCK TUBE — PRESSURE

CAPACITOR DISCHARGE
FLASH LAMPS
LASER FLASH — TEMPERATURE
MICROWAVE ABSORPTION

LIGHT ABSORPTION

FLUORESCENCE

OPTICAL ROTATION

CONDUCTANCE

A + B ⇌ C

VARIETY OF AGENCIES can be exploited for perturbing the equilibrium of a chemical reaction when employing the relaxation technique. Because chemical equilibria depend on temperature, pressure and strength of the electric field, the rates of fast reactions in solution can be studied by perturbing any one of these variables faster than the chemical system can respond. The illustration indicates some of the ways equilibria can be upset and methods for observing the speed with which equilibrium is reestablished.

action the single-molecule interconversion of two species A and C, the reciprocal of the relaxation time would have equaled the sum of the forward and backward rate constants. Here an independent measurement of the equilibrium constant would be required to evaluate the individual rate constants. More important, the relaxation time would have been independent of the concentrations of reactants [*see bottom illustration on page 217*]. Relaxation studies not only allow evaluation of rate constants but also discriminate among different possible mechanisms. If we had not known whether C formed directly from A or whether a two-molecule collision with another species B was involved, measurement of the relaxation time at different concentrations of A and B would provide the answer.

In multistep reactions a spectrum of relaxation times may be observed. The number of relaxation times expected depends on the number of independent reactions. If the reactions are sufficiently different, the individual relaxation times can be measured separately by simply changing the time base on an oscilloscope. If not, recourse must be taken to mathematical methods for their separation and interpretation.

The "temperature jump" instrument built by the author at Wesleyan University is illustrated on page 211. Technically it is among the simplest of the relaxation methods. It happens that it is also the most generally useful. In this instru-

ment a capacitor is charged to a high voltage. The energy stored in the capacitor is then discharged through the reaction cell by closing a variable spark gap. The reaction cell contains a neutral salt that does not enter into the reaction but carries the discharge current through the solution, which is rapidly heated by the passage of the current. Changes in the color or transparency of the solution in the cell are monitored by a spectrophotometer and recorded on an oscilloscope. The instrument produces a temperature jump of six to eight degrees Celsius in about five microseconds.

It is not necessary that the perturbation take the form of a simple step. Any perturbation that can be readily expressed mathematically can be used. For example, the pressure in a sound wave varies as a sine wave. Nor is it essential to perturb an equilibrium. Many reactions that go to completion, notably enzyme-catalyzed reactions, pass through a stationary, or quasi-equilibrium, state in which the concentrations of intermediate species are temporarily constant. The rate of formation and dissolution of those intermediates can be studied by rapidly perturbing the stationary state.

In the past decade Eigen, his co-worker Leo de Maeyer and their students have built instruments capable of measuring the fastest possible solution reactions. Instruments built at Göttingen and elsewhere have exploited a variety of agencies for rapidly perturbing chemical reactions [*see illustration above*]. For example, temperature perturbations can

be produced by the absorption of microwave energy or by the absorption of short bursts of light energy, either from flash lamps or from lasers. Pressure perturbations can be produced in liquid shock tubes as well as by sound waves.

The versatility of these approaches has been extended by coupling them to diverse detection and recording systems. In addition to ultraviolet and visible-light absorption spectroscopy, changes in fluorescence, in optical rotation and in electrical conductivity have been used to follow the progress of chemical reactions. Gordon G. Hammes of Cornell University has successfully coupled stopped-flow and relaxation methods to study the stationary state in an enzyme reaction.

The chemist is not content to know the results of a reaction, the proportions in which reactants combine and the rate of the overall process. He seeks to determine the detailed mechanisms by which chemical transformations occur. It is known that complex reactions involve a series of discrete steps. A variety of techniques allow the identification of reaction intermediates. More formidable is the study of the elementary steps themselves. In general, the transfer of protons (hydrogen ions), the association and dissociation of molecules, the transfer of electrons and the rearrangement of structures—any combination of which comprises the individual steps in complicated reaction mechanisms—are too fast to study by conventional methods.

Consider the association of two mole-

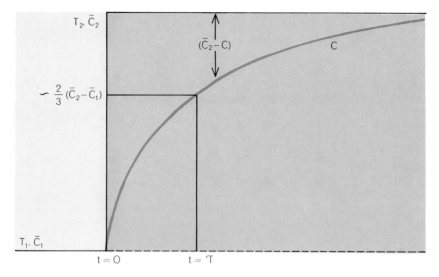

TEMPERATURE-JUMP TECHNIQUE provides data from which rate constants and equilibrium constants can be determined. The apparatus used is the one illustrated on page 31. At the outset, at the initial temperature T_1, product C is present at its equilibrium concentration $\bar{C}_1$. At time zero the temperature is rapidly increased to T_2. The exponential curve (*color*) shows how the instantaneous concentration of C approaches the new equilibrium value $\bar{C}_2$. The relaxation time τ (tau) is approximately equal to the time required for product C to travel two-thirds of the way toward the final equilibrium concentration $\bar{C}_2$.

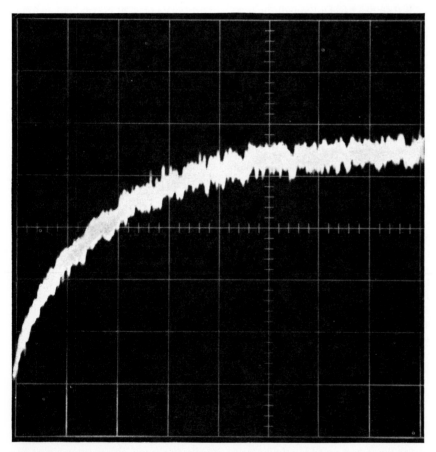

EXAMPLE OF CHEMICAL RELAXATION is shown in this oscilloscope trace made by the author. The reactants are the enzyme alpha-chymotrypsin and proflavin, a strongly colored substance that binds to the enzyme. When subjected to a temperature jump, the equilibrium amount of bound proflavin is abruptly altered, leading to a change in the amount of light at a wavelength of 460 nanometers absorbed by the solution. The trace shows the change in absorption in the 1.6 milliseconds following the temperature perturbation.

cules in water solution as forming some kind of identifiable intermediate complex. How fast can this reaction occur? Of course, since the rate depends on the concentrations of the reactants, it can be made as slow as one wishes by reducing the concentrations. Unfortunately one quickly reaches a point of diminishing returns. If the concentrations are made too small, the reaction is no longer observable. What one would like to estimate is the rate constant for association from which one could calculate the half-time of the reaction in solutions that contain detectable concentrations of the reactants. The theoretical calculation of rate constants is often prohibitively difficult. It is possible, however, to calculate the maximum rate constant for the association of two molecules by assuming that every collision results in the formation of a complex. The collision probability essentially depends on the mobility of the reacting molecules and their size. The faster the molecules move, the more probable a collision is. The bigger the molecules are, the greater the distance at which collision occurs. Both the thermal mobility and the size of the molecules can be measured by the methods of physical chemistry. For small molecules the calculated maximum rate constants range from about 10^9 to 10^{11} liters per mole per second. (A mole is the weight in grams equal to the molecular weight of a substance.) At the concentrations required for optical detection these values correspond to half-times in the microsecond range. Although such half-times are too short for continuous-flow and stopped-flow techniques, they can be measured by relaxation methods.

The study of proton transfer in water illustrates the use of relaxation methods to investigate a fast elementary process. This was one of the first reactions Eigen and his associates studied by chemical relaxation methods. They found that the rate constant for the bimolecular reaction in which protons (H^+) combine with hydroxyl ions (OH^-) to form water is 1.4×10^{11} liters per mole per second. This value was surprisingly higher than the theoretical maximum calculated by assuming that every collision leads to reaction and by using the measured mobilities of protons and hydroxyl ions. In making such a calculation one also has to assume a certain size for the reacting species. Earlier studies had suggested that a proton in water is associated with a water molecule, forming a hydronium ion (H_3O^+). Because it is much larger than a bare proton, H_3O^+ would collide more frequently with OH^-. Even this assumption, however, led to values smaller than the newly measured rate constant. It turned out that the measured value was in good agreement with the value calculated by assuming that the reacting species were the still larger ions $H_9O_4^+$ and $H_7O_4^-$.

The proton in water solution can be regarded as being hydrated by four water molecules and the hydroxyl ion as being hydrated by three water molecules [see top illustration on next page]. Water is a polar molecule, meaning that its electric charges are not evenly distributed. The oxygen has a partial negative charge; the protons (hydrogens) have a partial positive charge. The hydrated proton complex and the hydrated hydroxyl complex are stabilized by the attraction between the charged ions and the oppositely charged ends of the surrounding water molecules. These attractions and the weak interactions between polar water molecules themselves are called hydrogen bonds. The combination of a proton and a hydroxyl ion involves the diffusion together of the hydrated complexes $H_9O_4^+$ and $H_7O_4^-$. Once collision occurs, a hydrogen-bond bridge is formed, and the excess proton is rapidly transferred from $H_9O_4^+$ to $H_7O_4^-$. In ice all the water molecules are bridged to four others by hydrogen bonds. It has been determined from measurements of proton mobility in ice crystals that the mean time an excess proton remains associated with a particular water molecule in a hydrogen-bonded structure is 10^{-13} second. Therefore the slowest step in the combination of protons and hydroxyl ions in water is the random encounter of the hydrated ions. The measured rate constant corresponds to the probability of collision between $H_9O_4^+$ and $H_7O_4^-$.

The unique ability of protons to penetrate spheres of hydration has been confirmed by comparing the rates at which protons combine with negatively charged ions in water solution and the rates at which positively charged metal ions combine with negatively charged ligands (a ligand is anything that associates with metal ions). An example of the former process is the association of acetic acid, in which H^+ reacts with CH_3COO^- to form CH_3COOH. As anticipated, the rate constant for combination (4.5×10^{10} liters per mole per second) equals the calculated collision probability. Once the hydrated hydrogen and acetate ions have collided, pro-

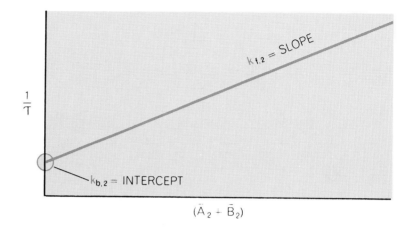

MECHANISM OF REACTION can be clarified by measuring the relaxation time, τ. If the reaction is of the form $A + B \rightleftharpoons C$, $1/\tau$ equals the forward rate constant times the sum of the final equilibrium concentrations of A and B (designated $\bar{A}_2$ and $\bar{B}_2$) plus the backward rate constant. When $1/\tau$ is plotted against $\bar{A}_2 + \bar{B}_2$, the slope of the curve gives the forward rate constant ($k_{f,2}$) and the point of interception gives the backward rate constant ($k_{b,2}$).

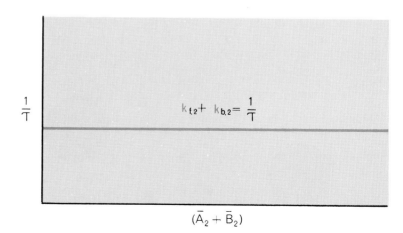

IF REACTION IS UNIMOLECULAR of the form $A \rightleftharpoons C$, it turns out that $1/\tau$ is independent of the concentrations of the reactants, so that the resulting curve is simply a horizontal line. In this case $1/\tau$ equals the sum of the forward and backward rate constants.

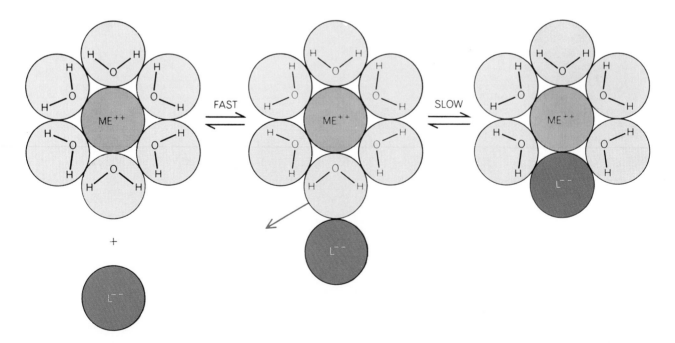

REACTION OF PROTONS AND HYDROXYL IONS in water can be thought of as taking place in two steps, the second much faster than the first. Relaxation experiments and other studies indicate that a proton (H^+) in water is normally hydrated, or surrounded, by four water molecules to form $H_9O_4^+$ and that the hydroxyl ion (OH^-) is hydrated by three water molecules to form $H_7O_4^-$ (*left*). When these complex ions collide, they neutralize each other and form uncharged water molecules. The neutralization actually takes place by a flow of negative charge density, which results in the flipping of weak hydrogen bonds and real bonds (*middle and right*). In the neutral structure that results, each oxygen is strongly bound to two hydrogen atoms and more weakly linked by hydrogen bonds (*broken lines*) to other water molecules. Dots on oxygen atoms represent the potential of forming other hydrogen bonds. This diagram represents in two dimensions structures that actually have three dimensions; thus the circles represent spheres.

FORMATION OF METAL COMPLEX can also be thought of as taking place in two steps, the first being faster than the second. In solution a metal ion (Me^{++}) is often hydrated by six water molecules. The ion with which it combines, called a ligand (L^{--}), is more loosely associated with water molecules (*not illustrated*). The metal ion and ligand come together swiftly (*middle*) but the step in which the ligand displaces a water molecule in the metal's inner hydration sphere takes place much more slowly (*right*).

ton transfer occurs rapidly along a hydrogen-bond bridge.

Metal complexes, for instance the combination of magnesium with sulfate ions, are not formed at a rate controlled by the rate of collision. Typically the rate constants for metal-complex formation are 10,000 to 100,000 times smaller than those for combination with protons. They are generally independent of the ligand but are correlated with the positive-charge density of the metal ion. The greater the ratio of positive charge to the size of the metal ion is, the more slowly the complex is formed. The explanation is that the metal ion and the ligand cannot penetrate each other's hydration spheres. Since the metal ion binds water more tightly, the rate is determined by the charge density of the metal ion. Before a metal-ion complex can form, a water molecule must dissociate out of the metal's inner hydration sphere [see bottom illustration on opposite page].

In addition to proton transfer and metal-complex formation, other elementary reactions have been studied by relaxation methods. The rates of electron transfer, of structural rearrangements and of association reactions have all been successfully measured. The use of relaxation methods to investigate processes involving a series of elementary steps is illustrated by the study of enzyme reactions.

Among the most challenging problems in biochemistry is the detailed explanation of the functioning of enzymes. It has been known for more than a century that enzymes are nature's catalysts. In their absence the myriad chemical reactions in living cells would proceed much too slowly to sustain life. In the past decade it has become clear that some enzymes also function as control units in metabolic pathways, accelerating or decelerating selected regulatory steps, depending on the abundance or deficiency of key metabolites.

Most of the enzymes involved in metabolism have now been identified, isolated and purified. The amino acid sequences of a dozen enzymes are now known. In the past four years the three-dimensional structures of five enzymes have been worked out by X-ray crystallographers. Early this year researchers at the Merck, Sharp & Dohme Research Laboratories and Rockefeller University independently completed the first chemical synthesis of an enzyme (ribonuclease) from its constituent amino acids.

The understanding of how an enzyme functions has proved to be more elusive. Detailed knowledge of the enzyme's structure is an essential first step. In fact, the X-ray analysis of the enzyme lysozyme has given a remarkably clear picture of how the enzyme binds itself to a simple molecule that mimics the structure of the cell wall of a bacterium, and clarifies the mechanism by which lysozyme catalyzes the wall's dissolution. Complete understanding of enzyme function, however, is ultimately a problem in kinetics. The role of enzymes is to regulate the speed of chemical reactions. A complete description of their functioning must therefore include both an identification of the reaction intermediates and an evaluation of the rate constants for the elementary steps in the reaction pathway. So far only three enzymes have been examined in detail by relaxation methods, but it is already apparent that such studies will play an important role in elucidating the detailed mechanisms of enzyme action.

In order to appreciate the power of relaxation methods for studying enzymatic catalysis, it is helpful to understand the limitations inherent in earlier methods. In 1913, 13 years before enzymes were shown to be proteins, Leonor Michaelis and Maud L. Menten explained why the rate of enzyme reactions does not increase indefinitely in linear fashion with substrate concentration but levels off at some maximum rate [see top illustration on next page]. They proposed that an enzyme-substrate intermediate is formed and that it may dissociate either before or after conversion of substrate to product. Since the former is usually much more likely, free enzyme and substrate are virtually in equilibrium with the intermediate complex. The rate of product formation is proportional to the concentration of enzyme-substrate intermediate. As the substrate concentration is increased the concentration of intermediate complex approaches the total enzyme concentration and the rate approaches a maximum value. At substrate concentrations insufficient to saturate the enzyme the rate is the maximum rate times the fraction of the enzyme complexed with substrate.

The difficulty with the Michaelis-Menten formulation of enzyme catalysis is that it postulates three rate constants (k_1, k_{-1} and k_2) but supplies explicit values for only one of them (k_2). The first constant, k_1, determines the rate of formation of the enzyme-intermediate complex. The second, k_{-1}, determines the rate of dissociation of the complex. The third, k_2, determines the rate at which product is formed from the complex. As shown in the top illustration on the next page, two constants appear in the derived rate expression: the maximum rate, V_m, and the Michaelis constant K_s. After obtaining V_m one can calculate k_2 from a knowledge of the total enzyme concentration (E_0). Called the catalytic rate constant, k_2 is a measure of the number of substrate molecules that an enzyme molecule can convert to product per unit of time. The Michaelis constant (K_s) is a measure of the fraction of enzyme complexed with substrate. It is equal to the substrate concentration required to reach half the maximum rate, or $V_m/2$. The Michaelis constant is a ratio of two rate constants: k_{-1} divided by k_1. Their individual values cannot be extracted in this approach.

The problem is that information is thrown away in the derivation of the rate equation. The enzyme and the substrate are not in equilibrium with the enzyme-substrate intermediate, nor is the intermediate complex in a stationary state throughout the course of the reaction. The derivation of the rate expression without simplifying assumptions requires the solution of a second-order differential equation. The resulting rate expression includes a transient term that corresponds to the buildup of a stationary concentration of the enzyme-substrate intermediate and provides the additional information needed to evaluate the individual rate constants. Generally the transient decays too rapidly to be measured by the older kinetic methods, but even if it could be evaluated, the problem is more complex.

The simple Michaelis-Menten mechanism is chemically unrealistic. It is clear on the basis of many kinds of studies that several functional groups in the enzyme must cooperate in catalyzing the chemical transformation of substrate into product. Any realistic mechanism for enzymatic catalysis must therefore include several identifiable intermediates. It does no good simply to rewrite the Michaelis-Menten equations to provide for an extra intermediate (or more than one), because V_m and K_s then are expressed as ratios of five (or more) different rate constants [see bottom illustration on next page]. The physical meaning of V_m and K_s is thus quite obscure.

The solution to this dilemma is to measure directly the rate of formation and disappearance of the reaction intermediates. Such measurements were seldom possible before the advent of re-

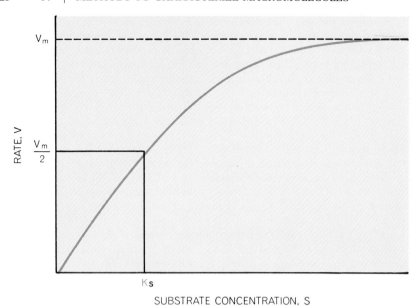

REACTION RATES OF ENZYMES were studied as early as 1913 by Leonor Michaelis and Maud L. Menten. They proposed that an enzyme and its substrate are in virtual equilibrium with an enzyme-substrate complex and that the complex breaks down, in turn, to release the enzyme and the product. The maximum rate of reaction, V_m, depends on the catalytic rate constant (k_2) and the total enzyme concentration (E_0). V_m is reached when the enzyme is saturated with substrate. The equilibrium, or Michaelis, constant (K_s) is equal to the substrate concentration needed to reach half the maximum rate, or $V_m/2$. Constant K_s is a ratio of the backward rate constant (k_{-1}) and forward rate constant (k_1).

$$E + S \underset{k_{-1}}{\overset{k_1}{\rightleftharpoons}} ES$$

ENZYME   SUBSTRATE   ENZYME-SUBSTRATE
                                    COMPLEX

$$ES \xrightarrow{k_2} E + P \text{ (PRODUCT)}$$

$$\text{MAXIMUM RATE } V_m = k_2 E_0$$

$$\text{EQUILIBRIUM CONSTANT } K_s = \frac{k_{-1}}{k_1} = \frac{E \cdot S}{ES}$$

$$E + S \underset{k_{-1}}{\overset{k_1}{\rightleftharpoons}} ES \underset{k_{-2}}{\overset{k_2}{\rightleftharpoons}} EP \xrightarrow{k_3} E + P$$

ENZYME   SUBSTRATE   ENZYME-   ENZYME-   ENZYME   PRODUCT
                          SUBSTRATE   PRODUCT
                          COMPLEX   COMPLEX

$$V'_m = \frac{k_2 \cdot k_3}{k_2 + k_{-2}}$$

$$K'_s = \frac{k_{-1} \cdot k_{-2}}{k_1(k_2 + k_{-2})}$$

ACTUAL ENZYME REACTIONS are more complicated than visualized by Michaelis and Menten. If another intermediate is included in the Michaelis-Menten scheme, V'_m and K'_s can still be determined by experiment, but five different rate constants now appear. V'_m and K'_s provide only ratios of various constants whose individual values remain unknown.

laxation methods. A single enzyme molecule can typically convert 1,000 substrate molecules to product per second. If enough enzyme were used to produce a detectable reaction, the reaction would be over before it could be measured by earlier methods. Relaxation methods now make such high-speed reactions accessible to study. The successful use of relaxation methods to explore the individual steps in an enzyme reaction is nicely illustrated by the study of the enzyme D-glyceraldehyde-3-phosphate dehydrogenase (GAPDH), which catalyzes the phosphorylation of D-glyceraldehyde-3-phosphate to form D-1,3-diphosphoglycerate, one step in the metabolism of sugars.

Kasper Kirschner of the Max Planck Institute in Göttingen has used the temperature-jump method to study how GAPDH is bound to the coenzyme, or cocatalyst, nicotinamide adenine dinucleotide (NAD$^+$). Each enzyme molecule consists of four subunits, and on each subunit is a catalytic site. The binding of the coenzyme to the enzyme is cooperative, that is, the coenzyme is bound more readily at high concentrations than it is at low ones. Since NAD$^+$ binds cooperatively, its availability can regulate the rate at which glyceraldehyde-3-phosphate is phosphorylated. When little NAD$^+$ is available, the phosphorylation goes slowly. When NAD$^+$ reaches a critical concentration, the reaction rate sharply increases. The coenzyme functions rather like an on-off switch.

Jacques Monod, Jeffries Wyman and Jean-Pierre Changeux of the Pasteur Institute in Paris have suggested a mechanism to explain cooperative binding. Because the structure of such a coenzyme, or other effector molecule, that binds cooperatively to the enzyme is distinctly different from that of the substrate, they call their proposal the allosteric model. ("Allo" is the Greek combining form for "other.") In the allosteric model a multiunit enzyme, T, is assumed to be in equilibrium with another form of the enzyme, R, which has a different three-dimensional structure [see illustration on opposite page]. Subunits in the R form are assumed to bind effector molecules more readily than the T form, but the T form is assumed to predominate when no effector molecules are bound. Now if the R form of the fully bound enzyme is favored, binding will be cooperative. At low effector concentrations most of the enzyme is present in the T form. The T form has little affinity for effector, so that most of it remains unbound. As the concentration of effector is increased, more of it is bound, and the equilibrium between R and T shifts toward R. At some point R predominates. Since R binds effector tightly, effector molecules rapidly saturate the enzyme binding sites. The binding process resembles a zipper. Getting it started is difficult, but once started it proceeds rapidly.

The allosteric model predicts as many as nine relaxation times for the binding of an effector to an enzyme of four subunits. Happily Kirschner found that there were only three readily separable relaxation times for the binding of NAD$^+$ to GAPDH. The two faster relaxation times correspond to two-molecule processes. The reciprocal of the fastest relaxation time depends linearly on the sum of the concentration of coenzyme and the concentration of free binding sites on the R form of the enzyme that binds NAD$^+$ more tightly. The other two-molecule process depends on the concentrations of NAD$^+$ and unoccupied T binding sites. Since only two bimolecular relaxation processes were observed, it could be concluded that the binding of effector to either the R or the T form of the enzyme is independent of the number of effector molecules already bound. The third relaxation time does not depend on enzyme concentration. It can therefore be associated with the conformational change between T and R. It does depend on the total concentration of NAD$^+$, reflecting the shift in the equilibrium between T and R with effector concentration, which one would expect if the binding is cooperative. Using the experimentally determined rate constants for the combination of effector with each conformation of the enzyme, and the rate constants for the interconversion of the two forms, Kirschner was able to construct an S-shaped binding curve that was in good agreement with the curve found experimentally.

The fact that the binding of NAD$^+$ to GAPDH can be described by the allosteric model does not mean that all regulatory enzymes conform to this model. It does, however, dramatically illustrate the potential importance of chemical relaxation studies to an understanding of enzyme function. The study of enzyme reactions by relaxation methods is in its infancy. Although the difficulties are formidable, such investigations promise to yield a deeper understanding of enzyme reactions and other complex biochemical processes.

T FORM

R FORM

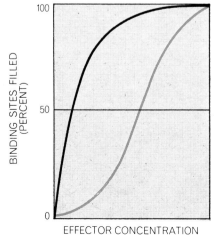

ENZYME MODEL depicts an enzyme with four subunits that responds to an effector molecule (E) in addition to its normal substrate. Such an effector is said to be allosteric. The enzyme can exist in two configurations: T form and R form. All four subunits are assumed to change shape together. The R form binds effector molecules more tightly than the T form does. When no effector molecules are bound, the T form is favored. As more and more effector molecules are bound, however, the equilibrium shifts to favor the saturated R form. This type of binding behavior is described as cooperative and is characterized by an S-shaped binding curve (*colored curve at left*). When the binding is noncooperative, the binding curve is hyperbolic (*black curve*).

BIBLIOGRAPHIES

I BASIC BIOMOLECULAR STRUCTURE

1. The Structure of Protein Molecules

COMPOUND HELICAL CONFIGURATIONS OF POLYPEPTIDE CHAINS: STRUCTURE OF PROTEINS OF THE ALPHA-KERATIN TYPE. Linus Pauling and Robert B. Corey. *Nature*, Vol. 171, No. 4341, pages 59–61; January 10, 1953.

PROTEINS. Joseph S. Fruton. *Scientific American*, Vol. 182, No. 6, pages 32–41; June, 1950.

THE STRUCTURE OF PROTEINS: TWO HYDROGEN-BONDED HELICAL CONFIGURATIONS OF THE POLYPEPTIDE CHAIN. Linus Pauling, Robert B. Corey, and H. R. Branson. *Proceedings of the National Academy of Sciences*, Vol. 37, No. 4, pages 205–211; April, 1951.

2. Proteins

THE NATIVE AND DENATURED STATES OF SOLUBLE COLLAGEN. Helga Boedtker and Paul Doty. *Journal of the American Chemical Society*, Vol. 78, No. 17, pages 4,267–4,280; September 5, 1956.

THE OPTICAL ROTATORY DISPERSION OF POLYPEPTIDES AND PROTEINS IN RELATION TO CONFIGURATION. Jen Tsi Yang and Paul Doty. *Journal of the American Chemical Society*, Vol. 79, No. 4, pages 761–775; February 27, 1957.

POLYPEPTIDES, VIII: MOLECULAR CONFIGURATIONS OF POLY-L-GLUTAMIC ACID IN WATER-DIOXANE SOLUTION. Paul Doty, A. Wada, Jen Tsi Yang, and E. R. Blout. *Journal of Polymer Science*, Vol. 23, No. 104. pages 851–861; February, 1957.

SYNTHETIC POLYPEPTIDES: PREPARATION, STRUCTURE, AND PROPERTIES. C. H. Bamford, A. Elliott, and W. E. Hanby. Academic Press, 1956.

3. The Three-Dimensional Structure of a Protein Molecule

BIOPHYSICAL CHEMISTRY. Edited by John T. Edsall and Jeffries Wyman. Academic Press, 1958.

THE CRYSTALLINE STATE. Edited by Sir W. H. Bragg and W. L. Bragg. G. Bell and Sons, 1933.

THE MOLECULAR BASIS OF EVOLUTION. Christian B. Anfinsen. Wiley, 1959.

X-RAY ANALYSIS AND PROTEIN STRUCTURE. F. H. C. Crick and J. C. Kendrew. *Advances in Protein Chemistry*, Vol. 12, pages 133–214; 1957.

4. The Hemoglobin Molecule

THE CHEMISTRY AND FUNCTION OF PROTEINS. Felix Haurowitz. Academic Press, 1963.

RELATION BETWEEN STRUCTURE AND SEQUENCE OF HAEMOGLOBIN. M. F. Perutz. *Nature*, Vol. 194, No. 4832, pages 914–918; June, 1962.

STRUCTURE OF HAEMOGLOBIN: A THREE-DIMENSIONAL FOURIER SYNTHESIS OF REDUCED HUMAN HAEMOGLOBIN AT 5.5 Å RESOLUTION. Hilary Muirhead and M. F. Perutz. *Nature*, Vol. 199, No. 4894, pages 633–639; August, 1963.

5. The Structure of the Hereditary Material

THE BIOCHEMISTRY OF THE NUCLEIC ACIDS. J. N. Davidson. Methuen, 1954.

HELICAL STRUCTURE OF DEOXYPENTOSE NUCLEIC ACID. M. H. F. Wilkins and others. *Nature*, Vol. 172, No. 4382, pages 759–762; October 24, 1953.

SYMPOSIUM PAPERS ON THE NUCLEIC ACIDS. Proceedings of the National Academy of Sciences, Vol. 40, No. 8, pages 747–772; August 15, 1954.

6. Hybrid Nucleic Acids

DISTRICT CISTRONS FOR THE TWO RIBOSOMAL RNA COMPONENTS. S. A. Yankofsky and S. Spiegelman. *Proceedings of the National Academy of Sciences*, Vol. 49, No. 4, pages 538–544; April, 1963.

ORIGIN AND BIOLOGIC INDIVIDUALITY OF THE GENETIC DICTIONARY. Dario Giacomoni and S. Spiegelman. *Science*, Vol. 138, No. 3547, pages 1328–1331; December, 1962.

STRAND SEPARATION AND SPECIFIC RECOMBINATION IN DEOXYRIBONUCLEIC ACIDS: BIOLOGICAL STUDIES. J. Marmur and D. Lane. *Proceedings of the National Academy of Sciences*, Vol. 46, No. 4, pages 453–461; April, 1960.

THERMAL RENATURATION OF DEOXYRIBONUCLEIC ACIDS. Julius Marmur and Paul Doty. *Journal of Molecular Biology*, Vol. 3, No. 5, pages 585–594; October, 1961.

II   MACROMOLECULAR AGGREGATES AND ORGANIZED STRUCTURES

7. Giant Molecules in Cells and Tissues

CONFERENCE ON TISSUE FINE STRUCTURE. *The Journal of Biophysical and Biochemical Cytology*, Vol. 2, No. 4, Part 2 (Supplement), pages 1–454; July 25, 1956.

MACROMOLECULAR INTERACTION PATTERNS IN BIOLOGICAL SYSTEMS. Francis O. Schmitt. *Proceedings of the American Philosophical Society*. Vol. 100, No. 5, pages 476–486; October 15, 1956.

DIE SUBMIKROSKOPISCHE STRUCKTUR DES CYTOPLASMAS. A. Frey-Wyssling. *Protoplasmatologia*, Band 2^{A2}. Springer-Verlag, 1955.

SYMPOSIA OF THE SOCIETY FOR EXPERIMENTAL BIOLOGY, No. IX: FIBROUS PROTEINS AND THEIR BIOLOGICAL SIGNIFICANCE. Cambridge University Press, 1955.

SYMPOSIUM ON BIOMOLECULAR ORGANIZATION AND LIFE-PROCESSES. Francis O. Schmitt, Paul Doty, Cecil E. Hall, Robley C. Williams, and Paul A. Weiss. *Proceedings of the National Academy of Sciences*, Vol. 42, No. 11, pages 789–830; November 15, 1956.

8. The Structure of Viruses

THE FINE STRUCTURE OF POLYOMA VIRUS. P. Wildy, M. G. P. Stoker, I. A. Macpherson, and R. W. Horne. *Virology*, Vol. 11, No. 2, pages 444–457; June, 1960.

A HELICAL STRUCTURE IN MUMPS, NEWCASTLE DISEASE AND SENDAI VIRUSES. R. W. Horne and A. P. Waterson. *Journal of Molecular Biology*, Vol. 2, No. 1, pages 75–77; April, 1960.

THE ICOSAHEDRAL FORM OF AN ADENOVIRUS, R. W. Horne, S. Brenner, A. P. Waterson, and P. Wildy. *Journal of Molecular Biology*, Vol. 1, No. 1, pages 84–86; April, 1959.

THE MORPHOLOGY OF HERPES VIRUS. P. Wildy, W. C. Russell, and R. W. Horne. *Virology*, Vol. 12, No. 2, pages 204–222; October, 1960.

THE STRUCTURE AND COMPOSITION OF THE MYXOVIRUSES. R. W. Horne, A. P. Waterson, P. Wildy, and A. E. Farnham. *Virology*, Vol. 11, No. 1, pages 79–98; May, 1960.

SYMMETRY IN VIRUS ARCHITECTURE. R. W. Horne and P. Wildy. *Virology*, Vol. 15, No. 3, pages 348–373; November, 1961.

9. The Structure of Cell Membranes

MEMBRANES OF MITOCHONDRIA AND CHLOROPLASTS. Edited by Efraim Racker. Van Nostrand Reinhold, 1969.

STRUCTURE AND FUNCTION OF BIOLOGICAL MEMBRANES. Edited by Lawrence I. Rothfield. Academic Press, 1971.

MEMBRANE MOLECULAR BIOLOGY. Edited by C. Fred Fox and Alec Keith. Sinauer Associates, Stamford, Conn., 1972.

10. Pumps in the Living Cell

SINGLE PROXIMAL TUBULES OF THE NECTURUS KIDNEY. [I:] METHODS FOR MICROPUNCTURE AND MICRO- PERFUSION. Joseph C. Shipp, Irwin B. Hanenson, Erich E. Windhager, Hans J. Schatzmann, Guillermo Whittembury, Hisato Yoshimura, and A. K. Solomon. *American Journal of Physiology*, Vol. 195, No. 3, pages 563–569; December, 1958. II: EFFECT OF 2,4-DINITROPHENOL AND OUABAIN ON WATER REABSORPTION. Hans J. Schatzmann, Erich E. Windhager, and A. K. Solomon. *American Journal of Physiology*, Vol. 195, No. 3, pages 570–574; December, 1958. III: DEPENDENCE OF H₂O MOVEMENT ON NaCl CONCENTRATION. Erich E. Windhager, Guillermo Whittembury, Donald E. Oken, Hans J. Schatzmann, and A. K. Solomon. *American Journal of Physiology*, Vol. 197, No. 2, pages 313–318; August, 1959. IV: DEPENDENCE OF H₂O MOVEMENT ON OSMOTIC GRADIENTS. Guillermo Whittembury, Donald E. Oken, Erich E. Windhager, and A. K. Solomon. *American Journal of Physiology*, Vol. 197, No. 5, pages 1121–1127; November, 1959.

11. The Mechanism of Muscular Contraction

A DISCUSSION OF THE PHYSICAL AND CHEMICAL BASIS OF MUSCULAR CONTRACTION. Organized by A. F. Huxley and H. E. Huxley. *Proceedings of the Royal Society*, Series B, Vol. 160, No. 981, pages 433–542; October 27, 1964. ELECTRON MICROSCOPE STUDIES ON THE STRUCTURE OF NATURAL AND SYNTHETIC PROTEIN FILAMENTS FROM STRIATED MUSCLE. H. E. Huxley. *Journal of Molecular Biology*, Vol. 7, No. 3, pages 281–308; September, 1963. FILAMENT LENGTHS IN STRIATED MUSCLE. Sally G. Page and H. E. Huxley. *Journal of Cell Biology*, Vol. 19, No. 2, pages 369–390; November, 1963. THE STRUCTURE OF F-ACTIN AND OF ACTIN FILAMENTS ISOLATED FROM MUSCLE. Jean Hanson and J. Lowy. *Journal of Molecular Biology*, Vol. 6, No. 1, pages 46–60; January, 1963. X-RAY DIFFRACTION FROM LIVING STRIATED MUSCLE DURING CONTRACTION. G. F. Elliott, J. Lowy, and B. M. Millman. *Nature*, Vol. 206, No. 4991, pages 1357–1358; June 26, 1965.

III   ENZYMES: MACROMOLECULAR STRUCTURES

12. The Enzyme-Substrate Complex

THE DEVELOPMENT OF ENZYME KINETICS. Harold L. Segal. In *The Enzymes*, Vol. I, 2d ed., edited by Paul Boyer, Henry Lardy, and Karl Myrbäck. Academic Press, 1959. ENZYME MECHANISMS IN LIVING CELLS. Britton Chance. In *A Symposium on the Mechanism of Enzyme Action*, edited by William D. McElroy and Bentley Glass. Johns Hopkins Press, 1954. ENZYME-SUBSTRATE COMPOUNDS. J. B. Neilands and Paul K. Stumpf. In *Outlines of Enzyme Chemistry*. Wiley, 1958. PROTEIN STRUCTURE AND ENZYME ACTIVITY. K. V. Linderstrøm-Lang and J. A. Schellman. In *The Enzymes*, Vol. I, 2d ed., edited by Paul Boyer, Henry Lardy, and Karl Myrbäck. Academic Press, 1959. THERMODYNAMIC STUDY OF AN ENZYME-SUBSTRATE COMPLEX OF CHYMOTRYPSIN: I. David G. Doherty and Fred Vaslow. *Journal of the American Chemical Society*, Vol. 74, No. 4, pages 931–936; February 20, 1952. THERMODYNAMIC STUDY OF SOME ENZYME-INHIBITOR COMPLEXES OF CHYMOTRYPSIN: II. Fred Vaslow and David G. Doherty. *Journal of the American Chemical Society*, Vol. 75, No. 4, pages 928–932; February 20, 1953.

13. Protein Shape and Biological Control

INTRACELLULAR REGULATORY MECHANISMS. H. E. Umbarger. *Science*, Vol. 145, No. 3633, pages 674–679; August 14, 1964. ON THE NATURE OF ALLOSTERIC TRANSITION: A PLAUSIBLE MODEL. J. Monod, J. Wyman, and J. P. Changeux. *Journal of Molecular Biology*, Vol. 12, pages 88–118; 1965. REGULATION OF ENZYME ACTIVITY. Daniel E. Atkinson. *Annual Review of Biochemistry*, Vol. 35, Part I, pages 85–124; 1966. THE CATALYTIC AND REGULATORY PROPERTIES OF ENZYMES. D. E. Koshland, Jr., and K. E. Neet. *Annual Review of Biochemistry*, Vol. 37, pages 359–410; 1968. THE MOLECULAR BASIS FOR ENZYME REGULATION. D. E. Koshland, Jr. *The Enzymes*, Vol. I, 3d ed., edited by Paul D. Boyer. Academic Press, 1970.

14. The Three-Dimensional Structure of an Enzyme Molecule

BIOSYNTHESIS OF MACROMOLECULES. Vernon M. Ingram. W. A. Benjamin, 1965.

INTRODUCTION TO MOLECULAR BIOLOGY. G. H. Haggis, D. Michie, A. R. Muir, K. B. Roberts, and P. M. B. Walker. Wiley, 1964.

THE MOLECULAR BIOLOGY OF THE GENE. J. D. Watson. W. A. Benjamin, 1965.

PROTEIN AND NUCLEIC ACIDS: STRUCTURE AND FUNCTION. M. F. Perutz. American Elsevier, 1962.

STRUCTURE OF HEN EGG-WHITE LYSOZYME: A THREE-DIMENSIONAL FOURIER SYNTHESIS AT 2 A. RESOLUTION. C. C. F. Blake, D. F. Koenig, G. A. Mair, A. C. T. North, D. C. Phillips, and V. R. Sarma. *Nature*, Vol. 206, No. 4986, pages 757–763; May 22, 1965.

15. Enzymes Bound to Artificial Matrixes

WATER-INSOLUBLE DERIVATIVES OF ENZYMES, ANTIGENS, AND ANTIBODIES. Israel H. Silman and Ephraim Katchalski. *Annual Review of Biochemistry*, Vol. XXXV, Part II; 1966.

SELECTIVE ENZYME PURIFICATION BY AFFINITY CHROMATOGRAPHY. Pedro Cuatrecasas, Meir Wilchek, and Christian B. Anfinsen. *Proceedings of the National Academy of Sciences*, Vol. 61, No. 2, pages 636–643; October, 1968.

MATRIX-BOUND ENZYMES, PART I: THE USE OF DIFFERENT ACRYLIC COPOLYMERS AND MATRICES. Klaus Mosbach. *Acta Chemica Scandinavica*, Vol. 24, No. 6, pages 2084–2092; 1970.

MATRIX-BOUND ENZYMES, PART II: STUDIES ON A MATRIX-BOUND TWO-ENZYME SYSTEM. Klaus Mosbach and Bo Mattiasson. *Acta Chemica Scandinavica*, Vol. 24, No. 6, pages 2093–2100; 1970.

IV METHODS TO CHARACTERIZE MOLECULES

16. How Giant Molecules Are Measured

LIGHT SCATTERING BY SMALL PARTICLES. H. C. van de Hulst. Wiley, 1957.

LIGHT SCATTERING IN PHYSICAL CHEMISTRY. K. A. Stacey. Academic Press, 1956.

LIGHT SCATTERING IN SOLUTIONS. P. Debye. *Journal of Applied Physics*, Vol. 15, No. 4, pages 338–342; April, 1944.

MOLECULAR-WEIGHT DETERMINATION BY LIGHT SCATTERING. P. Debye. *The Journal of Physical & Colloid Chemistry*, Vol. 51, No. 1, pages 18–32; January, 1947.

TABLES OF LIGHT-SCATTERING FUNCTIONS FOR SPHERICAL PARTICLES. R. O. Gumprecht and C. M. Sliepcevich. Engineering Research Institute, University of Michigan, 1951.

17. The Ultracentrifuge

THE ULTRACENTRIFUGE. Thé Svedberg and Kai O. Pedersen. Oxford University Press, 1940.

HIGH CENTRIFUGAL FIELDS. Jesse W. Beams. *Journal of the Washington Academy of Sciences*, Vol. 37, No. 7, pages 221–241; July 15, 1947.

18. Density Gradients

DENSITY GRADIENT CENTRIFUGATION. J. J. Hermans and H. A. Ende. In *Newer Methods of Polymer Characterization*, edited by B. Ke. Interscience, 1964.

DENSITY GRADIENT TECHNIQUES. Gerald Oster and Masahide Yamamoto. *Chemical Reviews*, Vol. 63, No. 3, pages 257–268; June, 1963.

TECHNIQUES FOR MASS ISOLATION OF CELLULAR COMPONENTS. Norman G. Anderson. In *Physical Techniques in Biological Research, Volume III: Cells and Tissues*, edited by Gerald Oster and Arthur W. Pollister. Academic Press, 1956.

19. Chromatography

PRINCIPLES AND PRACTICE OF CHROMATOGRAPHY. L. Zechmeister and L. Cholnoky. Wiley, 1941.

PARTITION CHROMATOGRAPHY. Biochemical Society Symposia No. 3. Cambridge University Press, 1949.

20. Electrophoresis

ELECTROPHORESIS. Dan H. Moore and H. A. Abramson. In *Medical Physics*, Vol. 2, edited by Otto Glasser. Year Book Publishers, 1950.

21. Relaxation Methods in Chemistry

RELAXATION METHODS. M. Eigen and L. de Maeyer. In *Technique of Organic Chemistry, Vol. VIII, Part II: Investigation of Rates and Mechanisms of Reactions*, edited by S. L. Friess, E. S. Lewis, and A. Weissberger. Interscience, 1963.

FAST REACTIONS IN SOLUTION. E. F. Caldin. Blackwell Scientific Publications, 1964.

RELAXATION SPECTROMETRY OF BIOLOGICAL SYSTEMS. Gordon G. Hammes. *Advances in Protein Chemistry*, Vol. 23, pages 1–57; 1968.

NAME INDEX

SUBJECT INDEX

Absorption of sound, 210
  kinetic studies, 210, 212
Actin, 75, 107, 113
Active transport, 95–97, 99–106
  potassium ion, 95, 97, 106
  potential gradients in, 102, 104
  sodium ion, 104–106
Adenosine triphosphate (ATP), 77, 107, 131,
  135, 136
Aggregates
  actin, 113
  in control systems, 78
  hierarchies, 74
  myosin, 112
  subunits, 138
Albumin
  bovine serum, 197
  egg, 7, 184
Allosteric effects, 5, 51, 220, 221
Allotropy, 155
Amide group, 11, 13
Amino acids, 8, 17–19, 25, 50
  nomenclature, 28, 50
  optical activity, 20
Amphipathic substances, 92
  directionality, 115
  structure, 113, 114
Analysis
  biochemical, 160
  enzyme electrodes, 160, 161
  enzyme separation, 162
Angstrom unit, 33, 54, 88
Anode, 205
Antiparallel pleated sheet, 146
Averages, 164, 172

Bacteriophage, T2, 61, 63 (See also Virus)
Biocatalysts. See Catalysts.
Blood serum
  globulin fractions, 203, 206
Boiling point elevation, 177
Buffers, 202

Capsids, 70, 82, 84
Capsomeres, viral, 84
Carbonic anhydrase, 121
Carotene, 196
Carboxypeptidase A, 129
Catalysts, 121
  action, 122
  biocatalysts, 158, 159
Cathode, 205
Centrifuge, 184

Characterization methods
  absolute, 164
  relative, 164
Chitin, 151
Cholinesterase, 126
Chromatography
  absorbants, 195
  affinity, 161, 162
  definition, 193
  displacement development, 199
  frontal analysis, 199
  in hybridization studies, 68
  ion exchange, 199
  partition, 197
  paper, 92, 94, 95, 125, 197–199
  separations in, 193–195, 197
  separations of amino acids, 98, 99
  solvents used, 195
  use in sterol chemistry, 196
Chromosomes, 53, 73, 78, 79
Chymotrypsin, 126, 161, 162, 216
Cilia, 75, 76
Clots, blood, 78
Codon, 150
Collagen, 7, 22–24, 72, 73–77
  banding of fibrils, 74, 75–77
  structure of fibrils, 76
Complementarity, 5 (see also DNA)
Complex, metal, relaxation methods in studies
  of, 218
Conformations in proteins, 149
Continuous flow, 210
Contractile polymers
  mechanisms, 77
Control mechanisms
  cyclic AMP, 130
  conformational changes, 138, 139
  definition, 128, 219
  feedback inhibition, 52
  genetic repressors, 139
  genetic inducers, 139
  hormones, 130
  sensory receptors, 138
  specificity, 140
Cores, viral, 84
Cortisol, 158, 160
Cortisone, 196
Cooperativity, 132, 167, 221
  positive and negative, 132, 133, 137
  conformational change, 132, 133, 189
  in enzyme binding, 220, 221
Crystallinity
  characterization, 189

CTP Synthetase, 134
Cytochrome C, 125

Dalton, 131
Δ-1, 2 dehydrogenase, 160
Denaturation
  DNA, 63
  proteins, 19, 20
Density gradients, 186
  centrifugation, 186
  preparation, 186, 190, 191
Dialysis
  in enzyme studies, 124
  equilibrium, 124, 125
Diffusion, 178, 182
  in density gradients, 188
  in electrophoresis, 192
  in enzyme kinetics, 189
  measurement, 183, 188, 191, 192
  in polymerizations, 189
  in separations, 189
  in the ultracentrifuge (see Sedimentation)
Diffusion Layer, 157
Diisopropylfluorphosphate (DFP), 125
  in enzyme research, 125, 126
Distribution coefficient, 194
DNA, 5, 53, 78, 79
  A and B forms, 57
  aggregation, 79
  in chromosomes, 79
  dimensions, 54
  function, 53, 54
  information content, 59
  replication, 57, 58, 79, 166
  single stranded, 88
  complementarity, 56, 59, 60
  in viruses, 81, 88
Drosophila, 79

Electrodes, enzyme, 160
Electron microscopy
  freeze-etch, 90, 92
  negative staining, 81, 110
  shadowing, 81, 111
  thin sectioning, 110
  of viruses, 81
Electrophoresis, 166, 201
  and Brownian motion, 201
  cells, 204
  in clinical medicine, 208, 209
  cylindrical lens method, 205, 208
  density gradient, 192
  fringe interference optics, 207